Seed Production Technology of Vegetables

Authors

Dr. Prabhkar Singh Associate Professor, Department of Horticulture, Indira Gandhi Krishi Vishwavidyalaya, Raipur (Chhattisgarh).

Dr. B. S. Asati Served for four year as Research Associate in ICAR Research Complex for NEH Region, Umiam (Meghalaya and worked as Technical Assistant, Department of Horticulture at Indira Gandhi Krishi Vishwavidyalaya, Raipur.

Seed Production Technology of Vegetables

Authors

Dr. Prabhakar Singh

Associate professor

Department of Horticulture

Indira Gandhi Krishi Vishwavidyalaya, Raipur (Chhattisgarh)

Dr. B.S. Asati

Lecturer (Horticulture)

Faculty of Agriculture

Mahatma Gandhi Chitrakoot Gramodaya Vishwavidyalaya

Chitrakoot, Satna (Madhya Pradesh)

2020

Daya Publishing House®

A Division of

Astral International Pvt. Ltd.

New Delhi - 110 002

First Published, 2003

Reprinted, 2020

ISBN: 978-81-7035-555-7 (HB)

Disclaimer:

Every possible effort has been made to ensure that the information contained in this book is accurate at the time of going to press, and the publisher and author cannot accept responsibility for any errors or omissions, however caused. No responsibility for loss or damage occasioned to any person acting, or refraining from action, as a result of the material in this publication can be accepted by the editor, the publisher or the author. The Publisher is not associated with any product or vendor mentioned in the book. The contents of this work are intended to further general scientific research, understanding and discussion only. Readers should consult with a specialist where appropriate.

Every effort has been made to trace the owners of copyright material used in this book, if any. The author and the publisher will be grateful for any omission brought to their notice for acknowledgement in the future editions of the book.

Published by : **Daya Publishing House®**
A Division of
Astral International Pvt. Ltd.
– ISO 9001:2015 Certified Company –
4736/23, Ansari Road, Darya Ganj
New Delhi-110 002
Ph. 011-43549197, 23278134
E-mail: info@astralint.com
Website: www.astralint.com

Digitally Printed at : Replika Press Pvt. Ltd.

Preface

Vegetables play the vital role in the nutritional security of the Indian populace and economy of the majority of small and marginal farmers. For success in vegetable growing enterprise, high quality seed is not only highly desirable but also a statutory requirement. To meet the requirement of the farmers for seed of good quality and genetic purity, it should be produced by using the standard agrotechniques under strict seed certification standards.

This book is intended as a reference for all concerned with the basic vegetable seed production technology–land requirements, field inspection, field and seed standards, brief cultural practices and important Varieties/Hybrids characteristics.

The book Seed Production Technology of Vegetables may be of great utility and intend to be a better guide to the students, teachers, research scientist, extension worker, policy makers, various seed companies, vegetable seeds producers and farming community as a whole who directly and indirectly are engaged in the production of quality seeds.

Our sincere thanks are due to our colleagues for the help rendered in various ways in the completion of this book.

Though every effort has been made to ensure the authenticity of the content of the book. The views and facts presented are solely those of the authors of various books journals consulted. The authors do not take any responsibility of the correctness, authenticity or views of the previous authors.

Prabhakar Singh
B.S. Asati

Contents

	Preface	*v*
1.	Introduction	1
2.	Classification of Vegetables	11
3.	History of Vegetables Seed Production	20
4.	Climatic Factors Affecting Vegetable Seed Production	33
5.	Botany and Floral Biology of Vegetable Crops	37
6.	Modes of Reproduction in Vegetable Crops	80
7.	Principles of Vegetable Seed Production	84
8.	Techniques for Hybrid Seed Production	97
9.	Seed Production of Vegetables	127
	Family–Solanaceae	
	1. Brinjal	
	2. Tomato	

3. Chilli and Bell Pepper

4. Potato

Family–Malvaceae

5. Okra

Family–Cruciferae

6. Radish

7. Cauliflower

8. Cabbage

9. Turnip

Family–Alliaceae

10. Onion

Family–Leguminosae

11. Pea

12. Fenugreek

13. French bean

14. Cowpea

Family–Chenopodiaceae

15. Spinach Beet (Palak)

Family–Umbelliferae

16. Coriander

17. Carrot

Family–Cucurbitaceae

18. Cucumber

19. Muskmelon

20. Watermelon

21. Bitter Gourd

Family–Compositae

22. Lettuce

10. Quality of Seeds 299

11. Seed Germination and Seed Dormancy 311

12. Economics of Hybrids Seed Production 323

13. Biotechnology in Vegetable Seed Production 337

14. National Seed Policy 359

15. Certification of Seeds 375

16. Seed Act and Seed Rules 384

Glossary 417

Appendices 455

References 515

Index 521

Chapter 1

Introduction

Vegetables play a vital role in the health and nutritional security of human beings in addition to improve the economy of the people of the country. India is next only to China in area and production of vegetables and occupies prime position in the production of cauliflower, second in onions and third in cabbage in the world. The total production of vegetables has been estimated at 108.20 million tonnes from an area of 7.05 million hectares and average productivity of 15.34 tonnes per hectare during 2005-06 (NHB, 2006).

Seed production in vegetables is the limiting factor for cultivation of vegetables in India. The vegetables require specific temperature and other climatic conditions for flowering and fruit setting. Some vegetables are grown in one part of the country but their seed production is restricted to another part. To reduce such microclimatic condition a protected environment is essential. Summer squash requires a mild climate for flowering, fruit setting and fruit development, and seed formation. Therefore, its seed production is only restricted to hilly region of north India in summer season. But nowadays seed production of summer squash 'Australian Green' and `Pusa Alankar' is also feasible in north Indian plains in a low-and medium-cost greenhouse. Similarly, seed production of highly remunerative crops–tomato, capsicum and cucumber–is also performed under protected environment. The maintenance of purity of different varieties/lines can be achieved by growing them under greenhouse without giving isolation distance particularly in cross-pollinated vegetables–onion, cauliflower and cabbage. To get proper pollination and fruit set in onion, summer squash, cucumber and bittergourd, the bee-hives are kept inside during flowering. Vegetable production of our country is still dominated by the locally available genotypes or open pollinated varieties. This is mainly due to farmer's ignorance and poor extension activities. Although, it is possible to produce vegetables in one or the other parts round the year due to ports being near and also good infrastructural facilities created, the production is presently limited in certain pockets for meeting the demands of the foreign markets.

Table 1: Area of Production of Important Crops

Crop	*Area of Production*
Okra	Nasik, Ozar, Saikheda, Dindhori, Kolhar, Naraingaon and Sholapur in Maharashtra.
Chilli	Pen, Alibaugh and Choul in Raigad district Dindhori, and Niphad and Itgatpuri Taluka in Nasik district of Maharashtra.
Watermelon	Panvel near Mumbai and Maharashtra state.
Bottle gourd and bitter gourd	Nasik and Pune district in Maharashtra.
Capsicum	Nasik, Pune and Satara district in Maharashtra.
French bean and cluster bean	Dindhori in Nasik district, Wai in Satara district and Naraingaon in Pune, Dhule and Ahmednagar.
Onion (big)	Nasik, Pune and Satara district in Maharashtra, Periyar and Coimbatore district in Tamil Nadu, Bhudaun in Uttar Pradesh and Patna, Biharsharif in Bihar, Bhavnagar and Rajkote in Gujarat.
Onion (small)	Kolar and Bangalore in Karnataka and Cuddapah in Andhra Pradesh.
Onion (multiplier)	Anna, Madurai, Salem and Coimbatore district of Tamil Nadu.
Garlic	Indore and Mandsaur in Madhya Pradesh, Ooty in Tamil Nadu, Jamnagar and Rajkot in Gujarat and Kullu in Himachal Pradesh.
Tomato	Nasik and Pune in Maharashtra and Bangalore in Karnataka.
Potato	Jallandher and Ludhiana in Punjab, Kurukshetra and Karnal in Haryana, Ooty in Tamil Nadu and Indore in Madhya Pradesh.

Need for Quality Seed Production

A statement " Good seed on good land yields abundant produce"–appearing in the ancient Hindu scriptures, the Manu Smriti provides an age-old recognition to the good quality seed in crop production. The use of good quality seed is indispensable for the successful production of any crop. The cost of seed is small fraction of the total cost of inputs involved in raising the crop to maturity. A good quality seed is good looking, viable and vigorous, genetically pure (true to type), bold and uniform size of the desired type, free from disease, insects-pest, weeds seeds, foreign matter which may impair the quality of seed, fairly priced, better longevity, good yielding ability and have wider adaptability. Quality seed production is a highly specialized activity and vegetable produce retained for seed can not be substituted for quality seed as it generally lack genetic vigour and has poor germination. The vegetable seed production comprising of complex interlocking operations that are necessary to ensure the regular supply of uniformity high quality seed to farmers. However, the quality seed production is only possible by maintaining proper isolation distance, timely removal of off types, volunteer plants and wild relatives is also important.

Seed Import and Export

The biggest seed trader (import + export) is Japan with 80 million $, followed by Korea with 23 million $, and Thailand 20 million $. China, and India seed sector are in full growth, and their export capabilities will presumably increase. All other countries are importing vegetable seeds only. It has been observed that even countries with a very strong vegetable seed industry, have high volumes of import, of which the value is compensated by the value of export. The high imports include seed for domestic market which is produced abroad, because, either for economic, or climatic reasons, the production is not possible locally. This practice of custom seed production abroad has become very common for developed seed companies. A lot of developing countries are concerned about their dependency on foreign seeds, and the loss in their trade balance. Some had even taken measures to restrict import of seed. The fact is that some countries with tropical climate will always need to import seed of certain crops such as cabbage and cauliflower for which seed cannot be produced locally. Moreover, countries like India, which has restricted import of seed for a long time, and revised the decision in 1988, realized the positive change it had brought to the vegetable industry. For this reason, vegetable seed import should not be restricted nor banned anywhere, so that the best material available can be made available to the farmers. These policies should be accompanied by the proper quarantine measures, in order to avoid the spread of unwanted diseases. India imported (in 1993) 170 tonnes of vegetable seeds, of which the main items are cabbage, capsicum radish, tomato, cucumber, squash, chili, and cauliflower. Most popular cabbage F_1 varieties is still imported from Japan. With the development of the seed sector in the past decade, India has been exporting officially or unofficially a lot of vegetable seeds to neighboring countries.

Although India is second largest vegetable producing country in the World, it has 8th ranks among the sixteen countries listed in seed trade the world over. In recent years, a lot of emphasis is being given to improve the quality of seed and planting material in view of globalization of the world trade which had opened the ample opportunities for the export of quality seed. Enjoying the diversity of agroclimatic conditions, strong seed production infrastructure and market opportunity, India holds significant promise for export of seeds. The export of fruits and vegetable seed during 2001-02 was 6188.67 metric tones amounting Rs. 6457.89 lakhs (Table 2). For our fruits and vegetable seeds USA, Japan, Netherlands, Italy and Pakistan were major market for the year 2001-02.

Table 2: Export of Seeds of Major Vegetables

Vegetable Seeds	*Rupees*		
	1998-99	*1999-00*	*2000-01*
Potato	39584130	15398192	14387935
Cabbage	5657697	484910	23494487
Cauliflower	3194145	591608	3511670
Onion	5660434	16141209	11384220
Pea	3845670	826891	160993
Radish	178650		69245
Tomato	162452647	110806580	87516676
Others	290601538	189881409	248583846

Source: APEDA, New Delhi.

Seed Production System

Most of vegetable seed industry is handled by private sector. Today all hybrid seed production of tomato, cucurbits, and okra is done on contract with farmers who are well trained by private operations, and the companies themselves. Public institutes, especially NSC (National Seed Corporation) or SFCI have concentrated mainly on high volume, low value crops such as okra and peas. The distribution is made by over 10,000 dealers spread all over the country, very well controlled by the private sector. Varieties/ hybrids released by public research institutes can be multiplied by private sector with no royalties. Seed and seed related government policies. In 1998, a new seed policy was promulgated, the motive of which was to ensure that farmers should have access to best material available. This opened the gates to the importation of seed. In theory, official registration or notification of new varieties is to be made and is a prerequisite for certification. Practically, the private sector receives their evaluation at the hands of the farmers. A draft act on Plant Variety Right (consequence of GATT) has been proposed but is being discussed and debated between public and private sector, which have different interests.

Which Seeds Can be Saved?

Vegetable seeds can be saved to sow new crops in the future, but not all seeds are suitable for saving. Varieties suitable for seed saving include local varieties that have been grown in one region for a very long time, self-pollinating crops (for example, beans and peas), and open-pollinated varieties of some cross-pollinating crops (for example, pepper, cucumber and carrot). Commercial F_1 hybrid varieties are popular among many vegetable growers today. However, the seed of hybrid fruits should not be saved, because the F_1 hybrid seeds were produced by crossing two different parent varieties. Seed saved from hybrids will either is sterile or the plants of the next generation may show wide variation in characters, uniformity and maturity.

Seed saving involves selecting suitable plants from which to save seeds, harvesting seeds at the right time, and storing them properly. The seed saving techniques of many common vegetables will be introduced in this manual. Before you start to save seeds, it is necessary to know a few things about there productive ability of plants. Many vegetable species produce flowers with the male part (anther) and the female part (stigma) in the same flower. These are called perfect flowers (Figure 1). However, in most varieties of the cucurbit family (cucumbers, melons, pumpkins, etc.), the anthers and the stigma are in the same plant but on different flowers. These are called imperfect flowers (Figure 2).

Figure 1: Perfect Flower of Eggplant: The Stigma (Green) is Surrounded by Anthers (Yellow)

Figure 2: Imperfect Flowers of Squash: Female Flower with Exposed Stigma (Left) and Male Flower with Exposed Anthers (Right)

Pollination occurs in plants when pollen from the anthers of the flower is deposited on the stigma. In some perfect flowers, self-pollination occurs. Lettuce, tomato, and okra have the stigma so close to the anthers such that the slightest wind movement can cause the pollen to drop onto the stigma with in the same flower. In peas and beans, self-pollination occurs even before the flower opens. Other types of perfect flowers require cross-pollination. An external pollinator such as an insect is necessary. Onion, carrot, cabbage, and radish, for example, belong to this type. Plants with imperfect flowers require wind or insects such as bees to transmit pollen from the anthers of the male flowers to the stigma of the female flowers. For example, cucurbits are cross-pollinated by bees.

How to Keep Seeds Pure?

Keep in mind that natural cross-pollination can always happen to some extent under a field situation, even in self-pollinating plants. It often occurs when pollen grains stick onto the bodies of insects visiting flowers, and then carry the pollen grains to the next flower they visit.

Isolation in Distance

Pure seeds can be produced by leaving enough distance between two or more varieties to prevent cross-pollination by insect or wind-blown pollen. How far apart differs among vegetables; this will be described for each vegetable in the following chapters.

Bagging

When only a small amount of seed is needed, cover the unopened flowers with a paper bag. This is applicable for crops with a high but not 100 per cent rate of self-pollination, such as pepper and eggplant. You can also bag the flowers of cucurbits (Figure 3); in this case, both male and female flower should be bagged, but hand-pollination is required.

Caging

Cages can be used for vegetables that flower over a long time or to prevent insects from transmitting pollen from two near by varieties of the same crop (Figure 4). You can use bamboo rods stuck in theground to make an arched tunnel and covered with nylon mesh. Because the cage will exclude all insects, you may need to hand pollinate the plant to ensure seed set, or you can introduce bees into the cage if they are cross-pollinated species.

Figure 3: Bagging Bitter Gourd Flower for Hand Pollination

Figure 4: Isolation of Pepper Selections in Nylon Net Tunnels

What is the Best Way to Store Seeds?

After saving your seeds, it is important to keep them alive for future use. Newly harvested seeds should not be immediately stored in a plastic bag because the moisture content of the seed is still high and will lead to deterioration. Before keeping seeds in the storage, seeds should be dried. Keep in mind that seeds are alive–but they breathe very slowly. To keep seed a live for a long time, keep them under low temperature and low humidityduring storage.

Humidity

Seeds will absorb moisture from the storage environment. High humidity levels cause seed to increase their respiration rate and use their stored energy. Make sure your seeds are dry enough (seed moisture content around 7–8 per cent) before storage, and keep them in an air-tight container, such as a screw-top jar (Figure 5).

Figure 5: Seeds Stored in Air-Tight Containers to Prevent them from Absorbing Moisture

Darkness

Exposure to sunlight will shorten the life of seeds. Use dark-color jars or non-transparent containers to protect seed from sunlight. If using clear jars, place them in paper bags to shield out sunlight.

Temperature

For most vegetable seeds, a temperature below 15 °C is ideal. You can keep the seeds in an air-tight container and place the container in the refrigerator. For short-term storage, keep the seeds in a cool and shady dry place.

Most vegetable seeds can be safely stored for at least three to five years. Place seeds in manila envelopes, cloth or mesh bags, plastic containers, or foil envelopes. The best containers are air-tight, such as a sealed glass jar, metal can, or foil envelope. Label each container carefully. Note the names of the line or variety, the year, and any othe rinformation you feel is valuable. Store seeds in a cool, dry place.

Conclusion

Vegetable production in India has increased significantly at the rate of 2.7 per cent per year, to meet the demand of the population in the region increasing at the rate 1.8 per cent annually. Despite the growing vegetable production have less vegetable available per capita than the dietary requirement. The total cropped area is almost reaching saturation in India, but in average the areas under vegetables have increased; similarly, but at a slower rate, the productivity has also improved. This achievement is a result of the recent interest of government programs in developing the vegetable industry. The public research started ground work in order to uplift the technical and managerial level of the farmers, by developing and disseminating (slowly) improved agricultural practices. The private seed sector has also contributed considerably to the overall growth of the vegetable industry, by developing and making available superior hybrids and quality seeds to the farmers, the demand for quality seed and hybrids is steadily increasing with the improving technological skills of the vegetable farmers. Some 10 to 15 years ago, the private vegetable seed sector was in active only in a limited number in India, and was developing hybrids adapted to local markets and exporting them to other countries. Today however, the vegetable seed sector development has expanded. The seed sector is under going fundamental changes, because local governments have realized the importance of the seed as major agricultural input for the development of the vegetable industry and the key role that private sector has to play in it. In future, and with adequate support from government policies, the vegetable seed sector is expected to be active and dynamic with hybrid varieties being developed locally for domestic markets and commercial farmers and superior open pollinated varieties produced for the benefit of marginal farmers and homestead gardens. Finally there is a great scope for development of the vegetable seed industry and hybrids in India. The market is immense, and the farmers are gradually picking up the technology. The time is also right to develop the seed sector and to start programs on hybrids through public and private sector.

Chapter 2

Classification of Vegetables

Botanical Classification

Botanical classification of vegetables is showing relationships, and gives an idea of the plant families represented as well as the important crops belonging to each of these families which is helpful for a vegetable breeder to know the botanical relationship of vegetable crops which he intend to work upon, viz; the morphological and cytological similarities and dissimilarities, their place of origin, crossability, floral biology and other details which are relevant in crop improvement. Many times, the common names are confusing even within the same geographical region, in such situation the botanical classification is useful in relation to the study of cultivation. Almost, all vegetable crops belong to sub community spermatophyta and division angiospermae, but spices like juniper belongs to division gymnospermae; family pinaceae, genus: *Juniperous* and species: *communis*. The division angiospermae is further divided into sub division or classes: monocotyledonae and dicotyledonae.

Monocotyledonae

Sl.No.	Family	Botanical Name	Common Name
1.	Graminae	*Zea mays* var. *rugosa*	Sweet corn
2.	Alliaceae	*Allium cepa*	Onion
		Allium cepa var. *aggregatum*	Multiplier onion
		Allium cepa var. *viviparum*	Top onion
		Allium porum	Leek
		Allium sativum	Garlic
		Allium fistulosum	Welsh onion
		Allium ascalonicum	Shallot
		Allium schoenoprasum	Chive
3.	Liliaceae	*Asparagus officinalis*	Asparagus
4.	Araceae	*Colocasia esculenta*	Taro or Arvi
		Aqconus calamus	Calamus
5.	Dioscoraceae	*Dioscorea alata*	Yam
6.	Orchidaceae	*Vanilla fragrans*	Vanilla
7.	Iridaceae	*Crocus sativus*	Saffron
8.	Zingiberaceae	*Alpinia galangal*	Galangal
		Zingiber officinale	Ginger
		Amomum subulatum	Large cardamom
		Elettaria cardamomum	Small cardamom
		Curcuma longa	Turmeric

II. Dicotyledonae

Sl.No.	Family	Botanical Name	Common Name
1.	Aizoaceae	*Tetragonia expansa*	New Zealand Spinach
2.	Chenopodiaceae	*Beta vulgaris*	Beetroot or Palak
		Beta vulgaris var. *cicla*	Chard
		Spinacia oleracea	Spinach
3.	Compositae	*Lactuca sativa*	Lettuce
		Lactuca sativa var. *capitata*	Head lettuce
		Cichorum intybus	Chicory
		Cichorum endivia	Endive
		Cynara scalymus	Artichoke
		Artemisia dracunculus	Tarragon

Contd...

Contd...

Sl.No.	Family	Botanical Name	Common Name
4.	Convolvulaceae	*Ipomoea batatas*	Sweet potato
5.	Cruciferae	*Brassica oleracea* var. *capitata*	Cabbage
		Brassica oleracea var. *botrytis*	Cauliflower
		Brassica oleracea var. *gemmifera*	Brussels sprouts
		Brassica oleracea var. *italica*	Sprouting broccoli
		Brassica oleracea var. *acephala*	Kale
		Brassica caulorapa	Kholrabi or knolkhol
		Brassica napus var. *napobrassica*	Rutabaga
		Brassica campestris var. *rapa*	Turnip
		Brassica chinensis	Chinese cabbage
		Barbarea vulgaris	Upland Cress
		Raphanus sativus	Radish
		Armoracia rusticana	Horse radish
		Raphanus sativus var. *longipinnatus*	Chinese radish
		Brassica juncea L.	Indian mustard or Brown mustard
		Brassica nigra	Black mustard or True mustard
		Brassica alba	White mustard or Sufai Rai
6.	Cucurbitaceae	*Cucurbita moschata*	Pumpkin
		Cucurbita pepo	Summer squash
		Cucurbita maxima	Winter squash
		Citrullus lanatus	Watermelon
		Citrullus lanatus var. *fistulosus*	Tinda
		Cucumis sativus	Cucumber
		Cucumis melo	Muskmelon
		Luffa acutangula	Ridge gourd
		Luffa cylindrica	Sponge gourd

Contd...

Contd...

Sl.No.	Family	Botanical Name	Common Name
		Lagenaria siceraria	Bottle gourd
		Momordica charantia	Bitter gourd
		Momordica dioica	Kakrol
		Trichosanthes dioica	Pointed gourd or parwal
		Trichosanthes anguina	Snake gourd
		Sechium edule	Chow-chow or chayote
7.	Euphorbiaceae	*Manihot esculenta*	Tapioca
8.	Leguminosae	*Pisum sativum*	Pea
		Phaseolus vulgaris	French bean
		Phaseolus lunatus	Lima bean
		Vicia faba	Broad bean
		Vigna unguiculata	Cowpea
		Vigna unguiculata var. *sesquipedalis*	Asparagus bean
		Cyamopsis tetragonolaba	Cluster bean or Guar
		Dolichos lablab	Sem
		Psophocarpus tetragonolobus	Winged bean
		Glycine max	Soybean
		Trigonella foenum graecum	Fenugreek
9.	Malvaceae	*Abelmoschus esculantus*	Bhindi or Okra
10.	Polygonaceae	*Rheum rhaponticum*	Rhubarb
		Rumex vesicarius	Sorrel
11.	Solanaceae	*Solanum tuberosum*	Potato
		Solanum melongena	Brinjal or Eggplant
		Lycopersicon esculentum	Tomato
		Capsicum fruitescence	Chilli
		Capsicum annum	Bell pepper
12	Umbelliferae	*Daucus carota*	Carrot
		Petroselinum crispum	Parsley
		Apium graveolens	Celery
		Pastinaca sativa	Parsnip

Contd...

Contd...

Sl.No.	Family	Botanical Name	Common Name
		Anthriscus cerefolium	Chervil
		Trachyspermum ammi	Ajowan
		Angelica archangelica	Angelica
		Pimpinella anisum	Aniseed
		Ferula asafetida	Asafoetida
		Carum carvi	Caraway
		Apium graveolens var. *rapaccum*	Celeriac
		Coriandrum sativum	Coriander
		Cuminum cyminum	Cumin
		Nigella sativa	Cumin Black
		Anethum sowa	Dill (Indian)
		A. graveolens	European
		Foeniculum vulgare	Fennel
		Levistcum officinale	Lovage
		Carum bylbocastanum	Kala zira
13.	Rutaceae	*Murraya koenigii*	Curry leaf
14.	Piperaceae	*Piper nigrum*	Black pepper
		Piper longum	Long pepper
15.	Myrtaceae	*Pimenta officinalis*	Spice
		Eugenia carryophyllus	Clove
16.	Myristicaceae	*Myristica fragrans*	Nutmeg
17.	Labiateae	*Mellisso officinalis*	Balm
		Ocimum basilicum	Basil
		Hyssopus officinalis	Hyssop
		Majorana hortensis	Marjoram
		Mentha arvensis	Mint
		Mentha piperita	Pepper mint
		Origanum vulgare	Origanum
		Rosemarinus officinalis	Rosemary
		Salvia officinalis	Sage
		Satureia hortensis	Savoy
		Thymus vulgaris	Thyme

Contd...

Contd...

Sl.No.	Family	Botanical Name	Common Name
18.	Guttuferae	*Garciana indica*	Kokam
19.	Lauraceae	*Laurus nobilis*	Bay leaf
		Cinnamomum aromaticum	Indian cassia
		Cinnamomus zeylanicum	Cinnamon

Classification Based on Nature of Pollination

Highly Cross-pollinated

1. *Wind-pollinated*: Amaranths, beet-leaf, spinach, garden beet, sugar beet.
2. *Insect-pollinated*: All cucurbits, all cole crops, radish, turnip, carrot, and onion

Often Cross-pollinated

Limabean, egg plant, okra, chilli, sweet pepper

Highly Self-pollinated

Asparagus bean, cluster bean, Dolichos bean, French bean, cowpea, garden pea, methi, lettuce, tomato

In groups b and c self-pollination occurs by reason of floral structures and any cross-pollination takes place is affected by insect-pollinators. The frequency of such cross-pollination is far greater in-group b.

Grouping Based on Planting Response

1.. *Easily survive transplanting*: All cole vegetables, Chinese cabbage, lettuce, tomato, brinjal, capsicum and chilies.

2.. *Needs care in transplanting*: Onion, leek, celery, parsely, amaranths, cardamom.

3.. *Not suited to transplanting*: Garden pea, turnip, radish, carrot, garden beet, spinach, gourds, bhindi, methi

4.. *Suits both transplanting and direct seeding*: Chinese cabbage, chillies, amaranths.

5. *Polypack planting*: All cucurbits mostly, cucumber, bitter gourd, squashes, melons, pumpkins, tomato hybrids.

Grouping Based on Nature of Vegetables

Annual

Tomato, brinjal, capsicum, chillies, pea, bhindi, colocasia, methi, amaranths, all cucurbits, celery, parsely.

Biennial

All cole crop, Chinese cabbage, onion, turnip, carrot, garden beet, beet leaf, leek, radish, spinach.

Perennial

Asparagus, chyote, ginger, cardamom, turmeric .

Grouping Based on the Area of Seed Production

Temperate

Cabbage, knolkhol, kale, Brussels sprout, sprouting broccoli, sugarbeet, chicory, potato, amaranths, temperate varieties of radish, turnip, carrot, and garden pea.

Sub-temperate

Late cauliflower, beet leaf, spinach, celery, parsley, asparagus, chyote, sub-temperate varieties of radish, turnip, carrot.

Tropical

Tomato, brinjal, okra, chillies, capsicum, all cucurbits, ginger, turmeric, colocasia, cardamom, sweet potato.

Classification Based on Hardiness in Hills

Hardy

Asparagus, sprouting broccoli, brussels sprout, cabbage, kale, leek, onion, garlic, parsely, garden peas, radish, spinach, turnip.

Half-hardy

Carrot, cauliflower, celery, garden beet, artichoke, lettuce, palak, parsnip, potato, chicory, chinese cabbage, endive.

Tender

Snap bean, sweet corn, tomato.

Very tender

Amaranths, bhindi, brinjal, capsicum, chilli, cluster bean, cucurbits, colocasia, sweet potato, yam, tapioca.

Grouping Based on Methods of Seed Production

1. *Seed to seed (in situ)*: Carrot, cauliflower, celery, garden beet, cabbage, garden pea, garlic tomato bhindi, brinjal, capsicum, chilli, cluster bean, cucurbits
2. *Replanting method (steckling/Head/curd/bulb to seed)*: Carrot, radish, garden beet, chinese cabbage, cabbage, sprouting broccoli, chicory, sugar beet.
3. Both methods: Carrot, radish, celery, Chinese cabbage, cabbage, turnip.

Classification Based on Breeding Systems

Sexual				*Asexual*
Predominantly Self-Pollinated	*Often-Cross-Pollinated*	*Largely Cross-Pollinated*	*Dioecious*	*Vegetatively Propagated*
Degree of selfing 90-100%	Adapted to self-pollination but degree of cross-pollination more than 4%	Degree of crossing can be 100%	Entirely cross pollinated	Colocasia, sweet potato, pollinated gourd, ginger, turmeric
Lettuce, garden pea, french bean, tomato, potato	Okra, chillies, capsicum, brinjal, lima bean	Cole crop, root vegetable, cucurbit, onion, amaranths, spinach beet, chenopodium	Spinach, asparagus, parwal	

Classification Based on Growing Season

Kharif Vegetables

Beans, bhindi, brinjal, bitter gourd, bottle gourd, cowpea, cucumber, muskmelon, chilli, capsicum, pumpkin, ridgegourd, summer squash, sweet potato, smooth gourd, snake gourd, tomato, watermelon, ginger.

Rabi Vegetable

Garden beet, cabbage, cauliflower, carrot, garlic, knolkhol, lettuce, broad bean, methi, onion, spinach beet leaf, radish.

Ziad Vegetable

Dwarf beans, dwarf tomato, cucumber (summer varieties), summer squash.

Classification Based on Soil Reaction (pH)

Slightly Tolerant (pH 6.8-6.0)

Asparagus, garden beet, sprouting broccoli, cabbage, cauliflower, celery, palak, spinach, Chinese cabbage.

Moderately Tolerant (pH 6.8-5.5)

Bean, brussels sprout, carrot, cucumber, brinjal, garlic, radish, knolkhol, parsely.

Very Tolerant (pH 6.8-5.0)

Chicory, endive, potato, rhubarb, sweet potato, watermelon.

Chapter 3

History of Vegetable Seed Production

In earlier times, that is, before world War II no one used to produce vegetable seeds on large scale from the business point of view except production was only for self use. Moreover, vegetables were produced on a very limited scale. Seed growing as business probably started in Europe as early as the first decade of 18th century. In 1719, vegetable seeds were being produced and exported to America from Europe. World War had a global effect on seed industry. Seed supplies were entirely cutoff from European countires. The USA and Canada were called upon to meet their seed requirements through indigenous seed production. After second World War, Jenson, abotanist from Kew, introduced cauliflower in the year 1822 and grew it at Saharanpur (Company Bagh). This all happened during those days, however Agriculture–Horticultural Society, Calcutta, also introduced seeds of English vegetables from South Africa during the year 1924. Later on, Jenson introduced seeds of several English vegetables such as cabbage, cauliflower, knolkhol, radish, turnips, garden beet and tomatoes, etc. from M/S Sutton and Sons, London. In India, however, the seeds of European vegetables were not produced in sufficient quantities and were imported from Europian countries. During 1860-80, the seeds were mostly supplied by M/s Sutton and Sons, Reading,London. It was perhaps the only company, which established its office in India in 1916 at Calcutta. The seed was imported from Europe and distributed in India by the company. This was perhaps the only seed company in India during those days. Thus, before World War II, India depended entirely for these vegetable seeds, on imports from abroad. During World War II, the supplies of seeds were cut-off, hence seed production of temperate vegetable/varieties was started in Quetta during 1942-43. Katrain and Kashmir valley during 1942-43. During this period vegetable seed industry made a rapid progress in comparison to field crops. By 1945, a few private pioneering seedsmen developed temperate vegetable seed production programmes/facilities in Quetta and Kashmir valley. Contacts with foreign seed companies were developed by them to put the Indian vegetable seeds in the International trade. The notable seed companies were M/s Suttons, Pochas, Palekars.

With the partition of the country, supplies of vegetable seeds were cut-off from Quetta and the seed industry of Kashmir was considerably affected by disturbances in1948. Therefore, during 1949, seed production programme was started at vegetable Breeding sub-station at Katrain Kullu valley by Government of India, with a view to standardize the quality seed production techniques.

Immediately after the partition, the Government of India set up a Central Vegetable Breeding Station in Kullu, that is first at Raison, then at Naggar and finally at Katrain. In 1955 the station was transferred to IARI, New Delhi with a view to intensifying the improvement work on temperate vegetables and was renamed as IARI, vegetable Research Station and finally as IARI Regional Station. The systematic breeding work on temperate vegetables was taken up and contributed greatly to the standardization of seed production technology. Indigenous production of temperate vegetable seeds is a unique example of import substitution leading to the development of a viable vegetable seed industry in HP. Till date the station has been making steady progress in the area of vegetable improvement and seed production resulting in development of large number of varieties and multiplication of their seeds to cater to the need of the entire country for breeder and foundation seed especially of temperate vegetables. Now the main mandate of the station is production of quality seeds of temperate vegetables and their distribution among the growers, development of new varieties, standardization of technology.

The CPRI was established in 1949; however, work on potato breeding was started as early as 1934. Now the main center at Shimla has regional and sub-centres in different parts of the country. In HP in the initial years, potato breeding and certification station at Kufri and foundation seed production station at Phagu in Shimla district with the objective of breeding new varieties, multiply their seed, certification before distribution among the various agencies and farmers in HP/country. During the early years, extensive world collections of potato varieties and species were built-up at this station, especially with the cooperation of the commonwealth potato expeditions. The collected germplasm was utilized in breeding programme. As a result the institute which is under the control of the council has developed a series of improved potato varieties replacing the old one in the state/country and also zone/region-wise. This institute has also developed, the seed plot techniques, which makes use of low aphid periods for growing disease free seed stocks of potato in the plains.

In the early sixties trials on sugar beet were started in HP at Solan, Dhaulakuan and Kalpa and Ribba in Kinnaur with the assistance of ICAR All India Coordinated Research Project main Centre Kalpa was sanctioned in 1972-73 on the basis of the exploratory trials. The main objectives of the sugarbeet scheme was to conduct research on seed production/multiplication on cultivators fields to be supplied in other parts of the country near sugarbeet diffusion plants for sugar extraction.

Snowball cauliflower seed production was taken up in Sproon valley in Solan on growers field in 1965 and by 1970 all imports of cauliflower seeds were stopped. The import of the seeds of other temperate vegetables was also stopped. Similarly seed production of garden beet, sugarbeet and chicory was taken up in Kinnaur district of HP. State seed certification Agency came into existence in 1978 and started registering vegetable crops for seeds including chicory and sugarbeet.

During the eighties, the importance of vegetable seed technology research was duly recognized to solve some of the burning problems which were often encountered by seed producers, certification staff at the field level and during seed processing and storage, considering the importance of seed technology research, ICAR started All India Coordinated Research Projects under the National Seed Programme. Seed Technology Research Units have been established in most of the states and SAU's. The role of NSC in this respect is highly appreciable. Presently the public sector is not involved in vegetable breeding, plant pathology and other research disciplines including cultivar assessment and quarantine. It is dependent upon Government Research Institutions and SAU's for the research and extension. The Project Directorate on Vegetable Research carries out the cultivar assessment and the Directorate of Plant Protection Advisor, Quarantine and Storage look after the quarantine. The public sector is involved in operations of multiplications, processing, marketing and to some extent in quality control through interval checks and extension.

For last few years vegetable breeders in India diverted their attention towards the development of F_1 hybrids. The hybrid varieties have been found to be high yielding more tolerant to diseases and pests and suitable to varied agro-climatic situations. A major breakthrough in the past few years has been realized in tomato, muskmelon, cabbage, okra, cucumber, capsicum, chilli, brinjal, carrot, cauliflower and onion. The cost of hybrid seeds is extremely high. There is a need to develop technologies for evolving cheaper vegetable hybrid seeds. Keeping in view the importance of hybrids in vegetables, ICAR has sanctioned a network project on " Promotion of Hybrid Research in Vegetable Crops" with a total of Rs 3.38 crores. The project was operative at 15 centers and covers nine vegetable crops.

Role of Various Agencies in Seed Production

Project Directorate of Vegetable Research (PDVR)

The Project Directorate of Vegetable Research is a constituent institution under the aegis of ICAR. An important milestone in the vegetable research was the establishment of All India Coordinated Vegetable Improvement Project (AICVIP) in 1971 at IARI, New Delhi. The project has been envisaged as a national network of multidisciplinary research on the improvement of the major vegetable crops at seven main centers and 10 sub-centres initially. To give a fillip to the research and to meet the future challenges of the country, the status of AICVIP was elevated to the level of PDVR in 1986 and subsequently it was delinked from IARI and shifted to its present location at Varanasi in 1992 by doing so the ICAR has promoted the establishment of a distinct institutional structure for strengthening.

Table 1: The Important Events in Indian Seed Industries

1928	Royal Commission on Agriculture submitted Report citing that facilities for increasing the supply of breeder seeds were inadequate, and that varietal purity of seeds was not maintained.
1957	All-India Coordinated Maize Improvement Project began.
1960	Similar projects on sorghum and pearl millet began. Training in seed technology started.
1961	Four maize hybrids released.
1963	National Seeds Corporation established.
1964	First sorghum hybrid released.
1965	First pearl millet hybrid released. 250 tonnes of dwarf.
1966	Seeds Act passed (came into force in 1969).
1963	Certified seed production increased from 30 ha to 35,000 ha.
1968-69	Seed review team report submitted.
1969	Tarai Development Corporation established with World Bank assistance primarily to produce certified wheat seed. Concept area development for seed production initiated.
1971	National Commission on Agriculture submitted its report, Government decided to recognize seed production in India. The Indian Society of Seed Technology (ISST) was born.
1971	ISST organized first All-India Seed Seminar.
1975	Project Report on National Seeds Programme (NSP) was submitted.
1981	First workshop on seed technology under NSP was held. Certified seed production increased from 137000 tonnes in 1974-75 to about 350000 tonnes in 1980-81.
1983	Seed Control Order envisage-constituted in October, the quality claimed on the label.
1987	An Expert Group on seed was constituted in October, the report was submitted in June, 1989.
1988	A new Seed Policy to encourage import of seed and planting material was announced in September
1989	Provision of major incentive and more liberation policy for vegetable seed industry.

National Bureau of Plant Genetic Resources (NBPGR)

It is one of the most important organizations centrally located and has close linkage within the country and the outside world regarding the procurement, supply and maintenance of germplasm of almost all vegetables. It has got a network of collection, evaluation, identification and malignance of the germplasm at all its centres. However, immediately after coming to existence the PDVR made an exhaustive effort in collaboration with other agencies mainly with NBPGR, in collecting the rich vegetable germplasms of our country. In this regard the cooperation and assistance extended by the NBPGR and other agencies, enabled PDVR to collect and maintain the germplasm of the following important vegetables and vegetable seed research in the country. At present the PDVR has seven main centres, 18 sub-centres and 31 voluntary centres to undertake the coordinated research project. At Varanasi head quarter, it has the activities of research seed production as well as coordinated unit.

Statutory Bodies and Agencies

1. Central Seed Committee (CBC)
2. Central Seed Certification Board (CSCB)
3. State Seed Certification Agencies (SSCA)
4. Central Seed Testing Laboratory (CSTL)
5. State Seed Testing Laboratories (SSTL)
6. Appellate Authority (AA)
7. Recognition of seed certification agencies of foreign countries
    (*a*) International Seed Testing Association (ISTA)
    (*b*) Association of Official Seed Analysis (AOSA)
    (*c*) The Society of Commercial Seed Technologist (SCST)
    (*d*) The Commercial Seed Analysts Association of Canada (CSAAC)

Financing Organization (Public Sector)

1. Central/State governments
2. Central/State departments of agriculture/horticulture
3. Indian Council of Agricultural Research
4. National Horticultural Board

5. National Agricultural Research Projects
6. All India Coordinated Research Project
7. National Agricultural Cooperative and Marketing Federation Ltd (NAFED)
8. National Cooperative Development Corporation (NCDC)
9. National Bank for Agriculture and Rural Development (NABARD)
10. NDOB
11. NERAMAC
12. Tarai Development Corporation Ltd (TDC)

Societies

1. Lahaul Potato Growers Cooperative Marketing Society Ltd. Manali
2. Indian Society of Seed Technology (ISST)
3. Indian Society of Vegetable Science (IAVS)

Universities

1. Sher-e-Kashmir University of Agricultural Sciences and Technology, Srinagar, Kashmir
2. Himachal Pradesh Krishi Vishvavidyalaya, Palampur (HP)
3. Dr. Y.S. Parmar University of Horticulture and Forestry, Solan (HP)
4. Himachal Pradesh University, Simla
5. Assam Agricultural University, Jorhat, Assam
6. North-East Hill University, Shillong, Meghalaya.

Institutes

1. IARI, Regional Vegetable Research Station, Katrain district Kullu, HP
2. Vivekananda Parvatiya Krishi Anusandhanshala, Almora, UP Hills
3. ICAR Research Complex for NEH Region, Shillong, Meghalaya with its centres in all the seven states of NEH Region

4. CPRI, Shimla, (HP) and its Regional Stations
   (*a*) CPRS, Kufri/Fagu, (HP)
   (*b*) CPRS, Darjeeling, West Bengal
   (*c*) CPRS, Shillong, Meghalaya
   (*d*) CPRS, Mukteshwar, UP

Semi-government Organizations

1. National Seeds Corporation Ltd.
2. Associated Agriculture Development Organizations
3. State Seed Corporation in each hill state

Various State Government Pilot Projects

1. Tribal Area Development Programmes in Lahaul and Spiti and Kinnaur
2. HPMC
3. SFDA in Sirmur district
4. Indo-German Project in Mandi and Kangra districts
5. Many others

Private Sector

1. Maharashtra Hybrid Co.
2. Century Seeds
3. Durga Seeds Co.
4. Sungrow Seeds Co.
5. Sutton and Sons
6. Pocha's Seed Co., etc.

Others

1. Ranichauri–GB Pant University of Agriculture and Technology, Regional Station, (UP)
2. Tejpur–Defence Research Laboratory, Assam
3. Ladakh–Defence Vegetable Farm, Ladakh

The various research schemes, project and voluntary centres are operating in the hilly areas sanctioned and funded by ICAR from time to time both for long and short period basis.

ICAR Funded Centres

1. Shalimar–Srinagar (Shere-Kashmir University of Agricultural Sciences and Technology)
2. Solan–(Y.S. Parmar University of Horticulture and Forestry)
3. Katrain (IARI, Regional Vegetable Research Station)

Voluntary Centres

1. Palampur-Himachal Pradesh Krishi Vishvavidyalaya, (HP)
2. Almora-Vivekananda Parvatiya Krishi Anusandhan-shala, UP
3. Shillong–ICAR Research Complex for NEH Region
4. Ranichauri–GB Pant University of Agriculture and Technology, Regional Station, (UP)
5. Tejpur–Defence Research Laboratory, Assam

National Seeds Corporation

The National Seeds Corporation Limited (NSC) is a Government of India undertaking under the administrative control of the Ministry of Agriculture and Irrigation, Government of India. It was set up in March 1963 and started functioning from July, 1963. The NSC was established to organize the development of a sound seed industry in India. The main objectives, interalia are:

1. To carry on in India the production of and to enter into contracts for distribution and selling of agricultural seeds including vegetables.
2. To carry on the business of establishing seed farms, and
3. To undertaken by inspection and other means, seed quality control measures in all phases of the seed business carried on by, on behalf of, or in cooperation with the corporation.

NSC has pioneered the seed industry. Best expertise is available with the corporation in the following fields:

1. Foundation seed production
2. Certified seed production
3. Custom service production
4. Processing
5. Quality control
6. Marketing
7. Information, communication, advertising and public relations
8. Training, and
9. Project formation and project management

NSC handles high quality certified seed of almost all vegetable Varieties/Hybrids grown when a farmer receives his bag of NSC quality seed, the label and seal on each seed bag assures him that he has just brought the best available vegetable variety seed. With the launching of the National Seed Project the NSC has assumed the role of a leader to develop the seed industry on sound lines. Specifically the NSC is contributing towards the state seed corporations share capital. NSC coordinates certified seed production programmes of several state seed corporations assesses seed demand and is responsible for interstate marketing of certified seed. It also conducts biennial surveys of seed demand, plans and organizes the production of foundation seed and breeder seed in consultation with ICAR. It provides technical assistance service to state seed corporations and private sector agencies, coordinates market research and sales promotion efforts, provides training facilities for staff participating in seed industry development, operates the reserve stock scheme, provides certification services for states not having established independent seed certification agencies and produces vegetable seeds for local and export market.

Major Conditions of Seed Production Under NSC

1. The reserved area shall be so located that the seed crop in that seed crop in that area will conform to the isolation requirements prescribed in the Indian Minimum Seed Certification Standards-March, 1971, as amended from time to time and or the NSC's prescribed standards as indicated in the annexure failing which the seed crop shall be liable to be rejected in part or full as may be necessary.

2. The NSC shall procure processed seeds conforming to the standards referred to at the fixed rate on rupees per quintal if seed is processed in the NSC's plant or in a plant designated by it and at the rate in rupees per quintal if seed is processed, with NSC's prior approval, in a different plant.
3. The procurement price shall be deemed to be an all-inclusive price and no other claim including tax if any, leviable shall be entertained. The NSC shall be entitled to recover the amount due from the amount payable to the grower.
4. For seed production of notified kinds/varieties, in regions where the NSC is the certification agency, the grower shall submit to the NSC an application for certification as prescribed in the Seeds Rules, 1968 as amended from time to time and remit Rs 25 by cash/MO/DD per application, as application fee at least 15 days prior to sowing. The application fee is not refundable. One application can include the entire area of all the seed crops of notified kinds/varieties grown in a season by one grower.
5. Seeds to be used for sowing in the reserved area shall invariably be obtained from the NSC only irrespective of whether or not the NSC is the certification agency in the region where the reserved area lies.
6. The grower shall pay to the NSC the seed cost-plus sales tax if applicable in advance in Rs per unit along with packing and forwarding charges. The cost of seed is not refundable except under the provision contained in agreement form.
7. If the seed obtained from the NSC is sown in different locations separated by 50 metres or more between them, prescribed inspection fee shall be charged for each location and any additional amount which will thus, become due should be paid before the first inspection. For sowings made with the NSC's, prior approval on areas other than or over and above the reserved area prescribed inspection fee should be paid before the first inspection. If use of carry-over seed with the grower is permitted the grower shall

remit the inspection fee at the prescribed rate before executing the agreement or before the first inspection.

8. In regions where the NSC is the certification agency the grower shall render all facilities to the NSC for conducting field inspection of the seed crops at a time and at any stage. A seed crop found not conforming to the standards, by the NSC shall be liable for rejection.
9. The grower shall arrange to transport in his gunny bags the harvested and threshed seed material to the processing plant assigned by the NSC or the state seed certification agency, as the case may be, or shall pay to the NSC the prevailing charges of transport, if the same is arranged by the NSC.
10. For analysis of seed samples the NSC will levy a fee of Rs 10 per sample irrespective of the nature of results of analysis except in respect of vegetable seeds produced in Jammu and Kashmir and Himachal Pradesh for which no testing fee is recoverable. No testing fee will be levied for vegetable seed except when the value of the lot concerned is Rs 1000 or more.

Need of the Seed Industry

In order to increase the production of vegetable seeds for home as well as export market, it is highly necessary that the seeds must be of the best possible quality, a high degree of genetic and physical purity and free from diseases. The prices of the seeds must be competitive, particularly in export market, hence, must be produced as cheaply as possible. In order to achieve these objectives the following facilities were created in the country:

1. Research on development of superior varieties and F_1 hybrids to cater to the needs of export market and also for home market.
2. Facilities for conducting research on important aspects of vegetable seed production such as minimum isolation distances, studies on the crossing of seed crops with local weeds, agronomic requirements, seed borne diseases and seed technology aspects concerning threshing, drying, processing, packing and storage of seeds.

3. Facilities for maintenance and purification of nucleus and stock seeds of varieties.
4. Facilities of isolation zones in case of highly cross-pollinated vegetable crops.
5. Creation of necessary facilities for processing, drying, storage and safe transport of vegetable seeds.
6. Creating agency for market survey for assessing the demand within the country and also for export.
7. Facilities for survey of prospective buyer countries for their demand of seeds of specific varieties.
8. Agency for strict quality control, particularly for the seeds to be produced for export market.
9. Training facilities for seed producers concerning seed production, processing, storage and marketing.

National Seeds Corporation has initiated the temperate vegetable seed in various crops such as cauliflower, chicory, sugar beet, cabbage, radish and turnip of temperate types in HP and Srinagar valley.

Chapter 4

Climatic Factors Affecting Vegetable Seed Production

Before seeds can be produced from vegetables it is necessary for the crop to flower. Flowering is required by the plant for sexual reproduction and maintenance of generation. The process of flowering is complex. Some plant species pass from vegetative phase to the reproductive phase with special requirement or stimulus, whereas in others such stimulus is not required. Species which have a special physiological requirement to pass from the vegetative phase to reproductive phase are generally either dependent on day length (photoperiod) or have a low temperature requirement (vernalization). Not only photoperiod and temperature but some other climatic factors, *e.g.* rainfall, wind etc. are also responsible for best yield of high quality seeds. The major environmental factors influencing vegetable seed production are :

1. Photoperiod
2. Temperature

3. Rainfall
4. Wind

Photoperiod

Photoperiod influences vegetable seed production by affecting photosynthesis and day length. Within the limits of each crop plant, the higher the light intensity, the higher will be the rate of photosynthesis required to manufacture the important constituents of the seed.

Different crop plants have different requirements of photoperiodism. According to Thomson (1979) the transition from vegetative to the reproductive phase in some vegetable crop plants can occur only at the season when the days are of particular length, and if the plants are kept in the wrong day length, they remain in vegetative phase for longer time. Crop species of temperate regions tend to flower in the long days of summer while tropical crop species require shorter days. Plants can be classified into three main groups according to the specific duration of light and dark requirement in each 24 hour period or cycle in order to initiate flower, *viz.* short-day plants, long-day plants and day-neutral plants.

Short-day Plants

This group includes species which will not flower unless the day light period is shorter than a particular critical time, which is between 10 and 12 hours, for example amaranthus sps, soybean, pepper (some varieties), kidney bean.

Long-day Plants

These include plants, which will flower only when the light period is greater than a critical time. The critical photoperiod for most long-day plants is between 12 and 14 hours, for example, spinach, radish, cabbage, cauliflower, broccoli, turnip etc.

Day-neutral Plants

This group does not have a specific day length requirement for flowering, for example, tomato, brinjal, lettuce, cucurbits, carrot etc.

Temperature

Some vegetable species do not initiate flowers until the plant has received a cold stimulus. The requirement of cold temperature stimulus for flowering is called vernalization, *e.g.* cabbage, Brussel's sprout, beet, biennial radish, carrot, onion etc. Depending upon the vernalization requirement, vegetable crops are grouped in to the following three classes :

Annual

Vegetable species which do not have vernalization requirements for flower initiation, such crops produce flower and seed in the first year, *e.g.* tropical radish, Batisak (*Brassica chinensis*).

Biennial

Biennial plants tend to remain in vegetative stage in the first year of growth and they flower and produce seed in the second year, *e.g.* beet, carrot, cabbage and biennial radish etc. Most of the biennial plants require vernalization for flower initiation.

Perennial

Perennial plants survive for several years producing flowers and seeds each year. Most vegetables are, however, annuals and biennials.

Temperature significantly influences the transition from the vegetative to the reproductive phase of the crop species having specific critical temperature requirements for flowering. Not only the flowering, the prevailing temperature during growth and developmental phases of plant determine the final seed yield.

Rainfall

An appropriate balance between sufficient rainfall for crop growth and establishment and sufficiently dry conditions for satisfactory pollination and seed ripening is the most important factor in the vegetable seed production. Sufficient soil moisture must be ensured during flowering stage when the crop is moisture sensitive, owing the reduced root growth. Seed viability can be seriously affected by high rainfall during the seed ripening period.

Wind

Excessive wind increases water loss from the crop and soil, prevents maximum activity of pollinating insects, carries wind-borne pollen over long distances and increases loss of seed by enhancing shattering during seed ripening (George, 1980). Strong winds during the reproductive phase can cause severe crop losses through lodging, shattering and shedding of seed. On the other hand gentle wind facilitates increased pollination in cross-pollinated crops.

Chapter 5

Botany and Floral Biology of Vegetable Crops

Family–Solanaceae

Brinjal

Plants are polymorphous, erect, aculeate or unarmed herb, woody at the base and 0.5-1.5 m in height. The crown is broad. Branches are light green or tinged with violet, rather thick, densely clothed with grey and stellate tomentum. Leaves are alternate, solitary or in pairs, on shorter or longer stalks, simple, ovate or oblong-ovate and elleptic. Base is often unequalsided and oblique, rounded, truncate or cordate, apex triangular, acuminate, obtuse or acute, irregularly undulate, angular or more or less deeply lobed, above dark green, on both sides dull, densely clothed with soft, grey, stellate tomentum, flaccid, 7-25 cm long, 5-15 cm wide and reticulate-veined. Veins are transparent. Petiole is robust, subterete, yellowish green, densely white stellate tomentose and 1.5-10 cm long.

Figure 6: Brinjal Plant with Fruits

Figure 7: Flowers of Brinjal

Flowers are rather large, stalked, lateral or leaf-opposed, the female ones solitary, the males in many-flowered inflorences. Pedicel is light green or strongly tinged with violet, terete, aculeate or not, densely stellate tomentose, patent or curved and 1.5-2 cm long. Calyx is tubular-campanulate, 5-lobed and less than half way down, on the outside greyish green, often strongly tinged with purple, on both sides densely stellate-tomentose and 1.5-2 cm long. Tube is ovoid-oblong, obtusely pentagonous, + 1 cm long and 0.5-0.7 cm diameter; segments narrowly oblong, acuminate, obtuse, obliquely erect and 0.5-1.5 cm long. Corolla is gamopetalous, deeply 5-lobed, stellately spreading, on the outside light violet, densely white stellate tomentose, within dark violet, glabrous, rugose and 2.5-4 cm diameter. Lobes are short, ovate-oblong and acute. Stamens are 5 erect, ovoid-oblong, vitelline, opening at the top by two pores and 0.5 cm long. Berries are most variable as to shape, colour and dimensions, globose or oblong, white, yellow or more or less intensely violet, glabrous, 2.5-15 cm long and 2.5-4 cm diameter. Fruiting calyx is campanulate, much enlarged, bursting on one or two sides, pale green or tinged with purple, thinly stellate tomentose, unarmed or beset with scattered prickles, 3-6 cm long, 3-5 cm diameter, fruiting pedicel much elongated and thickened towards the apex, unarmed or with scattered small prickles, pale green, nodding and 3-7 cm long. Seed is numerous.

Though brinjal is self-pollinated crop, but there is high degree of cross-pollination due to heteromorphic flower structure (called as heterostyly). The extent of cross-pollination has been reported as high as 29 per cent and hence, it is classified as often cross-pollinated

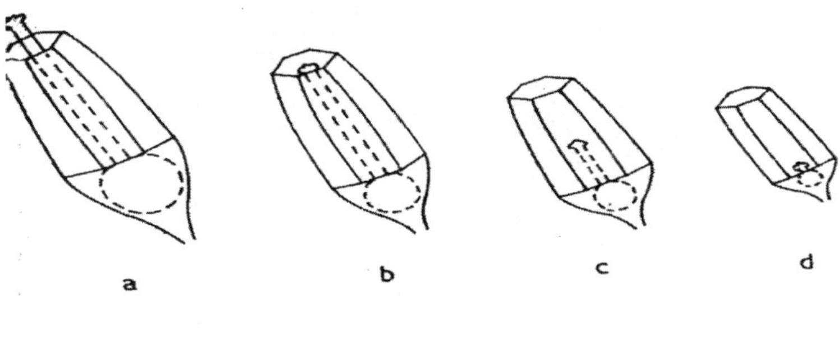

Figure 8: Types of Flowers in Brinjal: (a) long styled (b) medium styled (c) short styled and (d) pseudo short styled.

or facultative cross–pollinator. There are four types of flowers in brinjal like (a) long styled (b) medium styled (c) short styled and (d) pseudo short styled. Fruit setting occurs only in long and medium styled flowers. Flowers generally emerge 40-50 days after transplanting. Anthesis occurs at about 6-8 AM in August–September and usually between 9.30 to 11.15 AM during winter (December–January). Peak time for anthesis is 8.30 to 10.30 AM. The pollen dehiscence starts from 9.30 to 10.00 AM. It is observed that the daylight, temperature and humidity mainly influence anthesis and dehiscence. The receptivity of stigma can be observed from the pump and sticky appearance, which gradually turn brown with the loss of receptivity. Stigma receptivity is highest during anthesis i. e. flower opening. The period of effective receptivity ranges from a day prior to flower opening until about four days after opening. Pollen is most fertile immediately after the anther dehiscence. Opening of anthers is mostly by a pore or slit at or near the apex. Pollen usually remains viable for a day during summer and 2-3 days in winter under field conditions. According to Deshpande *et al.* (1978) maximum anthesis occurs at 6 AM in Arka Shirish and at 8 AM in Arka Sheel and Arka Kusumaker.

Tomato

Tomato produces a strong taproot but very often it is damaged at the time of transplanting, consequently, the lateral adventitious roots develop profusely. The adventitious roots develop at the rate of about 33 mm per day penetrating one-meter depth and attaining

Figure 9: Tomato Fruits in Plant

Figure 10: Flower Cluster of Tomato

a length of 1.5 meter (Ward, 1964). However, the root development largely depends upon the cultivar, the growing media and the moisture in it. The stem is soft, brittle and hairy when young and hard, woody, and conspicuously branched when mature. Its organization is in the sequence of two or three leaved sympodia. It is erect to semi-erect. Epidermal hairs with oil glands are present at the early plant growth stage. The leaf is alternate, petiolate and 2.15 inch wide. The leaflets are unequal odd-pinnate. The apex is narrowed or acuminate, acute and irregularly serrate. The number of leaflets varies from four to several and the length of leaf from 6 to 12 inch on the leaves. Hairs or trichomes are formed by the outer growth of the epidermal cells.

The hairs may be unicellular or may become multicellular by subsequent cell division. Inflorescence is the extra-axillary cymes with dicotomous or polycotomous branching. It is lateral and arises from the stem opposite to the leaf. It may be terminal or may continue with the vegetative shoot (Muller, 1940). The number of flowers per cluster varies from three to several. The early flowers that open in the cluster are large-sized as in typical cymes. The abscission of the flowers may take place due to high or low temperature or some other internal and external factors. The flower drop is due to the development of the abscission layer at the mid point of the pedicel. The plant produces bright yellow flowers. The flowers are pentamerous, bisexual, regular, complete, ebracteate, and hypogynous. The lateral section of the flower shows the details of sepals, petals, stamens, style ovary and ovules. The pistil has two to several carpels. The anther connate appearing in the throat of the corolla and they dehisce through the poricidal dehiscence. Usually, the style is shorter has anther cane. Therefore, a high degree of self–pollination. The fruit is soft berry, glandular hairs and glands are usually present on the small fruit, which degenerate in the advanced stage. The fruits of some varieties or species do not have glandular hair and glands. The seed is oval flattened buff to pale brown, 3-5 mm in length. It is derived from the anatrosperm are enclosed in the seed. The anthesis of flower starts in the morning at about 6 AM with maximum flower opening in the late morning. The maximum amount of anther dehiscence takes place from 8 to 11AM, depending upon the season. The stigma becomes receptive 16 hours before anthesis and remains thus for five days after anthesis. Pollen grain germination takes place approximately two hrs after pollen has been applied on the stigma. Ovules ripen over two days. Pollen can be stored upto three days under ordinary conditions. At 0 ^{0}C, it can store for 2–3 months. While, it remains viable for six months at 5 ^{0}C in a desicator. Pollen from one days-stored flower provides best seed set. The optimum temperature for pollen germination is 18–25 ^{0}C. Tomato pollen germinates best in a liquid solution of 20 per cent sucrose and 10 ppm boron.

Chilli

Plants are perennial, widely branched, erect and woody at the base. Plant height reaches up to 1.5 meters. Leaves are solitary or 2-3 close together, very unequal having shorter or larger stalks, broadly ovate or oblong-lanceolate, base broad and abruptly contracted into the petiole. The apex is narrowed or acuminate, acute or obtuse, coarsely crenate or entire, shortly elliate, light green, feebly shinning, sparingly hairy, especially beneath, 1.5-10 cm long, 0.5-5.0 cm wide and having 0.5-3.5 cm long petiole. Flowers are small, solitary in the leaf-axils or 2-3 or more together at the tip of branches and long pedicelled. Pedicels are erect, nodding at the tip, light green shortly pubescent, terete and 1.5-2.5 cm long. Calyx is minute campanulate, minutely abtusely 5-toothed or subtruncate, yellowish green, glabrous and not or hardy enlarged under the fruit. Corolla is gamopetalous, patent, flat, yellowish or greyish white deeply 5-partite and 0.5-1.0 cm in diameter. Segments are ovate-oblong, acute, 0.3-

Figure 11: Fruits of Chilli in Plant

0.35 cm long and 0.2-0.25 cm wide. Stamens are five and erect. Anthers are bronze-green. Berries are small, narrow, variable as to length, usually erect, rarely pendulous, long stalked, ovate-oblong, narrowed upwards obtuse, glabrous, when ripe shinning blood red, orange coloured or yellowish white many-seeded, 0.75-2.5 cm long and 0.3-1 cm in diameter.

Figure 12: Chilli Flower

The extent of natural crossing in chilli is reported to be about 16.5 per cent. Flowers remain open for 2-3 days. The flowering begins 1-2 months after planting and it takes again one month for fruiting. The flower opening takes place in the morning 5-6 AM. Anther dehiscence takes place from 9 to 11 AM. However, flower opening and anther dehiscence to a large extent depends upon the weather conditions. During cold as well as cloudy days, the opening of the flower is delayed. Maximum fruit setting takes place when pollination is done at the time of opening of flower. Since anthesis is completed by 8.00 AM and anther dehiscence starts from 9 AM, emasculation and pollination can be done simultaneously.

Bell Pepper

It is dicotyledonous plant behaving as a perennial in tropics and annual in temperate regions. Leaves are solitary or 2-3 close together, very variable as to size, on shorter or longer petioles, ovate-oblong or elliptic-lanceolate. Base is acute, decurrent along the petiole. Apex is narrowed, acute, coarsely crenate or entire, thin,

Figure 13: Fruits of Capsicum in Plant

above dark green, beneath yellowish green, on both sides dull and feebly shining, barring the nerves almost glabrous, 1.5-12 cm long and 0. 7-5 cm wide. Petiole is 0.5-2.5 cm long. Flowers are solitary in the leaf-axils, stalks long, patent, nodding or pendulous at the apex or rarely erect, yellowish green, shortly pubescent, longitudinally furrowed and 1-1.5 cm long. Calyx is small, campanulate, shortly 5-dentate, green glabrous, 0.2-0.25 cm long and much enlarged under the fruit. Corolla is gamopetalous, campanulate, white or yellowish white, rarely violet, deeply, 5-partite and 1-1.5 cm in diameter. Segments are spreading, oblong, acute, 0.5-0.7 cm long and 0.3-0.4 cm wide. Stamens are five, erect, anthers lilac or violet, afterwards turning bronze-green. Fruit shape may be pointed, blunt or shrunken at the blossom end. Ripe fruits are shining blood red, violet, orange coloured or even green and many seeded. In bell pepper, anthesis starts at 7.15 AM and continues up to 11.15 AM with the peak at 7.15 (Vijay *et al.*, 1979). Anther dehiscence occurs after 30 minutes of anthesis. The stigma becomes receptive from the day of anthesis and remains receptive up to two days after anthesis. Fruit set (80 per cent) has been found maximum on the day of anthesis and pollen

Figure 14: Capsicum Flower

fertility around 96 per cent. However, it has been observed that there is considerable variability exists for time of anthesis and dehiscence. Fruit set in flower buds pollinated a day prior to anthesis has been observed to be good. The extent of natural crossing in peppers has been reported to vary from 7.6 to 36.8 per cent in the field with a mean of 16.5 per cent. The extent of natural cross pollination is estimated by planting a variety having recessive characters with another are having counterpart dominant traits. The seed set on recessive parent are grown and crossed ones are identified by the dominant phenotype. In an open field, the minimum safe isolation distance for bell pepper is estimated at 600 feet.

Potato

The potato is perennial but as a crop, it is treated as an annual. It is vegetatively propagated by means of tubers. The tuber is an

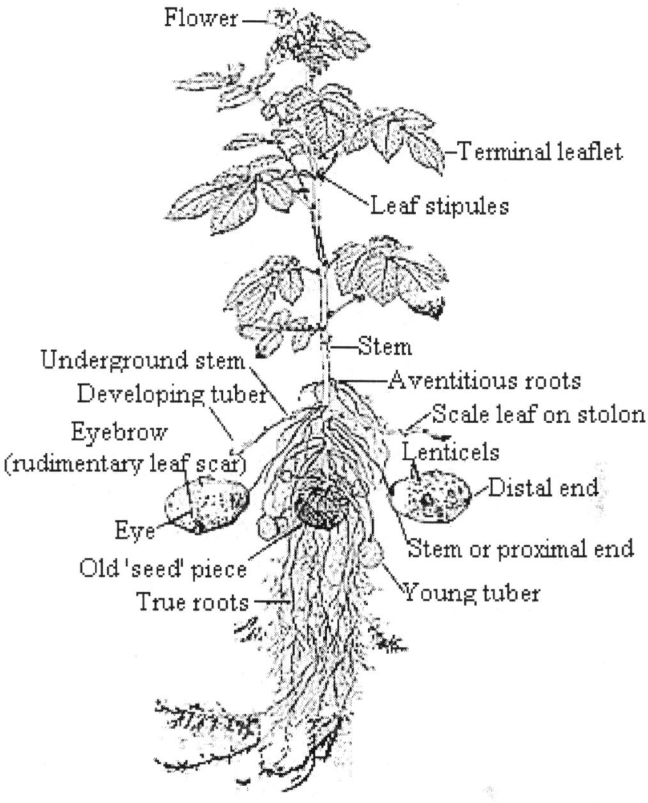

Figure 15: Potato Plant

enlarged underground stem produced on the end of a stolon and not on the roots proper. These tubers are morphologically stems. They possess 'eyes' that is bud. The tubers also contain lenticels (respiratory structure) like stems of other plants. The upper part of the sprout develops into the aerial stem. The aerial stems are round or angular, pubescent or glabrous, green or purplish. The stem is of a branching type. It is erect when young, and spreads as the plant grows. The leaves of the potato are alternate and compound (occasionally bi-compound). The leaves arise along the stem in a spiral arrangement. There are three to four pairs of leaflets arising in succession along the rachis. The leaflets are more or less opposite. There is a large terminal old leaflet. The root of potato is adventitious,

Figure 16: Potato Flowers, Leaves and Tubers

arising from the base of a sprout. The root growth is usually restricted to top layers, at a depth of about 20-25 cm from soil surface. In rich soils roots of some varieties may reach up to a depth of 90-100 cm. The flowers of the potato plant are in terminal clusters.

Each flower normally has five stamens, two-celled pistil, five sepals and five petals united for about half their length. Most varieties of potato bear infertile pollens and hence fruits or berries are not generally formed. In some of the varieties fruits or berries are formed. One inflorescence can contain a variable number of flowers (1 to 40) depending on the cultivar and time of the flowering season. Usually there are more flowers in a cluster but maintained six to eight flowers are recommended. Five or six well-developed fruits from a successful cross will give an abundance of seed. Berries are too numerous, they do not grow evenly. The stigma is receptive 1-2 days prior to dehiscence; therefore, the best time for pollination is just before the flower buds open, or shortly thereafter. Older flowers may become self-pollinated. Buds that are too young or too small are more difficult to manipulate before pollination, the anther need to be removed and the corolla trimmed to facilitate pollen application to the stigma, however, if pollination is done early enough, anther removal is not necessary because the numerous germinating pollen grains of the first pollination prevent additional fertilization from taking place. Pollen should be shaken and collected from flowers of the male parent plants, and this can be done by hand or with a mechanical device that will vibrate the anthers. Most pollen is shed one-three days after flower opening. It is convient to palce the pollen in a capsule or glass tube at a depth of 5 mm from the top so that the stigma can be easily dipped in it. Special care should be taken to pollinate abundantly, since the number of seeds in a berry (one fruit can contain up to 300) can be limited by the number of pollen grains applied. The fresh pollen should be used, maintained at 2-5 ^{0}C up to one month, or stored in a freez (-20 ^{0}C). Desiccated pollen can be stored at–20 ^{0}C for at least one year.

The pollinated inflorescences should be tagged. The fruit are ready for harvest 3-4 weeks after pollination. Up to that time, they usually stick firmly on the plant and do not require bagging, if fruits are left on for a longer period, they should be bagged in case of fruit drop. One-month-old fruits can be placed in a cool room to mature further, turning soft and possibly yellowish. However, the seed may be extracted as early as one month the time of pollination. A convient way to collect the seed is to squeeze a berry in warm water and decant the water, pulp and undeveloped seed. Plump, viable seeds are more dense and sink to the bottom of the container. Following seed extraction, the dried seeds can be treated with gibberellic acid (1500-2000 ppm) for 48 hours breaking dormancy. Fresh seeds are dormant for several months. Seed that is stored at low humidity and room temperature can retain viability for at least 10 years.

Family–Malvaceae

Okra

Plant is an erect, herbaceous annual and 1-2 meter tall. Stem is green or with reddish tinge. Leaves are alternate, broadly cordate, palmately 3-7 lobed, hirsute and serrate. Flowers are solitary, axillary with about 2 cm long peduncle. Epicalyx is up to 10, narrow hairy bracteoles that fall before the fruit reaches maturity. The calyx splits longitudinally as flower open. Petals are 5, yellow with crimson spot on claw and 5-7 cm long. Staminal column is united to the base of petals with numerous stamens. Ovary is superior, stigma 5-9 with deep red. Fruit is a capsule, light green, or sometimes red in colour pyramidal oblong, beaked, longitudinally narrowed, 10-30 cm long and dehiscing longitudinally when ripe. Seeds are green to dark brown rounded.

The flower bud appears in the axil of each leaf above 6th to 8th leaf depending upon the cultivar. The crown of the stem at this time bears 3-4 underdeveloped flowers but later on during the period of

Figure 17: Okra Plant with Flower and Pod

profuse flowering of the plant there may be as many as 10 under developed flowers on a single crown. As the stem elongates, the lower most flower buds open into flowers. There may be a period of 2-3 or more days between the times of development of each flower but never does more than one flower appear on a single stem. The buds measuring 2.82 x 1.55 cm open into flowers (Purewal and Randhawa, 1947). Flower bud initiation takes place 22-26 days after sowing. The first flowering occurs 20-22 days after initiation. Flowering continues for 40-60 days. Edible fruit is formed with in 3-6 days of flower opening. Stigma is the most receptive on the day of flower opening, anther dehiscence and dusting of the stigma with pollen being almost simultaneous. Anthesis is observed between 9 and 10 hr and is preceded by maximum anther dehiscence between 8 and 9 hr. The stigma remains receptive on day of anthesis. Protogyny has been reported in okra, which is mainly responsible for allogamy of the crop. According to Srivastava (1982), second to fourth positioned flowers carry more anthers (63 anthers/flower) while only 26 in the first flower and 10 in the 10th flower. Anther dehiscence commences 15-20 minutes after anthesis and is completed in 10-15 minutes. The dehiscence on anthers is transverse

(Purewal and Randhawa, 1947). Pollen out put/anther also differs by flower position. The flower numbers two to four are the best producers (Srivastava, 1982).

Pollen grains of okra are polysiphonous (Nair *et al.*, 1974). All the tubes except one stop elongation and that one is the functional tube. In other tubes, the protoplasmic activity is arrested by the formation of coagulated protoplasmic plug (callose plugs) at the tips. The middle part of the staminal column from the flower collected in the middle phase of flowering possesses the most balanced pollen biomass and therefore, the most desired for breeding. (Srivastava, 1982). The pollen remains viable in storage at 50 per cent RH for 55 days (Dubey and Singh, 1968). Pollen fertility is maximum during the period between one hr. before and one hr. after opening of the flower. It takes two–six hr. for fertilization after pollination. Studies on influence of ovule position on pollen tube entry and seed set revealed those ovule positions 7 to 14 received the maximum number of pollen tubes *i.e.* about 90 per cent (Chandra and Bhatnagar, 1975). It is basically a self–pollinated crop, however, cross–pollination through insects can be as high as 19 per cent, and hence, it is classified as often cross–pollinated vegetable crop.

Family–Cruciferae

Radish

Radish is an annual or biennial herb depending on types. The rosette leaves are lyrate-pinnatified herb and may vary from 10 to 15 cm in some small-rooted cultivars to as much as 45 cm long in some large rooted cultivars. The leaves are usually covered with stiff bristles. The edible portion of radish root develops from both the primary root and the hypocotyl. Radish roots vary greatly in size, shape and other external characters, as well as the length depending on cultivars. The shape varies from oblate to long tapering and the exterior colour may be white or various shades of scarlet. Some of the red cultivars are white tipped. However, the flesh colour of all cultivars is white.

The inflorescence of radish is a typical terminal raceme of cruciferae. The flowers are small, white, rose or lilac in colour with purple veins in bractless racemes. Sepals are erect and clawed petals. Fruit indehiscent, 3-7 cm long and upto 1.5 cm in diameter with 6-12 seeds and long conical beak. Radish seeds are globose, about 3 mm

Figure 18: Radish Roots

Figure 19: Radish Flowers and Pods

in diameter. Radish seed when mature, is at first yellowish but turning reddish brown with age. There will be 120 seeds in 1 g of seeds. The seed take 7-14 days for germination under optimum environmental conditions and remain viable for 5 years.

Cauliflower

The stem of the vegetative cauliflower plant is rather short, and it thickness to about the some extent that of cabbage. The leaves are large, generally oblong, the younger ones being nearly always sessile. In contrast to situation in other crops, buds do not usually arise in the leaf axils. During transition to the generative phase, which, in contrast to the situation in most other cole crops, is accomplished in

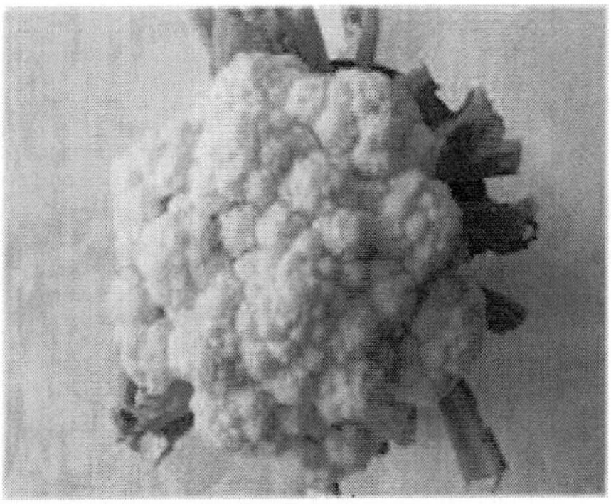

Figure 20: Cauliflower Curd and its Inflorenscence

many types of cauliflower in the first year, the peduncles in the axils of the bracts formed by the main growing point branch repeatedly, so that branches even of the fifth order can arise. At first the numerous peduncles do not grow length wise but become thick and fleshy. Thus, a colourless, roughly spherical, terminal compact head arises of which the upper surface consists of vast numbers of naked apical meristems. The peduncles are composed of thin walled parenchyma and vascular bundles, which do not become woody. The young curd is at first entirely covered by the foliage, on becoming visible it already has a diameter of over 5 cm after sometimes the flower stems elongate and a number of the apices develop into normal flowers. The inflorescence may attain a length of 1-2 m, but the slender pedicels are only 1.5–2 cm long. There are 4 sepals, 4 petals, 6 stamens and two carpals. The carpel form a superior ovary with false septum and two rows of campylotoropus ovules. The androecium is tetradynamous *i.e.* there are two short and four long stamens. The pollen grains are 30-40 u in diameter and have three germinating pores. The bright yellow petals become 15-25 mm long about 10 mm wide. The sepals are erect. The buds open under the pressure of the rapidly growing petals. This process starts in the after noon, and usually the flowers become fully expanded during the following morning. The anthers open a few hours later, the flowers being slightly protogynous. Insects, particularly bees, which collect pollen and nectar, pollinate the flowers. Two nectarines situated between the basis of the short stamens and the ovary secretes the nectar. Situated out side the basis of the pairs of long stamens is also two nectarines, but these are not active. Flowers are borne in racemes on the main stem and its branches. The fruits are glabrous siliques, 4-5 mm wide and sometimes over 10 cm long, with two rows of seeds lying along the edges of the replum (false septum, an outgrowth of the placentae). A silique contains from ten to thirty seeds. Three to four weeks after the opening of the flower from which it is formed, the silique two valves breaking away from below upwards, leaving the seeds attached to the placentas.

Anthers are being emasculated by physically removing the anthers for controlled hybridization. Anthers rarely dehisce before the flower is fully open and emasculation can safely be left until a day before anthesis, even though the stigma may be level with the anthers within the flower bud.

Cabbage

Plants are polymorphous, annual or biennial, erect, 30-150 cm high and much branched in the upper part. Stem is obtusely quadrangular or subterete, pruinose and glabrous. Leaves are very variable at the base of the stem 2-10 cm long, petioled, lyrate or obovate, subentire or undulate, more or less deeply and irregularly lobed, terminal to orbicular or ovate, apex rounded, higher leaves mostly sessile, semi-amplexicaul, ovate-oblong or linear-lanceolate, entire or subentire. All leaves are variable as to shape, colour and size, green, yellowish green or red, fleshy, thick, glabrous and more or less distinctly dotted. Inflorescence is terminal, elongate, lax, racemiform or paniculiform, 10-50 cm long and racemes 20-40 flowered. Flowers are rather large, buds raised far above the expanded flowers and pedicels 0.5-0.8 cm long. Sepals are erect, obtuse, 1 cm long, the outer ones ovate-oblong and narrow, saccate at the base, the inner ones more ovate or broadly ovate oblong. Petals are oblong-spathulate, rounded at the tips, light yellow and 1.5-2 cm long. Stamens are subequal, erect and 1-1.3 cm long. Pods on stalks are

Figure 21: Cabbage Head

Figure 22: Honeybee Pollinating Cabbage Flower

1.5-2 cm long, more or less patent, thin, cylindrical or linear elongate, with a long beak, many seeded, 5-9.5 cm long and 0.3-0.5 cm thick. Seeds are seriate, globuse and smooth.

Turnip

The fleshy thickened underground portion of turnip is actually the hypocotyl, the colour and shape of which vary depending mainly on the cultivar. Its shape may vary from flat through globular to top-shaped and long. The below ground colour may be white or yellow, while the skin colour of above ground portion may be red, purple, white yellow or green. A distinct taproot and secondary roots arise from the lower part of the swollen hypocotyl. A report from Japan shows that the elongation of turnip hypocotyl stops by the 6th leaf stage in all cultivars tested, whereas the beginning at the 4th leaf stage, while in others, it is not evident till the 6th leaf stage. Thickening begins in the central part of the hypocotyl followed by the upper and lower parts (Iwasaki and Takeda, 1977). Generally, the edible maturity of roots reaches in 40-80 days, depending on cultivars and cultural condition. Leaves and petioles of turnip are hairy and coarse to the touch and yellowish green in colour. Leaves are oblong to oval and may be entire, serrate or even-pinnate depending on cultivars and time of development. Leaves formed early are less likely to be pinnate. The leaves developing on the inflorescence are alternate, glaucous, oblong or lanceolate and entire or dentate. The inflorescence of turnip is a terminal raceme on the main stem. The

Figure 23: Turnip Plant and Inflorenscence

flowers in structure and arrangement are practically identical with other cruciferous crops.

Family–Alliaceae

Onion

Onion is bulbous biennial or perennial herbs, which give off a distinctive and pungent odour when the tissues are crushed. The onion smell and taste are important diagnostic features of the genus *Allium*. Other characteristics include the presence of bulb formed by the attachment of swollen leaf bases to the underground part of the stem, inflorescence in the form of umbels with numerous small flowers, a single spathe which split into segments as the young

Figure 24: Insects Pollinating Onion Flowers

flowers emerge and a superior ovary above the six stamens and petals like perianth.

Flowers are borne in simple umbels at the apex of the floral stem, which is commonly hollow when mature, round in cross section and somewhat swollen at the middle or near the base. The number of seed stems varies from 1-20 or more depending before it is fully expanded, is enclosed in a papery spathe, consisting of 2-3 bracts, which are split open by the pressure of developing flowering buds. The number of flowers may vary from 50-2000, depending upon the species and cultivars, time of planting, size and storage conditions of mother bulbs. The umbels are aggregates of cymes of 5-10 flowers each and the flower open in definite sequence. The flowers are white to bluish in colour. The perianth segments are six in two whorls spreading, reflexed, free and ovate. The stamens are also six in two whorls, anthers bilocular, ovary superior, three locules with nectary at base, each locule has two ovules. The anthers of inner whorls dehisce first and usually it occurs between 9 AM and 5 AM. The style, which is approximately one mm long, when the flowers first open, is not receptive, until it elongates to a length of 5 mm. This requires a day or two days after anthers have all dehisced. An inflorescence may continue opening for two weeks or more and a plant may be in bloom for more than 30 days. The flowers are protandrous in nature and generally cross-pollinated. It is pollinated chiefly by honey bees, which visit the nectaries. Though, onion is cross-pollinated crop but selfing also takes place as the insect visit many flowers in a single umbel and also umbels of the same plant before leaving for other plant. In onion, anthesis start early in morning (6 AM to 7 AM). An artificial pollination for breeding purpose, selfing is accompanied by bagging all the umbels of the same plant together and by shaking or by introducing houseflies.

Family–Leguminosae

Pea

The plant is a short lived, herbaceous annual, glaucous, which climbs by leaflet tendrils. The cultivar may be dwarf, semi dwarf or tall. The root system is not strongly developed except taproot. The stem is slender, circular and weak. The leaves are pinnate with upto three pairs of leaflets and terminal branched tendril. Stipules are large and leaf like ovate; stipule absents; leaflet ovate; elliptic upto

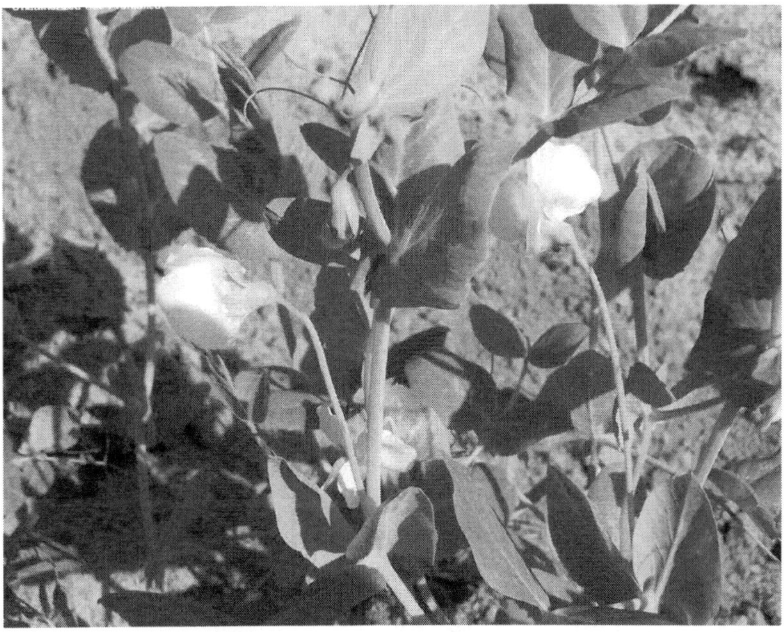

Figure 25: Pea Plant with Flowers

six in numbers entire with undulate margin. The flowers are solitary or up to three flowers per raceme. Bracts are very small. Calyx is oblique and lobes unequal. Corolla is white, pink or purple. Keel is short, incurved and obtuse. Stamens are diadelphous, filaments broad and anthers uniform. Style is falcate, flattened and bearded on inner surface. Stigma is minute and terminal. Pods are swollen or compressed, straight or curved on short stalk upto 12-15 cm with as many as 10 seeds. Seeds are angular or globose, 4-10 in numbers, smooth or wrinkled, no endosperm, green, grey or brown, sometimes mottled, seed weight upto 250 mg (Smartt, 1976).

A developing floral bud could be distinguished from a developing vegetative by its highly compressed sepals forming a cone. A period of 7-9 days are generally taken from this stage to bud initiation (indistinguishable whether floral or vegetative) till the first stage, having thus 12-14 days for competition of the development. Anthesis starts about 5.45 AM with the unfolding of axilla and

Figure 26: Pea Pods with Seeds

wings, exposing the keel. This process continues till 6.30 to 7.30 AM at stage III or IV of a bud. Dehiscence takes place by longitudinal splitting of anther lobes. Three longer stamens dehisce first followed by others. The stigma is receptive 48 hours before anthesis and remained 24 hours after it. The stigma retained good receptivity for only one day. Temperature is a factor, which affects the duration of dehiscence. It come down to 88 per cent at anthesis and further reduces to 50-53 per cent in 24 hours anthesis. Optimum temperature for pollen tube growth is about 25 ^{0}C. Exposure to low temperature of about ^{0}C has no injurious effect on subsequent development of pollen tubes. Germination percentage of pollen grains is mostly 75 to 80 per cent at anthesis and it gradually comes down to zero after 24 hours.

Fenugreek

Its yellowish flowers 12–18 mm long (1–2) are in the leaf axils. Flowers sessile, solitary or in pairs. Calyx hairy, the teeth subulate. The scented flowers are hermaphrodite (have both male and female organs) and are pollinated by insects. It is quick growing and produces erect shoot to a height of 40-70 cm. It has light to dark green leaves with or without pink margin and produces 2-3 cm small white flowers at the base of each leaf. The pods are slender and straw coloured when ripe, beak shaped and about 8-10 cm long with 8-15 yellowish brown coloured having smooth surfaced seeds. The seeds are about 0.3-0.5 cm long.

Figure 27: Fenugreek Leaves, Flowers and Seeds

French Bean

It is highly polymorphic species; annual herb, erect and bushy, 20–60 cm tall, or twining with stems 2–3 m long; with a taproot and nitrogenous nodules; (although the germinating bean has a tap root,

Figure 28: French Bean Pods and Seeds

adventitious roots usually emerge 1–2 days after germination, to dominate the tap root which remains 10–15 cm long); leaves alternate, green or purple, trifoliolate, stipulate, petiolate, a marked pulvinus at base; leaflets ovate, entire; acuminate, 6–15 cm long, 3–11 cm wide; flowers in lax, axillary few-flowered (12) racemes, zygomorphic, variegated, white, pink, or purplish, ca 1 cm long; pods slender, green, yellow, black, or purple, cylindrical or flat, 8–

20 cm long, 1–1.5 cm wide; seeds 4–6-(12), usually glabrous, sometimes puberulent, beak prominent; seeds white, red, tan, purple, grey or black, often variegated, reniform, oblong or globose, up to 1.5 cm long, endosperm absent; 100 seeds weight 10–67 g, depending on cv; germination phanerocotylar.

Cowpea

It is an annual, sinistrorse, erect or decumbent herb and 0.3-4 cm high or long. Leaves are long or petioled, pinnately, trifoliate. Petioles are glabrous and 3.5-10 cm long. Stipules are far above the

Figure 29: Cowpea Plant and Pods

acute base, ovate, obtuse and 1-1.8 cm long. Leaflets are shortly stalked, ovate-rhomboid, base broadly luneate or rounded, apex gradually narrowed, acute, entire or fainty lobed, glabrous or thinly pubescent, 6.5-16 cm long, pedicels rather thick. Calyx is more or less plicate-wrinkled, glabrous, with small triangular acute teeth and 1cm long. Standard is 2-3 cm in diameter, blush violet within or sometimes sordidly yellow. Wings are violet, truncate at the base. Pods are thin linear with a thick, obtuse beak, 10-85 cm long and 0.8-1.0 cm broad. Seeds are numerous, very variable as to size and colour.

The flowers open early in the morning, closes before noon and falls in the same day. The extra floral nectaries at the base of the corolla attract ants, flies and bees but a heavy insect is required to depress the wing petals and expose the stamen and the stigma. Optimum times for emasculation is at 18.20 hr to 18.30 hr and hand pollination at 18.20-18.40 hr in the next day. Male sterility has also been reported in cowpea. Cowpea flowers bloom between 6.30 AM and 7.00 AM. The process of opening of corolla takes 45-60 minutes.

Figure 30: Cowpea Seeds

Anthers mature and start dehiscing about 7-8 hrs before the time of opening of corolla. For hybridization of cowpea, emasculation can only be effective if it is completed 20 hrs before flower opening. The safest time for emasculation may be during the morning proceeding the day of blooming. Pollens remain viable for 24 hr at room temperature of 28.5 °C and relative humidity 91 per cent when stored in a refrigerator at 9 °C and 0 per cent relative humidity. The period of pollen viability increased from 12 hr before blooming to 6 hr after blooming.

Family–Chenopodiaceae

Spinach Beet (Palak)

Plant of spinach beet is a herbaceous annual for the edible leaf production, while it is biennial for seed production. In the early stages of the plant growth (at vegetative phase), it produces rosette,

Figure 31: Flower Stalk of Spinach Beet

succulent tender edible leaves on a small thick stem. Later, at reproductive phase the stem elongates after about 75 days of sowing which results in bolting. Plant has taproot system along with number of feeder roots. Leaves are long with entire margin and long petioles. Nath (1969) reported that the inflorescence in spinach beet is racemes with bracts of cymose type and emerges either in the axil of the leaf or from a spurious terminal bud from the main shoots as well as from the lateral shoots. Flowers are small sessile, bisexual (hermaphrodite) perfect, bracteate (leafy) and are borne in the axil of leaf in a group of 2-3, with 5 stamens opposite to perianth lobes on a fleshy disc within unilocular perigynous ovary.

Flowers produce abundant small and light pollen grains, which are carried by wind leading to wind pollination. Hence the crop is highly cross-pollinated. The whole phase of flower bud development takes 35 days to come in bloom. There are six developmental stages through which each bud passes before it opens into a flower. The anthesis takes place between 7 AM and 5 PM with peak period between 11.00 PM and 1.00 PM.The flowers open mostly during mid day. In an individual flower, anthesis is completed within two hours and hastened by high temperature and low humidity. Anther dehiscence takes place between 8.30 AM and 6.30 PM with the peak period between 12.30 and 2.30 PM depending upon the temperature and humidity. Stigma receptivity begins 8 hours before anthesis, reaches the maximum just after anthesis and continues for another 10 hours (Purohit and Nath, 1970). The fruit is actually a seed ball containing two to three seeds.

Family–Umbelliferae

Coriander

Coriander is thin-stemmed, small, bushy herb, 25 to 50 cm in height, with much branches, tops or umbels. Leaves are alternate, compound and petiole has a pair of stipules, which sheath the stem at the base. The compound leaves become segmented and linear as they reach the upper extremities. Both leaves and stems, may the whole plants have a strong pleasant aroma for which they are loved and used in cooking for flavouring and garnishing dishes. The infloresence is a compound umbel and usually comprises about 5 smaller umbels. The corolla is made up of five petals. Flowers are both bisexual and unisexual, the latter being mostly staminate. The

Figure 32: Coriander Leaves

Figure 33: Flowers and Seeds of Coriander

flower possesses five stamens and the ovary is inferior and bilocular. The fruit is globular, round and small. Its surface has longitudinal fine grooves and ribs alternately. When pressed, the fruit break into two locules, convex outside and concave inside, in each of which, there is one seed. The fruit are mild, delicately fragnant due to presence of volatile oil. Seeds are pale white to light brown in colour.

Carrot

The carrot is an annual or biennial herb, with an erect to much branched stem, 30-120 cm high arising from a thick, fleshy taproot and 5-30 cm long. Usually, the stem elongates and produces round, hispid branches during the second year. Leaves are pinnately compound with long petioles expended into a stalk at the base. The

Figure 34: Carrot Roots

edible portion of carrot root is actually enlarged flashy taproot. It consists primarily of phloem or cortex and core or xylem. Good quality carrot has a maximum of cortex and a minimum of core. In the so-called 'coreless' cultivars, the core is small and deeply pigmented, so that, the cortex and core are evenly coloured.

The carrot umbel is a compound inflorescence. The development of the umbel begins with a broadening of the floral axis and internodes

Figure 35: Umble of Carrot

elongation (Borthwick *et al.*, 1931). A primary umbel can contain over 1000 flowers at maturity, whereas, secondary, tertiary and quaternary umbels bear successively fewer flowers. Usually several floral stalks develop from a single plant. Flowers are arranged in umbellets and umbellets into an umbel. Within umbellets and umbels, floral development is centripetal and arrangement is spiral. Thus, the first mature flowers are those on the outer edges of the outer umbellets. The primary umbel consists mainly of bisexual flowers, but male flowers can occur frequently (between the edge and center of an umbellet in subsequent umbels). The pollen from flowers at the center of an umbellet is larger and more frequently fertile than that from peripheral flowers. Carrot is potardrous. The petals separate and the filaments begin unrolling to release the anthers at anthesis, although the anthers in one flower may not extrude simultaneously. After the filament is straightened, the pollen is shed and the stamen is quickly abscised. The petals then open fully and the style elongates. The carrot has a split style, which separates when the flower is receptive to pollination. The petals of male fertile plants fall soon after the split stigma is receptive. It is interesting that petals of petaloid, but not brown anther, male sterile plants are persistent until the seed ripens. Carrot flowers are epigynous with five small sepals, five petals, five stamens and two carpels. The mature flowers and consequent mature fruits are approximately two mm long. Early in development, each carpel bears two ovule primordia, but only the lower one continues to grow. The carrot embryo sac is monosporic (developing from the chalazal macrospore) and 8-nucleate. Nectaries on the ovary wall are important in insect attraction for pollination. The flowers are perfect with small petals, usually white or yellowish. The calyx is entire. There are five stamens. The ovary is inferior and consists of two locules, each with a single ovule. On the upper surface of the ovary, there is a swollen nectar-secreting disk, which support the style and stigma. The carrot fruit is a bilocular schizocarp, which dries and splits upon maturity to yield two mesocarp (achenes) with one seed each. The mericarp are covered with spines, which must be removed for easy of handling. The carrot seed is actually a fruit, an indehiscent mericarp, containing a single seed. Infact, two mericarp pair to form the schizocarp, the true carrot fruit, which develop from a two-loculed ovary. The mericarp or the seed is somewhat flattened on one side and opposite side has longitudinal ribs with bristly hairs, the fruit is oblong and about 3 mm long. It produces the first order umbel called king umbel, second order and tertiary umbels.

Pollen dehiscence may lasts one to two days per floret and six to seven days after anthesis and may take about 4-6 weeks for complete blooming. The different order of umbels flower at intervals 8-12 days from each other. Anthesis in a single umbel is completed in 7-9 days. The studies on carrot of variety Imperator revealed that umbels of the first two order set seeds in 100 per cent umbels and the third, fourth and onwards orders set in 82.7, 6.29 and 2.78 per cent umbels, respectively. The setting of seed in late formed umbels is extremely poor. The stigma fertility starts when dehiscence is completed in whole umbel. First king umbel comes in flower and it gets pollen from secondary umbel. The secondary gets pollen from tertiary umbel. The temperate types start bolting by third week of April. The king umbels start flowering by the end of May. The secondary umbels or tertiary and other umbels flower at intervals of 8-12 days from each other. Anthesis in a single umbel is completed in 7-9 days. Flowering in first, second and third order umbels are almost completed by end of June. The fourth and onwards umbels flower during July and August. Over 95 per cent of cross-pollination has been observed in carrot.

Family–Cucurbitaceae

Cucumber

Plant is annual, tendril bearing, and creeping or climbing, sparingly branched and scabrous plant. Stem is robust, obtusely quinquangular, densely hispidulous, yellowish green and 50-250 cm long. Tendrils are inserted beside the leaves, robust, patent, peduncled, simple, hispid, yellowish green and 10-30 cm long. Leaves are distichous, long petioled, simple, in outline broadly ovate and angular. Base is deeply cordate. Apex is acuminate acute, distinctly 3-7 lobed, with unequal glandular teeth, hispid along the margins, above light or dark green, beneath pale green or yellowish green, dull on both surfaces, densely patently hispid, very scabrid, 5-7 nerved, the lowest main-nerves branched on the outer side, lateral nerves white, transculent 7-18 cm long and 7-15 cm wide. Lobes are triangular, acuminate, acute and dentate. Petiole is robust, bractate, furrowed on the upper side, clothed with numerous, patent, very stiff hairs, yellowish green and 3-20 cm long. Flowers are axillary,

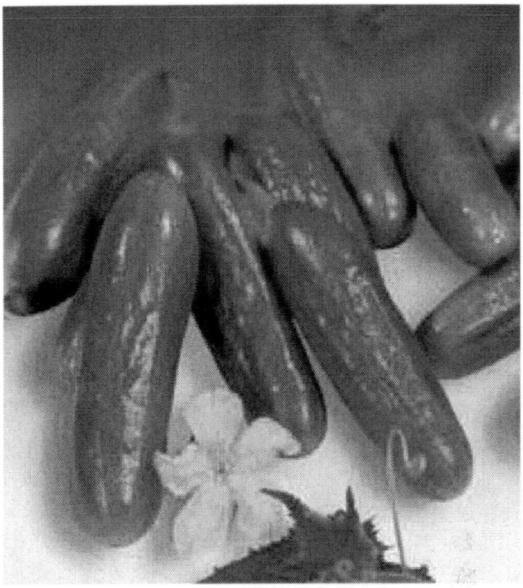

Figure 36: Cucumber Fruits and Flower

shortly stalked, unisexual-monoecious, rather large, densely hispid, 3-4 cm diameter, for the greater part male, the female ones solitary, and the male ones in fascicles of 3-7. Stalk of the male flowers is thin, lengthening after the anthesis, clothed with patent hairs with much swollen and ovoid base, 0.3-0.5 cm long and thick. Receptacle is campanulate or tubular, 0.8-1 cm high and densely tomentose. Calyx is segmented on the margin of the receptacle, 5-partite more than half-way down, with widely patent segments, vitalline or sulphureous, paler on the outside, with numerous longitudinal ribs or nerves, which are clothed with scattered long, stiff hairs, between the nerves densely and very shortly hispidulous, wrinkled on the inside, densely clothed with minute gland-tipped hairs and 2-3 cm diameter. Segments are ovate-oblong, lanceolate or obovate, acute, veined, 1.5-2 cm long and 1-1.5 cm wide. Male flowers are with a broadly ovoid, yellowish white, 0.5-0.7 cm long column of anthers, which are supported by 2-3 very short, free and white filaments. Female flowers are with an ellipsoid or fusiform, pruinose ovary, clothed with scattered hairs with swollen, ovoid or wart-like bases. Style is simple and stigmas 3-5 and thick. Fruits are pendulous on robust stalks, very variable as to shape and size, usually oblong, lanceolate or linear-cylindrical, small or large, straight or curved, at first green or whitish green, pruinose, with 5-10 more or less conspicuous longitudinal, white or yellowish white streaks, especially towards the apex and often with scattered spinous tubercles and warts, sordidly yellow or sordidly orange-coloured and glabrous when ripe, succulent, many-seeded, 10-20 cm long or longer and 3-7 cm diameter. Flesh is pale green. Fruit-stalk is much lengthened, robust, 2-5 cm long and hispid. Fruits of hermaphroditic flowers *i.e.* of perigynous flowers are more round than the fruits of epigynous flowers. Staminate, pistillate and hermaphrodite flowers can easily be distinguished from each other visually. Their pedicels can recognize the staminate flowers. Since hermaphroditic flowers must afternoon for anthesis as well as pistil, their column at the top of the ovary is enlarged in comparison with the pistillate flowers and can therefore, easily be distinguished from them. Seeds are flat, ovate-oblong, and rather acute at ends, sharply margined, 0.8-1 cm long and 0.3-0.5 cm broad. Choudhury and Phatak (1959) studied floral biology in cucumber. The whole developmental process from the initial bud stage to the stage when the flower is detached from the pedicel is divided into 8 stages (stage 1 to stage 9). The sunrise and sunset mainly influence opening and closing of the male flowers, which is by light and the time of the day. Anther in all varieties dehisced between temperature range of 20.5-21.5 ^{0}C. Pollen fertility has been considerable up to noon and by afternoon (2 PM), fertility

Figure 37: Female and Male Flower

was greatly reduced and it was negligible by the evening. Stigmatic receptivity was very short duration and pollination is carried out within two hours after anthesis. Rise in temperature causes early drying of stigmatic secretion. Different floral abnormalities like mixed inflorescence, hermaphroditism, fusion, dimorphic female flowers, reduction and increase in the floral parts were observed. A greater tendency was observed in exotic collections towards these abnormalities.

Muskmelon

Muskmelon belongs to, genus *Cucumis.* The characteristics of genus *Cucumis* are fruit fleshy, many seeded pepo, flower solitary, flowers lemon yellow to deep orange, leaves deep or shallowly lobed not pinnatified, corolla rotate deeply parled and small flowers. It is a polymorphic species where most of the cultivars are andromonoecious (hermaphrodite and staminate flowers) gynoecious (only pistillate flowers) or monoecious (pistillate and staminate flowers), monoecious and andromonoecious are most common. Stem is soft–hairy to glabrous, striate or angled, leaves orbiculate to ovate to reinform, usually five angled, sometimes shallowly three to seven–lobed, hairy or somewhat scabrous, 3-5 inches across. The staminate flowers are clustered. The pistillate flowers are solitary on stout pedicels (Whitaker and Davis, 1962). There are various horticultural forms within *C. melo* based on fruit characteristics, namely cantaloupes, nutmeg muskmelon, winter melons, white skinned melons, snake melons, oriental pickling melons, mango melons, pomegranate melons which hybridize readily with each other and there is apparently very little sterility even among progenies from crosses involving variant types. Under open natural system, it is partly crossed and partly self–pollinated by bees but there are reports of obtaining vigorous lines by inbreeding up to 25 generations (Robinson *et al*; 1976). Muskmelon is highly cross-pollinated crop and, honeybees are pollinators. Nandpuri and Brar (1966) reported time of anthesis is 5.30-6.30 hr, dehiscence 5.00-6.00 hr, duration of pollen fertility 5.00-14.00 hr and stigma receptivity 2 hr before to 2-3 hr after anthesis.

Watermelon

The plant of watermelon is typically a trailing (vine) annual, but recently dwarf forms, which may be described as "bush", have

Figure 38: Fruits and Seeds of Muskmelon

been located and are used to develop cultivars suitable for growth in a limited space. Dwarf may have a total diameter of as little as 2 fit, whereas, the more rampant vine cultivars may be 30 fit or more in diameter. Dwarfing is primarily related to shortened internodes (Mohr, 1956). In the typical vine, there is a delay in the development of lateral branches, with a dominant single "runner" reaching

Figure 39: Flower of Watermelon

several feet in length before branching, is initiated and maintaining dominance for sometime thereafter. In some of them dwarf types, multiple branching occur simultaneously from the crown of the plant when it is quite young, which provides more potential bearing areas concentrated fruit set.

The root system of watermelon is very extensive but shallow, consisting of a taproot and many lateral roots within the top 2 fit of soil. Early destruction of the tap root (as in transplanting) may be advantageous in getting high yields. Fruit shape varies, ranging from long, cylindrical to spherical fruits with the melons of most cultivars intermediate in shape. Rind colour varies from white various shades of green, with some cultivars having striped rind and some mottled rind. Rind ranges from brittle to tough and flexible, from very thin to excessively thick, and form very soft to very hard. Flesh texture varies from fine grained and 'melting' to firm coarse, and fibrous, seeds range in size from very small (size of tomato seed) to large size (size of pumpkin seed). Seeds usually range in colour from white to black, but red, green, brown and mottled colours also occur. The greater part of the fruit is the edible flesh, which is mostly derived from the placental, whereas, the cantaloupe and squash are largely pericarp. The pericarp of the watermelon is its hard outer rind.

Honeybees generally pollinate watermelon. Monoecious and andromonoecious types of flowering habits are found, of which the former is most common. The ratio of staminate to pistillate flowers under field conditions varies between 7:1 and 15:1. Staminate flowers usually appear earlier than pistillate ones. The first female and

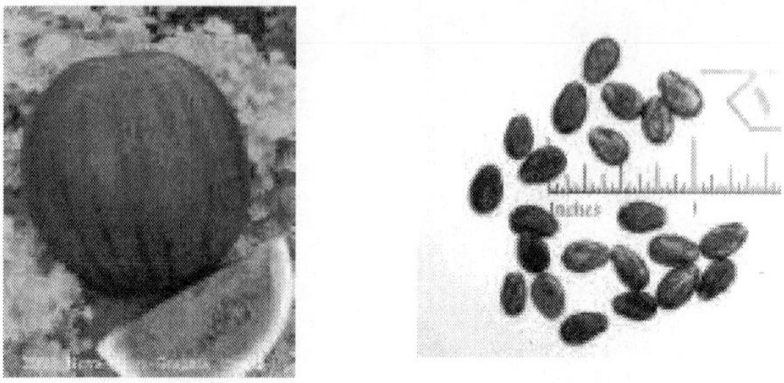

Figure 40: Fruits and Seeds of Watermelon

hermaphrodite flowers are formed on the main vine. The male flowers open first in the axil of 4th to 12th leaf and a few days later female and hermaphrodite flower open, those of early cultivars in the axil of 5th to 7th leaf and in late cultivars in the axil of 25th to 30th leaf. More female flowers are formed in spring than in autumn. The stigma of the watermelon is of the 'wet type' and consists of multicellular papillae invested by copious extracellular secretions (Sedgley, 1981 and 1982). Although, exudate production and secretion occur during the two days prior to anthesis, cells have the capacity to produce further exudates in response to pollination. Pollination of the stigma at anthesis results in a rapid pollination. After 30 minutes, the papillae and pollen grains are submerged in the secretion. The anthesis of flower in watermelon can be observed from 6 AM to 7.30 AM, with a peak between 6.30 and 7.30 AM. The anther dehiscence starts one hour before anthesis and continues until 6.30 AM. At high temperature stigmatic fluid starts drying and the stigma becomes non-receptive (Sidhu *et al.*, 1980). Pollen from hermaphrodite flowers germinates rather slowly than that form female flowers, but is equally effective for fertilization. Bees are the pollinators. In watermelon crosses, higher seed set has been observed when pollination occurred in the early morning immediately after flower opening. Seed set has been found lower when pollination is carried out in the afternoon. Pollen kept at room conditions for 5 hours gave a lower seed set than freshly collected pollen and after 30 hours completely lost viability.

Bitter Gourd

It is a slender, annual, tendril-bearing climber, somewhat fetid when bruised and 1-5 m long. Stem is thin, acutely quinquangular, with broad longitudinal furrows, light green and thinly white pubescent. Tendrils are inserted beside the leaves, on long, robust stalks, simple or cleft, faintly 5-angular, yellowish green and 15-30 cm long. Leaves are alternate, long stalked, in outline orbicular or broadly ovate. Base is deeply cordate, apex shortly acutely acuminate, deeply pedately and 5-9 lobed. Lobes are more or less deeply lobed, all of them narrowed at the base, shortly mucronate, on both surfaces appressed pubescent on the nerves, afterwards becoming almost glabrous, above dark green, beneath light green or yellowish green, dull or feebly shining, 3-5 nerved or pedately 5-9 nerved, with transparent veins, densely pellucidely dotted and 5-17 cm diameter. Petioles are robust, half terate, furrowed above, thinly pubescent, light green, 3-10 cm long and stipule absent. Flowers are unisexual–monoecious and solitary in the leaf-axils. Pedicels are long, filiform, thinly pubescent or subglabrous and 5-15 cm long. Bract is below the middle of the pedicel, often near its base, sessile, large, orbicular-cordate or reniform with an obtuse or rounded, often emarginate or shortly cuspidate apex, entire, thinly pubescent, shortly ciliate, green 0.5-2 cm diameter. Male flowers are much resembling the female ones but on shorter and thinner pedicels. Calyx is campanulate, deeply 5-fid and 1-1.25 cm long. Tube is short, thick, yellowish green, densely appressed pubescent and wartty. Segments are patent, elliptic or lanceolate, acute, light yellowish green, straight, densely shortly pubescent on surfaces, 0.4-0.85 cm long and 0.2-0.4 cm wide. Petals are 5 in numbers inserted in the throat of the calyx-

Figure 41: Fruits of Bittter Gourd

Figure 42: Bitter Gourd Seeds

tube, oblong lanceolate or elliptic with an obtuse or rounded, often-emarginated apex, entire, sulphureous, densely pubescent, 1.5-2 cm long and 0.75-1.2 cm wide. Stamens are three in number, filaments short, thick and white. Anthers are 3, conduplicate and cohering. Ovary in the female flowers is fusiform, rostrate, warty, dark green and 2.5-3 cm long. Calyx is small, 0.75-1 cm in diameter. Sepals are 5, oblong lanceolate, acute, yellowish green, shortly pubescent, 0.25-0.35 cm long and 0.1 cm wide. Petals are 5, patent, elliptic or oblong obovate, rounded at the apex, entire, citrine, densely shortly pubescent, 1.5-1.75 cm long and 1 cm wide. Staminodes are 5, tubercles-shaped; pale yellow, 4 of them approximate in pairs. Style is short and pale green. Stigmas are 5-6 and erect. Fruits are long peduncled, pendulous, and fusiform, beset with many irregular warts, very variable as to size, orange-coloured when ripe, 3-valvate at the top, 7-30 cm long and flesh blood-red. Seeds are numerous, flattened, ovate-oblong, warty, 1-1.5 cm long and 0.6-0.9 cm broad.

Bitter gourd is a monoecious crop, where male and female flowers are borned on the same plant. Varieties differ in days to flower, time of anthesis and pollen viability (Deshpande *et al.*, 1979). Anthesis occurs between 3.30 and 7.30 AM and anther dehiscence between midnight and complete opening of the flower (7.30 AM). Pollen viability has been high in all varieties (90-99 per cent), while pollen germination ranged from 66 to 68 per cent. The stigma remained receptive for 24 hrs before and after anthesis. Pollination is brought by bees. The appearance of first pistillate flower at a lower node number is an indication of the earliness of the variety. On the other hand, the higher the node number of the position of the first pistillate flower, the greater would be the production of pistillate flowers and a shorter interval between the appearance of the first staminate flower and the first pistillate flower indicates shorter plant life.

Family–Compositae

Lettuce

It produces taproot, which grows on the surface of the soil. Large number of leaves are produced of which outer leave are usually green to dark green and inner leaves are light green to whitish which are tender and crisp. Inner tender leaves have better quality than outer leaves. Inflorescence is head or capitulum consisting of a large number of flowers. Numerous small flowers remain closely together on a receptacle. Floret with a single ligulate yellow petal, 5 stamens and 2 carpels. There is very high degree of self-pollination. Pollen and ovule are usually highly fertile. Anthesis starts at sunshine and flower fully opens by late morning. Pollination takes place from 9.0-9.15 AM (Pal, 1975). Fruit is a typical achene and seed exalbuminous.

Figure 43: Plant, Flower Stalk and Seeds of Lettuce

Chapter 6

Modes of Reproduction in Vegetable Crops

Successful seed production of a crop depends on a thorough knowledge of the reproductive process of a particular crop. The techniques of seed production must take into account several features of reproduction: whether it is sexual, asexual, or a combination of the two, the nature of floral structures, the amount of pollen transfer, the degree and means of self-incompatibility, and the effect of inbreeding on vigour. For most purposes the important consideration to the seed producers is the extent of cross-pollination. It is, therefore, necessary that before engaging himself in seed production, a producer must acquaint himself with the following details of reproduction in the particular crop.

Methods of Reproduction

The method of reproduction in vegetable crop plants may be broadly grouped in to two categories, asexual/vegetative and sexual.

Asexual/Vegetative Reproduction

Asexual reproduction does not involve fusion of male and female gametes. New plants are developed from vegetative parts of the plant or may arise from embryos that develop without fertilization. In nature, a new plant develops from a portion of the plant body. This may occur through modified underground and sub-aerial stems, bulbils, cuttings and grafting. As for example, taro, pointed gourd, potato etc. are reproduced through asexual means.

Sexual Reproduction

Sexual reproduction involves fusion of male and female gametes to form a zygote and the zygote develops into a new plant. The main difference between asexual and sexual reproduction is that in asexual reproduction one parent or sex is concerned, while in sexual reproduction both parental sexes are concerned. All self–and cross-pollinated vegetables are in this group.

Modes of Pollination

Pollination refers to the transfer of pollen grains from anther to stigma. Pollen from an anther may fall on stigma of the same flower leading to self-pollination or autogamy. When pollen from flowers of one plant is transmitted to stigma of another plant, it is known as cross-pollination or allogamy. A third situation, geitonogamy results when pollen from a flower of one plant falls on the stigmas of the other flowers of the same plant, *e.g.* in maize.

Many cultivated vegetable species reproduce by self-pollination. These species, as a rule, have hermaphrodite flowers. But in most of these species, self-pollination is not exclusive and cross-pollination may occur from 5 to 50 per cent. The degree of cross-pollination in self-pollinated species is affected by several factors, namely variety, environmental conditions like temperature, humidity, wind, location and abundance of pollinating insects. However, the essential conditions for self-pollination are :

Bisexuality

In this both male and female reproductive organs are present in the same flower. Without this condition self-pollination is never possible.

Homogamy

This is the condition in which the anthers and the stigmas of a bisexual flower mature at the same time resulting in self-pollination.

Cleistogamy

In this condition the bisexual flowers never open and therefore, the self-pollination is only the way of pollination, *e.g.* lettuce.

Partial List of Self-pollinated Vegetables

Tomato (*Lycopersicon esculentum*), Lettuce (*Lactuca sativa*), Parsnip (*Pastinaca sativa*), Peas (*Pisum sativum*), Dwarf bean (*Phaseolus vulgaris*)

Cross-pollinated Vegetable Crops

The majority of the cultivated vegetable species are cross-pollinated crops. Cross–pollination may occur both in bisexual and unisexual flowers but it is the rule in unisexual flowers. Nature favours cross-pollination and the agencies which help in bringing cross-pollination in flowers are:

1. Air (Anaemophily), *e.g.* Amaranthus, spinach, beet
2. Insects (Entomophily): All cucurbits, all brassica, onion, carrot.

There are many reasons for which cross-pollination takes place in the crops, some of which are listed below:

Decliny

Decliny or unisexuality is a condition in which the flowers either staminate (male) or pistillate (female) occur on the same plant or on different plants, e. g. cucumber, watermelon, pumpkin, squash, asparagus etc.

Dichogamy

The stamens and pistils of hermaphrodite flowers may mature at different times facilitating cross-pollination, *e.g.* sweet corn, sugar beet etc.

Self-incompatibility

It refers to the failure of pollen from a flower to fertilize the same flower or other flowers on the same plant, *e.g.* cabbage, cauliflower, mustard, cole crops, root crops etc.

Male Sterility

Male sterility refers to the absence of functional pollen grains in otherwise hermaphrodite flowers. Male sterility is not very common in natural populations but is of great value in production of hybrid seed.

Partial List of Cross-pollinated Vegetable Crops

Cabbage (*Brassica oleracea* var. capitata), Cauliflower (*Brassica oleracea* var. botrytis), Broccoli (*Brassica oleracea* var. botrytis), Brussel's sprouts (*Brassica oleracea* var. gemmifera, Knol khol (*Brassica oleracea* var. caulorapa), Carrot (*Daucas carota*), Radish (*Raphanus sativus*), Beet (*Beta vulgaris*), Spinach (*Spinacea oleracea*), Cucumber (*Cucumis sativus*), Muskmelon (*Cucumis melo*), Watermelon (*Citrullus lanatus*), Pumpkin (*Cucurbita moschata*), Squash (*Cucurbita pepo*), Other cucurbits (bittergourd, bottle gourd, ridge gourd, sponge gourd, snake gourd, point-gourd, ash gourd etc), Amaranths (*Amaranthus tricolor, A.caudatus, A.spinosus*), Onion (*Allium cepa*)

Often Cross-pollinated Vegetables

In some vegetable crops, cross-pollination often exceeds 5 per cent and may reach up to 50 per cent. Such species are generally known as often cross-pollinated species, *e.g.* brinjal, okra, chilli, sweet pepper.

Cross Compatibility Among Cruciferous Vegetable Crops

Safe guarding the seed plants from foreign contamination through extraneous pollen transmitted by insects or transported by wind is essential for the production of pure seeds. Therefore, one should know the crossing patterns of different vegetables.

Chapter 7

Principles of Vegetable Seed Production

Production of genetically pure and quality pedigree seed requires high technical skill and specialization. Seed production must be carried out under standard and well-organized conditions. The producer should be familiar with genetic and agronomic principles of seed production.

Genetic Principles

During the course of seed production, it is necessary to ensure that the product is true-to–type. Genetic purity of a variety can deteriorate due to several factors during production cycle. The important factors of apparent and real deterioration of varieties are as follows as suggested by Kadam (1942):

1. Developmental variations
2. Mechanical mixtures
3. Mutations

4. Natural crossing
5. Minor genetic variation
6. Selective influence of diseases
7. Technique of the plant breeder

Of these, mechanical mixture, natural crossing and selective influence of diseases are perhaps the most important reasons of genetic deterioration of varieties during seed production followed by raising the seed crops in areas outside their adoption which may cause developmental variations and genetic shifts in varieties.

Maintenance of Genetic Purity

For the maintenance of varietal purity various methods have been suggested by Horne (1953) and Hartman and Kester (1968). The important safeguards for maintaining genetic purity during seed production are:

Control of Seed Source

The use of seed of an appropriate class and from an approved source is necessary for raising the seed crop. Four classes of seeds, namely, breeder's, foundation, registered and certified seeds have been defined by the Association of Official Seed Certification Agencies (AOSCA):

Breeder Seed

Breeder seed plays an important and key rôle in the pedigree system of seed production. Breeder seed is seed or vegetative propagating material directly controlled by the originating or sponsoring plant breeder of the breeding programme or institution and/or seed whose production is personally supervised by a qualified plant breeder and which provides the source for the initial and recurring increase of foundation seed. Breeder seed shall be genetically so pure as to guarantee that in the subsequent generation, *i.e.*, certified foundation seed class shall conform to the prescribed standards of genetic purity. The breeder seed production is undertaken by a qualified plant breeders/sponsored plant breeder.

Normally, the breeder seed is produced by the SAU's and ICAR institutes. In some cases, CSFCI and NSC also produce breeder seed. The requirement of breeder seed is received by the Seed Division,

Ministry of Agriculture, and Govt. of India. The seed indents are compiled by the Ministry and sent to the ICAR for organizing the production. The indents are then sent to the respective crop-coordinators for planning the production programme with the help of the concerned breeder in the Annual Workshop of the All India Coordinated Research Project. The breeder seed production is constantly monitored by sending the information in five different forms. Breeder seed crop is monitored by a joint inspection team consisting of the crop breeder, the producing breeder, representatives of the Crop Coordinator (CC), State Seed Certification Authority (SSCA) and the NSC to ensure the seed quality. Breeder seed approved by this team is eligible for its further multiplication as foundation seed.

Foundation Seed

Certified foundation seed shall be the progeny of breeder seed, or be produced from foundation seed, which can be clearly traced to breeder seed. Production of foundation seed shall be supervised and approved by the certification agency and be so handled as to maintain specific genetic identity and genetic purity and shall be required to conform to certification standards specific for the crop/ variety being certified.

During the production of certified foundation seed, the following guidelines shall be observed:

1. Certified foundation seed (CFS) produced directly from breeder seed shall be designated as foundation seed stage-I;
2. CFS produced from foundation seed stage-I shall be designated as foundation seed stage-II;
3. CFS stage–II will not be used for further increase of foundation seed and shall be used only for production of certified seed class;
4. Minimum seed certification standards shall be the same for both foundation seed stage-I and II, unless otherwise prescribed;
5. Certification tag shall be white colour for both foundation seed stage-I and II, and shall contain the information as to its stage; and

6. Production of foundation seed stage-II shall ordinarily be adopted in respect of such crop varieties, when it is felt by the certification agency that breeder seed is in short supply, *i.e.*, vegetatively propagated crops, apomictically reproduced crops and self-pollinated crops, often cross-pollinated and cross-pollinated crops, these being gene-pools, should not lose their genetic identity and purity if measures to safeguard the same are adequately taken composits and synthetics, parental line increase of hybrids.

Registered Seed

Registered seed is the progeny of foundation seed that is so handled as to maintain satisfactory genetic identity and purity, and that has been approved and certified by a certifying agency. This class of seed should be of a quality suitable for production of certified seed.

Certified Seed

Certified seed shall be the progeny of foundation seed and its production shall be so handled as to maintain specific genetic identity and purity according to standards prescribed for the crop being certified. Certified seed may be the progeny of certified seed provided this reproduction does not exceed three generations beyond foundation seed stage-I and it is determined by the certification agency that genetic identity and genetic purity will not be significantly altered and when the seed certification agency is satisfied that there is genuine shortage of foundation seed despite all the reasonable efforts made by the seed producer.

Certified seed produced from certified seed shall not be eligible for further seed increase under certification. Certification tags for such production which is not eligible for further seed increase under certification shall be subscribed with not eligible for further seed increase under certification. Certification tag shall not be blue color. Certified seed is produced by NSC, CSFCI, SSC's and private seed companies.

Crop Rotation

Satisfactory intervals between related or similar crops is required to minimize the risk of plant material or dormant seeds remaining from the previous crops, which are likely to cross-pollinate or make admixture with the planned seed crop. In addition to these the reasons for crop rotation include plant nutrition, maintenance of soil physical condition and minimizing the risk of soil-borne pests and diseases. In practice, therefore, attention must be paid to the numbers of years since a related crop was grown in the same soil.

Isolation

One major factor during the course of seed production is to ensure that the possibility of cross-pollination between different cross-compatible plots or fields is minimized. In addition to the question cross-pollination, adequate isolation also assists in avoiding admixture during harvesting and the transmission of pests and pathogens from alternative host crops. Vegetable seed crops can be isolated by time and by distance.

Isolation by Time

This type of isolation is possible within individual farms of multiplication stations. In this system seed production is arranged in such a manner that the cross-compatible varieties are grown in successive years or seasons provided the rules regarding rotation are applied.

Isolation by Distance

When isolation by time is not possible, then isolation by distance is to be followed. Isolation distance primarily depends on the nature of pollination of the crop. In general, highly cross-pollinated (by insects) vegetable crops like onion, radish, cabbage, cauliflower and cucurbits require isolation distance of 800-1000 meter while wind pollinated vegetables like spinach, beet require isolation distance of about 2000 meters. Isolation distance also varies according to category of seeds like foundation and certified seeds.

Roguing of Seed Crop

The existence of off-type plants in the seed crop is a potential source of genetic contamination. The removal of such plants is termed as roguing. Not only the off-types but the diseased and abnormal plants are also to be removed. The number of roguing required for the seed crop will vary with the kind of vegetables, purity of the seeds sown, nature of the previous crop etc. Roguing may be done at the following stages as soon as the off-types are recognizable:

1. Vegetative stage
2. Flowering stage
3. Maturity stage

In the seed crop, off-type plants should be rogued out at different times of the day by walking in different directions of the plot. In general the cross-pollinated vegetable crop for seed production should be thoroughly rogued before flowering. Regular supervision by trained manpower is important.

Seed Certification

The genetic purity in the commercial seed production is maintained through a system of seed certification. Seed certification implies that the crop and seed lot have been duly inspected and that they meet requirement of good quality pedigree seeds.

To achieve this purpose, qualified and well-trained personnel of seed certification agencies carry out field inspection at appropriate stages of crop growth. They also make seed inspections to verify that the seed lot is of the requisite genetic purity and quality. In addition to inspections, seed certification agencies also lay down the field and seed standards which the seed crop and seed lot respectively must conform to get approval as certified seed. The field standards include land requirements, isolation requirements, maximum permissible off-types etc.

Grow-out Tests

Varieties being grown for seed production should periodically be tested for genetic purity by grow-out tests to make sure that they are being maintained in their true form.

Agronomic Principles

Besides genetic principles of seed production, there involves the application of the following agronomic principles for the production of good quality seeds.

Site Selection

1. The site should have a good drainage.

2. The site should be flat with sufficient slope to avoid accumulation of water.
3. The soil should be fertile. Organic manures and fertilizers should be applied to the exact necessity.
4. The site must be free from soil which is chemically contaminated due to residues, etc, from the previous crop, harmful diseases and insect-pests.
5. A suitable rotation should be followed. The seed crop should not be raised year after year on the same field.
6. In case of cross-pollinated vegetables, a proper isolation distance from other varieties and crops crossable to main seed crop should be maintained.

Soil and its Preparation

1. Light loam is the best soil for vegetable seed production, sandy loam and loam soils are also good.
2. To ensure good tilth to the soil, sufficient number of ploughings should be given.
3. To facilitate proper irrigation, soil should be smoothly well leveled.
4. The soil should be tested well in advance to know the fertility status.
5. If the soil is deficient in nutrients, they may be supplemented with heavy doses of manures and fertilizers and micronutrients.

Manures and Fertilizers

1. Most of the vegetables are heavy feeders and in general 20-30 tonnes of well decomposed farmyard manure or compost/leaf mould should be added per hectare.
2. Leguminous vegetables such as garden peas, beans, etc, needs less manure doses and Cole crops, potato, tomato, etc, require heavy manuring.
3. The bulb, tuber and root crops, the application of potassic fertilizers at the rate of 80 kg/ha is quite essential.
4. The nitrogenous fertilizer at the rate of 100-kgN/ha applications is necessary for vegetative growth, which leads to higher seed yields.

5. Application of 80 kgK_2O/ha is essential for fruiting, seed setting of vegetables.
6. The whole quantity of manure, P_2O_5 and K_2O + 50 per cent N should be applied before sowing/planting vegetables and rest half N should be top dressed in two equal splits.
7. While applying commercial fertilizers, care should be taken to prevent seeds, seedling and young plants from coming in direct contact.

Selection of Varieties

1. A large number of vegetable, varieties hybrids have been developed in each vegetable for different locations.
2. The vegetable variety/hybrid widely adopted and seed of which is in great demand should be considered for seed commercial.
3. The varietal purity may be maintained and should be grown in a particular belt where the seed can be raised easily.
4. The area should be divided into zones to avoid genetical and physical admixture of seeds.
5. High seed yielding and disease/insect resistant varieties should be selected.
6. Preference may be given to hybrid varieties.

Seed Gowing

1. The vegetable such as garden peas, cowpea, beans, cucurbits, radish, carrot, spinach, turnip, etc, are directly sown in the field. For this, the land should be well prepared according to the need for planting the seed at different depths.
2. Seed sowing depth depends on the soil type, moisture content, season of planting and kind of seed.
3. In general, depth of three-four times the diameter or thickness of seed is considered appropriate.
4. Sowing should be done in rows as it facilitates in subsequent operations.

5. Pure good seed having high vitality, improved type, free from diseases, suitability for growing conditions and free from foreign matter should be used.
6. Healthy and fresh seeds have 75-85 per cent germination usually germinate more quickly and uniformly.
7. The seed should be obtained from a reliable sources *viz.* NSC or Vegetable Research Institution/Universities or some well reputed seed firm.
8. The seed may be treated before sowing.
9. Use breeder / foundation/certified seeds.

Nursery and its Preparations and Sowing

1. Some vegetables such as cabbage, cauliflower, knolkhol, brinjal, tomato, chillies, lettuce, etc, are first raised in nursery and transplanted later in the main field.
2. In case of kitchen garden of small-scale small wooden boxes (50 x 30 x 10cm) should be used for raising nursery.
3. In case of large-scale vegetable seed production out door nurseries (300 x 20 x 15cm) should be preferred.
4. For raising early and healthy seedlings, polyhouse or green house may be used based on recommended technology in each zone.
5. Normally five-six nursery beds will be sufficient for a hectare of seed crop.
6. The soil mixture used should be sterilized for killing the disease organisms and insect-pests (one part formaline in 100 parts water at the rate of five litres/sq metre to saturate fully up to 15 cm depth).
7. The seed should be sown in five cm apart in the raised beds.
8. Thick sowing should be avoided to get healthy seedlings.
9. If the seeds are very small a little sand should be mixed with them for facilitating uniform seed distribution.
10. The seeds should be covered with a film of soil mixture and water with the sprinkler to maintain optimum moisture in the soil.

11. The soil is kept moist till the germination sets in.
12. Light shade should be provided to the seed beds by detachable thatch during hot days and rains.

Transplanting

1. When the seedlings attain four-five leaf stage in four-six weeks should be transplanted.
2. Before transplanting, the seed bed should be water soaked as this facilitates the removal of seedlings with least root injury.
3. The seedlings should always be lifted up with little soil attached with roots.
4. The seedlings should be transplanted in rows little deeper in the soil.
5. The soil around the roots should be made firm with the help of hands and forefingers.
6. Soon after transplanting the field should be given irrigation.
7. Transplanting should be done preferably either in the evening or during cloudy weather.
8. Apply folidol or aldrin dust at the rate of 20-25 kg/ha against cut worms.
9. Gap filling should be done immediately to maintain uniform plant population.

Irrigation

1. A regular and assure water supply should be maintained for successful vegetable seed culture.
2. During summer, more frequent watering should be applied during dry spells while in rainy season and winter less watering.
3. Light and frequent irrigations are more beneficial than heavy irrigations given after a long interval.
4. Seed crop should never be allowed water, especially at flowering fruit/seed setting stage.

Interculture

1. The seed crop should be kept clean through frequent weeding.
2. Shallow cultivation should be given to keep down the weed population and to promote aeration.
3. Deep cultivation particularly when the plants have attained a fair amount of growth should be avoided as it may prove injurious as it prunes the roots.
4. One earthing up should be given to the seed crop at preflowering stage.
5. In case of pole type vegetable Varieties/Hybrids suitable staking may be provided.
6. In some vegetables thinning is needed which may be done at right stage and time.
7. Mulching may also be done to increase temperature and conserve moisture.

Methods of Seed Production

1. For commercial seed production, the seed to seed method should be followed.
2. In seed to seed method, only breeder or foundation seed must be used.
3. Root to seed or retransplanting method should be used for production of nucleus, breeder or foundation seed.
4. Optimum spacing combination may be followed to obtain high quality seed yield/hectare.

Isolation Distance

1. The isolation distance should be provided all round a seed field to separate it from fields of other variety and also from fields of the same variety not confirming to the verietal standard.
2. In case of garden peas and bean the isolation distance should be 10 metres whereas in lima bean it should be 50 metre.

3. Cole crops, beet, carrot, radish and turnip, it should be 1000 m and in case of cucurbits and onion it may be kept 400 m for commercial seed production.
4. In chillies, capsicum and bhindi, it should be 200 m and in case of tomato and lettuce it should be 25m.
5. However, in potato it should be five m.

Rouging and Inspection

1. Any plant which is outside the acceptable limit of varietal variation should be removed.
2. Rouging should be done to all undesirable and diseased crop plants whenever observed.
3. The number of rouging should be three-four times depending on the rouging vegetable crop and purity of seed desired.
4. Field inspection of seed crops should be done at least twice or thrice as desired or recommended.
5. Inspection makes the seed crop eligible for certification.
6. The time when inspection should be made verges with individual seed crops.
7. Normally in first inspection of all seed general standards should be checked-isolation, cultural practices, spacing, etc.
8. Second inspection should be done before flowering and third at flowering stage seed/fruit setting stage.

Harvesting, Threshing and Grading

1. The seed crop should be harvested as and when it is matured.
2. In some cases the seed crop can be harvested in one lot, for example, garden peas, beans etc, while in others harvesting is completed in two-three lots as the crops do not mature at a time.
3. In some cases the seed crop after harvesting needs curing while others can be threshed directly after sun drying.

4. Certain vegetables are ready when they are a bit over ripe, others need extra time to dry out even after they are picked and some need attentation at the height of ripeness.
5. Those that contain the seeds inside of them such as peppers, melons tomato, should be cut open and the seed are simply spooned out.
6. Seeds that are picked directly off the plants such as lettuce seeds can be cleaned by shaking them around on a fine screen used as a sieve. The unwanted debris will remain on the top.
7. Lima bean garden peas, etc. of course are simply removed from their pods.
8. After proper cleaning and grading, the large seeds are dried before storing for five-six days in the open and small seed for three-four days when the temperature is 30-35^{0}C.
9. Finally the seed packs should be well labeled and sent to different seed depots or units for further transmission to various parts of the country or even outside.
10. Others as pumpkin and squash seeds can be washed from the pulp.

Insect-pests, Diseases and Weeds

1. The insect-pests, diseases, weeds as and when they appear should be carefully examined and got inspected by the nearest scientists/extension staff, available and when studied carefully must be sprayed immediately for its control.
2. For future the insecticide, fungicide and weedicides should be kept in the ready stock.
3. Spraying should be done always on a sunny day with calm atmosphere for effective control.
4. Instructions and precautions mentioned on the containers should be strictly followed.
5. Spraying/dusting should not be done in the morning or at a time when the pollinators are busy in the seed field for pollination work.
6. Always spray safe chemicals so that the pollinators are not killed.

Chapter 8

Techniques for Hybrid Seed Production

Hybrid is the first generation cross-produced through crossing between two genetically different plants, lines or varieties. The seeds produced in this way have hybrid vigour, which serve as a key for increasing the yield of many self and cross-fertilized crops. With limited possibilities of horizontal expansion in area and to meet the ever-increasing demand of quality vegetables by the growing populations, the exploitation of hybrid vigour is one of the best options for increasing both the productivity and quality. Vegetable hybrids under optimum crop production and protection management, produce economically more yield than improved open pollinated varieties. The uniform size, early maturity, improved nutritional quality, adaptability to adverse environments and resistance to various biotic and abiotic stresses are certain additional advantages of hybrids resulting in better monetary returns to the vegetable growers. Development of hybrids in vegetables brought about a near revolution but still India has not been able to harvest full potential of hybrid technology as compared to countries like USA, Japan, Canada, Europe, Israel etc. Only about 10-12 per cent of the total vegetable area is under hybrids in India as against 75-90 per cent in Japan or USA. In India among different vegetables only tomato and cabbage are the leading crops occupying about 30-35 percent area under hybrids while rest of the vegetables covering less than 15 percent, which is very less. The hybrid seed production and their cultivation therefore has to be popularized on war footing to achieve intended vegetable production targets. Many hybrids have been developed both by public and private organizations especially the private companies, who are competing among then selves to produce best hybrids to compete in the world market and are selling seeds on very exhauburant rates ranging from Rs. 5000-50000 per kg seed. Though they are making very huge profits but their contribution in promotion of hybrid technology is exemplary and cannot be over looked.

Recently a major break through has been realized in cabbage, tomato and brinjal production using hybrids. The entire cabbage and tomato production in Maharastra, Karnataka, Tamil Nadu and West Bengal is under F_1 hybrids and is being extended even in other states like Punjab, Gujarat Delhi, Haryana, J and K, U.P. Rajasthan etc. Growing of these hybrid varieties not only fetching higher income to growers but also have better price in the market for their uniform quality produce as compared to the commonly grown varieties. The average yield of hybrid tomato in these states is about 75-80 t/ha as against 20-25 t/ha of the other varieties. Generally, the vegetables are sold at a premium price, about 1 or 2 rupees more per kg than others. For cultivation of hybrids varieties, advance production technology with adequate inputs is a necessary requirement. Without this, it would not be possible to exploit the full potential of hybrid vigour and obtain the maximum return to cover the extra expenses in production. Unfortunately in our country, the progress in developing and popularizing hybrids varieties has been slow and insignificant mainly due to high cost of hybrid seed and their non-availability. An average vegetable grower is not in a position to afford greater investment on buying hybrid seeds at such a high price.

The high cost of hybrid seeds which is mainly because of expensive manual operations involved in its production. However, it is false belief as the cost of the seeds is less than ten percent of the total expenditure incurred on the cultivation of one-hectare hybrid crop. The increased price of hybrid seeds is compensated by the higher yields, 3-4 times than the normal. In the coming years, strategies have to be developed for large scale production of hybrids seeds at low cost.

For this, hybrid production based on male sterility, self-incompatibility, gynoecious or monoecious lines in important crops will have to be developed. Use of growth substances like MH, etherl, gibberellic acid or silver nitrate may also be important in economizing hybrid seed production after proper standardization. Tapetum based stable transgenic male sterile system for production of hybrid seed and mass propagation of parental lines especially self-incompatible, male sterile and gynoecious lines *in vitro* are some of the important biotechnological areas, which can play a vital role to make the hybrid technology a success. To produce larger quantity of hybrid seed at a reasonable cost different strategies involving natural and man made mechanisms, which ease the hybrid production, have to be considered and exploited.

Hybrid Seed Production System

Public institutes and private sector may also cooperate indirectly in ways of developing the crop technological skills of the farmers and to produce quality seeds. With the present hectarage of vegetables grown in India, the vegetable seeds requirements are around 7981.34 tonnes. The majority of seed presently used is form open pollinated varieties, whereas the demand of hybrid seeds is gradually increasing.

Moreover, hybrids are available, have been developed by the private seed sector. Seed companies able to breed, to produce, and to market quality and hybrid seed in India are limited in number. In India the activities of the private vegetable seed industry is limited to seed trade/(import), seed production of local vegetables (mainly open pollinated varieties), and procurement of seeds from local sources. The quality of the seed distributed by those companies is often doubtful. Experience has shown that varieties (OP or F_1) of vegetable crops developed locally for the domestic market have a better success and acceptability with farmers. In view of this, and considering a long-term strategy, it is of interest to all private and public seed sector in India to strengthen their vegetable seed industry. It is also likely that the private sector will play the key role in this development.

Table 1: Estimated Seed Requirement of Vegetable Hybrids for 2002–2003

Crops	*Area Under Cultivation (ha)*	*Seeds Requirements in (tones)*
Tomato	299670	89.91
Brinjal	108420	43.37
Chilli	42100	21.05
Cabbage	156430	78.22
Cauliflower	46930	23.47
Gourds	147345	736.73
Melons	43550	130.65
Okra	95575	955.76
Total	940015	2079.13

Table 2: Major Private Vegetable Hybrid Seed Companies in India

Company Name	*Major Field of Activity*
Indo American Hybrid Seed	Tomato, okra, muskmelon and watermelon
Maharashtra Hybrid Seed Company Ltd	Eggplant, gourds and okra
Hoechst India Ltd	Eggplant and okra
Bejo Sheetal Seed Co Ltd	Tomato
Century Seeds Co Ltd	Okra and eggplant
Nath Seed Co Ltd	Watermelon
Pioneer Seed Co Ltd	Okra and tomato
Sungro Seeds	Tomato and eggplant
Namdhari Seeds Pvt, Ltd	Watermelon and tomato
Sandoz (India) Ltd	Tomato, chilli and cabbage

Types of Hybrids Used for Commercial Cultivation

The various types of hybrids utilized in the hybrid seed programme of vegetable crops are the single crosses, double crosses, double top crosses, triple crosses etc. A summarized account for these hybrids is given in Table 3.

Table 3: Commercial Hybrids Used in Some Vegetable Crops

Sl.No.	*Hybrids*	*Symbolic Representation*	*Crops*
1.	Single crosses	(A x B)	Brinjal, carrot, chillies, cucurbits, onion, tomato and watermelon
2.	Double crosses	(A x B) x (C x D)	Sugerbeet
3.	Triple crosses	[(A x B) x (C x D) x (E x F)]	Cabbage and kale
4.	All types of hybrids	–	Potato

A, B, C, D, E and F are constituent inbred lines of the hybrids.

Types of Crossing

The five different types of crossing used in the production of seed of F_1 hybrids are as follows:

1. Both emasculation and pollination have done by hand: Tomato, brinjal, capsicum and muskmelon (andromonoecious varieties).
2. Artificial pollination alone required:

   Cucurbits: Cucumber, watermelon, muskmelon, pumpkin and squash.

   Tomato: using male sterile lines.

   Cucumber: using pistillate lines.
3. Emasculation and use of wind pollination:
4. Sweet corn by detasselling.

   Spinach: Emasculation by pulling out male plants.
5. Emasculation or removal of male flowers and use of insect-pest pollination:

   (*a*) Cucurbits: Cucumber, watermelon, pumpkin, squash and bottle gourd.

   (*b*) Free Insect-pollination:

6. Using self-incompatibility: Cabbage, Chinese cabbage, radish, turnip, Brussels sprouts, broccoli and cauliflower.
7. Using male sterility: Onion, carrot and radish.

Procedures of Hybrid Seed Production

Heterosis breeding involves three important stages, namely production of inbred lines, testing of combining ability of inbred lines, and production of seed in bulk of the promising and desirable F_1 hybrids.

Inbreeding and Production of Inbred Lines

Inbreeding is the mating of closely related individual which results in a large number of lethal weak types and a general decline in their vigour, fertility, and fecundity. Recessive, lethal, and deleterious genes are usually concealed in a heterozygous system, which are increased in the homozygous balance after inbreeding. Inbreeding increases homozygosity and reduces the proportion of heterozygosity in the population, thus reducing the vigour of plants. For inbreeding, usually selfing and sibmating are practiced, for selfing in cole crops, bud pollination is practiced. For quick and efficient selfing, electrically aided pollination and thermally aided pollination is followed. In self–pollinated crops, i.e; tomatoes, eggplants, chillies and beans, pure lines maintained in the collection are used as inbred lines. In case of cross–pollinated crops like onion, cabbage, cauliflower, Brussels sprouts, radish, turnip, carrot and beet, it is necessary to produce inbred lines. However, in cucurbits like cucumber, pumpkin, squash, bottle gourd, watermelon and muskmelon in which varieties can be maintained in pure form by inbreeding without loss of vogour. It is not necessary to produce inbred lines. In cucurbitaceous crops, the homozygous varieties can be used directly in place of inbred lines as done in the self–pollinated crops like tomato, brinjal and capsicum. When inbreds are to be produced, selection is done within and between the inbred lines to obtain promising inbreds. These inbred lines are usually used in single crosses in onion, cucurbits, cabbage, cauliflower, broccoli, Brussels sprouts, carrot and spinach, in double crosses in cabbage and sweet corn or in three–way crosses in sweet corn and onion. For achieving uniformity in the hybrids, it is essential to inbreed the parents. The degree of inbreeding required in the parents will be determined by the extent of uniformity desired. The generations of artificial inbreeding of selfing depends upon the degree of inbreeding depression in crop. A high degree of inbreeding has been observed in Brussels sprouts, marrow–stem kale, cabbage and broccoli and a moderate to low degree in radish, carrot, onion and beet. In cole crops, there is a high degree of inbreeding depression, therefore, it requires five generations of inbreeding, but carrot and onion require only three to four generations of inbreeding, as they show moderate degree of inbreeding depression. In cucurbits, inbreeding can be practiced to obtain homozygous true–breeding varieties without any loss of vigour. Uniformity may not be desired so much in the hybrids in which disease resistance is an important consideration as heterogeneity may offer greater insurance against diseases, particularly to a new pathogen or physiological race. Therefore, testing of uniformity of parents is essential as the uniformity of hybrids depends entirely upon the homozygosity of the parents. When there is more homozygosity, there is a problem of maintenance of inbred lines. Proper maintenance of parental lines is an integral part of hybrid seed production.

Testing the Combining Ability of Inbred Line

The successful exploitation of hybrid vigour depends upon combining abilities (gca and sca) of the parental lines and the techniques used for economical hybrid seed production. It has been observed that some varieties have better ability to transmit high yield and quality to their progeny. Some times there are also differences in reciprocal hybrids. Generally, sca value of the cross gives better predictive information than gca of the parents. Besides, sca is also more reliable for the selection of parents for hybrid production. However, parents can also be selected on the basis of genetic diversity or crossing widely different varieties in a number of vegetable crops, which produces potential hybrids. The combining ability of inbred lines can be tested in top cross, single cross or paired cross, diallele cross and poly cross. The top cross and poly cross tests are helpful in determining the general combining ability, while, by single or pair crosses, the specific combining ability can be determined. The diallele cross, which includes a complete set of all possible combination of crosses, determine both the general and specific combining abilities. It is convenient to follow top cross in sweet corn and beet, poly cross in cabbage, Brussels sprouts, and cauliflower in which inbred lines can be easily maintained by vegetative propagation, and single and diallel crosses in onion, radish, turnip, cabbage, cauliflower, Brussels sprouts, broccoli, cucumber, watermelon, muskmelon, pumpkin, squash, spinach, tomato and brinjal. Earlier, Nieuwhof (1969) had also suggested the use of paired cross, diallel cross and poly cross in cole crops.

Production of Hybrid Seed

Shull (1909) was the first to produce a single–cross hybrid in maize involving two inbred lines. The practical utilization of a single–cross hybrid suffered from the limitation of its higher cost. It was produced from inbreds, which were poor yielder, thus, less quantity of hybrid seed was produced. This was the reason that single–cross hybrids were uneconomical and did not attain commercial importance. However, in tomato, pepper, cucumber, muskmelon, and in some cole crops, single–cross hybrids are being commercially used for hybrid seed production. The utilization of hybrid in maize came into practice only when Jones (1918) produced a double cross hybrid by crossing two single hybrids, and this was economical because of a heterotic hybrid was taken as the seed parent. In maize, such hybrids are very popular in maize growing countries. Production of a double–cross hybrids have also been suggested earlier in cabbage (Odlandand and Knoll, 1950) and marrow stem–kale (Thompson, 1959). In cole crops, particularly in kale, a triple–cross hybrid has been suggested by Thompson (1959) and in carrot, it has been reported by Erickson *et al* (1982). Single cross and triple cross hybrids have been reported by Thompson (1951) in spinach. Although double cross and triple cross hybrids are economical, they lack uniformity. In pointed–headed cabbage, the single cross hybrid has been found superior to double cross (Nieuwhof and Garretsen, 1975). Therefore, in vegetable crops, single–cross hybrids are generally preferred to double and triple hybrids. Thus, single cross is the best alternative for horticultural quality and value.

It is comparatively easy to produce the seed of F_1 hybrids in dioecious crops like spinach and asparagus, monoecious vegetables like cucurbits and sweet corn and self–incompatible crops like cabbage, Brussels sprouts, broccoli, cauliflower, radish and turnip than in the self–fertile species having hermaphrodite flowers, such as tomato, brinjal and capsicum. Three main mechanisms utilizing natural pollination have facilitated the procedure of crossing. These include the sex expression, dioecism in spinach and asparagus and monoecious in cucurbits and sweet corn; self-incompatibility in cabbage, Chinese cabbage, Brussels sprouts, broccoli, cauliflower, radish and turnip; and sterility–both genetically male sterility as in tomato, brinjal, capsicum, onion, pumpkin and squash, and triploidy in seedless watermelon. In self–fertile species possessing hermaphrodite flowers, emasculation is usually done either by hand or genetically with the help of male–sterility genes.

Factors Affecting Production and Popularization of Hybrid Seeds

In India, neither the actual seed grower nor the vegetable grower harvesting the full benefits. It has been generally seen that the maximum benefits have been shared by the middlemen and the companies. We have to develop such systems where prices of hybrids seed are reasonably low so that vegetable grower has no difficulties in purchasing the seed of high yielding hybrids and simultaneously the seed producer should get a maximum share and profit for the hybrid he is producing. Once, seed producer and vegetable growers both are satisfied, the system of seed production will be economically viable and lost till the balance is in equilibrium. Successful hybrid seed production programme depends upon a large number of factors such as:

1. Cost of hybrid seed production.
2. Hybrid seed demand of a particular crop
3. Hybrid seed productivity per unit area.
4. Share of profit on sale of hybrid seed by the seed producer.
5. Economic status, growing knowledge and expertise of the farmer.
6. Sale price and quality of the hybrid seed.
7. Additional productivity of vegetables to the grower through hybrid seed.
8. Additional returns from vegetables by use of hybrid seed.

Hybrid Varieties Testing

For commercial utilization, the hybrids are released for commercial cultivation by either the central or State Varieties Release

Committee. The release of a hybrid for commercial cultivation is based on conclusive demonstration of its superiority over the best existing hybrids (included as check in the evaluation trials) in yielding ability or in some other feature of economic importance *e.g.* abiotic and biotic stress etc. Due to above reason, the hybrids are extensively evaluated for their performance, diseases and insects resistance and quality etc. in multilocation trials. The multilocation trials are conducted under All India Coordinated Vegetable Improvement Projects, which play a key role in testing, identification and release of new hybrids. Before release of hybrids, three activities and operations are involved. These are:

1. Evaluation
2. Identification
3. Release and notification

Evaluation

It consists of following seven different types of trials / tests conducted during evaluation. In all these trials, the best existing hybrid is included as a check for comparison. In general, more than one (often two or three) hybrid is included as check.

1. Station trials
2. Multilocation trials
3. National trials
4. Adoptive trials
5. Minikit trials
6. Trials on disease and insects
7. Trials or quality test

Station Trials

The breeder who has developed the hybrids, conducts it. It may be conducted for one year or two years and is often called as preliminary yield trial. The objective of station trial is to make sure that the hybrid/hybrids developed by breeder are superior in performance (at that location) to the best available hybrid for the region before they are included in the trials of All India Coordinated Vegetable Improvement Project. This is of great importance because in the trials of coordinated project, the hybrid/hybrids would not only have to compete with the best existing hybrid, but also with the hybrids developed by other breeders. The salient points in station trial are as:

1. To avoid disappointment and rejection in the early stages of testing, the breeder should himself make sure to the superiority of the hybrids, he has isolated.
2. The best existing hybrids should be used as checks.
3. Standard package and practices should be adopted for cultivation.
4. Disease and insect relation with hybrid should also be seen.
5. The station trial may be repeated once make sure that new hybrids are superior to the existing hybrids.

Multilocation Trials

These trials are conducted under the All India Coordinated Vegetable Improvement Project. The objective of these trials is to evaluate the performance of newly developed hybrids at several locations distributed over the region. Since the soil and climatic conditions show a large variation from one region of the country to the other, the country has been divided into several agro-climatic zones (Table 4). Each zone consists of areas having similar agro-climatic conditions. It should be noted that zones have not been created on the basis of political boundaries, but on the basis of agro-climatic conditions, prevailing in the area. Thus it may be expected that the agro-climatic conditions at any two locations within a zone would be more comparable than those at two locations belonging to any two different zones.

The trials conducted under coordinated projects are grouped in two categories:

1. Initial evaluation trials
2. Advanced hybrid trials

Initial Evaluation Trials and Advanced Hybrid Trials

The initial evaluation trial is conducted at 10-12 locations within a zone and it is conducted under irrigated conduction. The hybrids are tested in IET is for two years only. These are promoted for AHT trials if their performance are outstanding, otherwise, they are rejected. The objective of IET is to eliminate relatively inferior entries, so that their number is reduced to a manageable size for AHT evaluations. Both in IET and IHT, the crops are raised under good environmental conditions. The plot size and spacing in both IET and AHT trials varies from crop to crop as given in Table 5.

Table 4: Eight Agro-climatic Zones and States Under Each Zone

	Agro Climatic Zones	*States*
I.	Humid Western Himalayan region	Jammu and Kashmir (J&K), Himachal Pradesh (HP) and Uttranchal.
II.	Humid Bengal-Assam Basin	West Bengal and Assam.
III.	Humid Eastern Himalayan and Bay Islands	Sikkim, Meghalaya, Manipur, Nagaland, Mizoram, Tripura, Arunachal Pradesh and Andaman and Nicobar Islands.
IV.	Sub-Humid Sutlej Ganga Alluvial Plain	Punjab, Uttar Pradesh and Bihar
V.	Sub-Humid to Humid Eastern South Eastern Uplands	Eastern Madhya Pradesh, Orissa and Andhra Pradesh
VI.	Arid Western Plains	Rajasthan, Gujarat, Haryana and Delhi
VII.	Semi Arid lava Plateau and Central High Lands	Madhya Pradesh (excluding eastern area) and Maharashtra
VIII.	Humid to Semi Arid Western Ghats and Karnataka Plateau	Karnataka, Tamil Nadu and Kerala

National Trials

The national trials are conducted throughout the country in all the zones. The entries in national trials consist of one entry from each zone that ranked first in that zone in uniformity Regional trials of the previous year. The national trials serve as IET for an entry in zones than that in which it was developed and tested in Uniformity Regional Trials. The purpose of national trials is to evaluate outstanding entries of one zone in the other ago-climatic zones to see if they perform well in other zones as well. It has been suggested that national trials should include IET entries of the entire zone so as to evaluate all the entries on a national level.

Table 5: Plot Size and Spacing for IET and AHT Trials

Sl.No.	Names of Crops	Plot Size (m)	Spacing (cm)	Design	Replication
1.	Brinjal	4x 3.6	60x45	RBD	3
2	Tomato	3.6x 3	60x45	RBD	3
3.	Chilli	4.3x 3	50x45	RBD	3
4.	Capsicum	4.3x 3	50x45	RBD	3
5.	Okra	3x 2.7	60x30	RBD	3
6.	Cucumber	4x 1.8	100x 30	RBD	3
7.	Bottle gourd	10x3	300x50	RBD	3
8.	Bitter gourd	5.5x 2.5	150x300	RBD	3
9.	Watermelon	2x 5.4	150x 90	RBD	3
10	Ridgegourd	6x 2.5	300x50	RBD	3
11.	Cauliflower	4.5x 3.2	45x40	RBD	3
12.	Cabbage	4.5x 3	45x30	RBD	3
13.	Broccoli	4.5x3	45x 30	RBD	3

Adoptive Research Trials

These trials are conducted on research stations or farms of state governments. The hybrids identified by the workshop in All India Coordinated Vegetable Improvement are included in these trials. Trials are conducted in without replication in very large area with agronomic practices as same for which the concerned entry has been identified. The data from these trials are considered for release of the identified entry as a new hybrid.

Minikit Trials

The minikit trials are conducted in the farmer's fields. These trials are conducted along with the adoptive trials in the following crop season after the entry has been identified for release. The objective of minikit trials is to popularize the new hybrid among farmers of the zone. It also serves another purposes; the seed of a good new hybrid reaches the farmers.

Diseases and Insects-test

Entries are evaluated for diseases and insects resistance throughout the period of testing, that is, during IET as well as AHT.

Diseases and insects resistances are tested both under natural epidemic as well as artificial epiphytotic (epidemic) conditions. The disease reaction tests for different diseases of various crops are carried out at different places where epidemics of the concerned disease occur regularly. The places where a disease occurs at a high intensity, that is, the disease occurs in an epidemic form, every year are known as hot spots for that disease. Serious diseases and insects occur in vegetable crops, required attentions are given in Table 6.

Table 6: Major Diseases and Insects of Vegetable Crops

Sl.No.	*Name of Crops*	*Diseases*	*Insects*
1.	Tomato	Fusarium wilt, verticilium wilt, late blight, septoria leaf spot, anthracnose, bacterial wilt, bacterial canker, leaf curl virus, tomato mosaic and root knot nematodes	Fruit borer and white fly
2.	Brinjal	Bacterial wilt, phomopsis blight, little leaf and root knot nematodes	Shoot and fruit borer
3.	Bell pepper and chilli	Fruit rot, cercospora leaf spot, powdery mildew, bacterial leaf spot, phytophthora root rot, root knot nematode and common TMV	Thrips, mite, aphid, and fruit borer
4.	Okra	Yellow vein mosaic virus, fusarium wilt, cercospora leaf spot and fruit rot	Fruit and shoot borer, jassids and whitefly
5.	Cauliflower and Cabbage	Black rot, sclerotinia rot, alternaria blight, irwinia rot and cabbage yellows	Cabbage butterfly aterpillar, aphids, diamond back moth.
6.	Carrot	Alternaria blight and cercospora leaf spot	
7.	Onion	Purple blotch, basal rot, stemphyllium blight and bacterial storage rot	Thrips
8.	Cucurbits	Powdery mildew, downey mildew, anthracnose, cucumber mosaic virus	Red pumpkin beetle, fruit fly and aphids

Quality Tests

Quality tests are conducted to determine the suitability of an entry for the various use of its produce, the quality in vegetable crops is not simple characters, and is often not easy to determine. Therefore, a number of different tests have to be conducted for which sophisticated equipments may be needed for speedy and reliable results. In each crop, a different set of quality tests has been developed in order to determine the suitability of new hybrids of the crops for various uses. The quality parameters, which requires in vegetable hybrids are given below in Table 7.

Table 7: Quality Parameters in Vegetable Crops

Sl.No.	*Name of Crops*	*Quality Parameters*
1.	Tomato	Large round fruit with adequate firmness and shelf life, uniform fruit size, shape, red colour and freedom from external blemishes or abnormalities for fresh market. Large fruit size, high fruit quality and continuous production for home garden tomatoes. Deep uniformally red coloured tomatoes, pH below 4.4, high total soluble solids (4.5-7 per cent) and high alcohol insoluble solids (AIS) in processing tomatoes.
2.	Brinjal	Fruit shape, size and colour as per consumer preference with soft flesh, low proportion of seed and lower olenine content.
3.	Bell pepper and chilli	Oblete or round fruit in bell pepper and long fruit in chilli. Pleasing flavour, high sugar/acid ratio, high pigment content and vitamin C in bell pepper and high capsaicin ($C_{18}H_{27}NO_3$) and high oleoresin in chilli
4.	Okra	Dark green, tender, thin, medium long, smooth 4-5 ridged pods at marketable stage. Pods should be free from conspicuous hairs and optimum seed setting ability.
5.	Pea	Long, attractive green pods with more seeds/pod. Sweetness in pods with high shelling percentage.
6.	Cauliflower	Non-ricey, compact bract free protected curds with retentive cream/white colour and bettter seeding quality.
7.	Cabbage	Longer staying capacity in field after formation, narrow, short and soft core with short stem.
8.	Knol-khol	Spherical swollen bulb like stem and it should be free from bolting.
9.	Radish	White, long/stump roots with thin tap root and non-branching habit, nonpithy roots with slow bolting habit. Pungency of roots should be as per consumer preference.

Contd...

Table 7–Contd...

Sl.No.	*Name of Crops*	*Quality Parameters*
10.	Carrot	Scarlet/orange colour roots with uniformity in shape and size. Smooth root surface, thick flesh roots, and thin and self-coloured core in roots. Broad shouldered, cylindrical, uniformly tapering or stump rooted with non-branching habit. Early rooting with free from cracking. High sugar and dry matter content in roots.
11.	Turnip	Root colour as per consumer's preference, white, purple types more liked in India that Golden Ball types. Stump rooted hybrids with thin taproot and non-branching habit. Slow bolting habit and appropriate dry matter (8-9 per cent) in roots.
12.	Beet root	Dark red and uniformly coloured roots along with uniform root shape. Absence of internal white rings in roots. Slow bolting with manogerm seeds.
13.	Onion	High TSS is important for dehydration industry producing onion chips and powder.
14.	Cucumber	Attractive green or dark green fruits with smooth surface and without prominent spines or prickles. Uniform long cylindrical shape without crookneck. Fruits should be free from carpel separation without hollow spots with free from bitterness along with less seeds at edible maturity.
15.	Muskmelon	Fruits should be attractive round/spherical in shape with thick flesh and attractive in colour. Small seed cavity, sweet, juicy musky flavoursome fruits. TSS should not less than 10 per cent with tough netted skin of fruit.
16.	Watermelon	Tough skinned fruits for long distance transportation. TSS content not less than 10 per cent. Fruits with smaller and fewer seeds with attractive deep red flesh and firm flesh.
17.	Squash and Pumpkin	Thick fruit flesh and small seed cavity. Round/oblong/flat round fruit shape. Orange flesh colour, rich in β carotene, a precursor of vitamin A.
18.	Bottlegourd	Attractive green fruit of round/long/club shaped with sparse hairs persisting on skin and nonfibrous flesh at edible stage. Fruits should be nonbitter.
19.	Bittergourd	Less ridged fruit surface with thick fruits particularly suitable for stuffing. Fruit size variation as per consumer's preference (Small 7.5-10 cm long, medium long 10-15 cm, long 15-20 cm and extra long 20-40 cm). Immature seeds for longer period during green edible stage.

Hybrid Testing Centres in India

Following organizations in various states of India are actively engaged for testing of vegetable hybrids under All India Coordinated Vegetable Improvement Project (Table 8).

Table 8: List of Organizations and their Respective Crops

Sl.No.	*Name of Organisations*	*Crops*
1.	Indian Institute of Vegetable Research, 1, Gandhi Nagar (Naria), Varanasi – 221 005, (Uttar Pradesh)	Brinjal, tomato, chilli, capsicum, okra, cucumber, bottle gourd, bitter gourd, cauliflower, watermelon, ridge gourd, broccoli, cabbage and onion
2.	Department of Vegatable Crops, Indian Agricultural Research Institute, Pusa – 110 012 (New Delhi)	Brinjal, tomato, chilli, okra, cucumber, bottle gourd, bitter gourd, cauliflower, ridge gourd, broccoli, cabbage and onion
3.	Division of Vegetable Crops, Indian Institute of Horticultural Research, Hessaraghatta, Lake Post, Bangalore – 560 089	Tomato, chilli and okra
4.	Indian Agricultural Research Institute, Research Station, Katrain – 175 129, Kullu Valley (Himachal Pradesh)	Capsicum, cucumber, broccoli and cabbage
5.	Vivekananda Parvitya Krishi Anusandhan Shala, Almora – 263 601 (Uttaranchal)	Tomato, okra and broccoli
6.	Horticulture and Agroforestry Research Programme, Namkum, Ranchi – 834 010 (Jharkhand)	Brinjal, tomato, onion and ridge gourd
7.	Department of Horticulture, Horticulture College and Research Institute, Tamil Nadu Agricultural University, Coimbatore – 641 003 (Tamil Nadu)	Brinjal, chilli, okra and onion
8.	Department of Vegetable Crops, Punjab Agricultural University, Ludhiana – 141 004 (Punjab)	Brinjal, tomato, chilli, okra, bitter gourd, watermelon, broccoli, cabbage and onion
9.	Department of Horticulture, Mahatma Phule Krishi Vishwavidyalaya, Rahuri – 413 722 (Maharashtra)	Brinjal, chilli, capsicum, cucumber, bitter gourd, cauliflower, watermelon, ridge gourd and onion

Contd...

Table 8–Contd...

Sl.No.	Name of Organisations	Crops
10.	Department of Horticulture, Mahatma Phule Krishi Vishawavidyalaya, Rahuri – 413 722 (Maharashtra)	Brinjal, chilli, capsicum, cucumber, bitter gourd, cauliflower, watermelon, ridge gourd and onion
11.	Department of Horticulture, College of Agriculture, Orissa University of Agriculture and Technology, Bhubaneswar – 751 003 (Orissa)	Okra
12.	Department of Horticulture, University of Agricultural Sciences, Dharwad – 580 005 (Karnataka)	Chilli, capsicum, watermelon, ridgegourd and onion
13.	Department of Horticulture, Agricultral Research Station, Durgapura, Jaipur – 30 2018 (Rajasthan)	Tomato, watermelon, ridge gourd and broccoli
14.	Department of Vegetable Crops, Narendra Dev University of Agriculture and Technology, Narendra Nagar (Kumarganj), Faizabad – 224 229 (Uttar Pradesh)	Tomato and chilli
15.	Department of Vegetable Crops, Choudhury Charan Singh Haryana Agricultural University, Hisar – 125004 (Haryana)	Chilli, bitter gourd and onion
16.	Agricultural Research Institute, Andhra Pradesh Agriculture University, Rajendra Nagar, Hyderabad – 500 030 (Andhra Pradesh)	Okra
17.	Department of Vegetable Crops, College of Agriculture, Jawaharlal Nehru Krishi Vishawavidyalaya, Jabalpur – 482 004 (Madhya Pradesh)	Tomato and okra
18.	Department of Horticulture, Assam Agricultural University, Jorhat – 785 013 (Assam)	Okra
19.	Vegetable Research Station, Gujarat Agricultural University, Junagadh – 362 001 (Gujarat)	Okra and ridge gourd
20.	Department of Horticulture, Vegetable Research Station, Chandra Sekhar University of Agriculture and Technology, Kalyanpur, Kanpur – 208 024 (Uttar Pradesh)	Chilli, okra, bottle gourd, bitter gourd, cauliflower, ridge gourd and onion

Contd...

Table 8–Contd...

Sl.No.	Name of Organisations	Crops
21.	Department of Horticulture, Bidhan Chandra Krishi Vishawavidyalaya, Distt. Nadia, Kalyani – 741 235 (West Bengal)	Tomato, chilli, watermelon and onion
22.	ANG Research Agriculture Research Station, Ranga Agriculture University, Lam.	Tomato, chilli, okra, and ridge gourd
23.	Department of Vegetable Crops, Govind Ballabh Pant University of Agriculture and Technology, Udham Singh Nagar, Pantnagar – 263 145 (Uttar Pradesh)	Tomato, cauliflower, bottle gourd, bittergourd, cauliflower and ridge gourd.
24.	Department of Vegetable Crops, College of Horticulture, Dr. Y.S. Parmar University of Horticulture and Forestry, Solan – 173 230 (Himachal Pradesh)	Tomato, capsicum, cucumber, broccoli, and cabbage
25.	Division of Olericulture, Sher-E-Kashmir University of Agriculture and Technology, Shalimar, Srinagar – 191121 (Jammu and Kashmir)	Brinjal and capsicum
26.	Department of Olericulture, College of Horticulture, Kerala Agriculture University, Vellanikkara – 68 0654 (Kerala)	Bitter gourd
27.	Department of Botany and Plant Breeding, B.A. College of Agriculture, Gujarat Agriculture University, Anand – 388 110 (Gujarat)	Brinjal, tomato, chilli, capsicum, okra and cauliflower
28.	Department of Vegetable Crops, Himachal Pradesh Krishi Vishawavidyalaya, Palampur, Distt. Kangra, Himachal Pradesh	Cabbage
29.	College of Forestry and Hill Agriculture, Hill Campus, Ranichauri – 249 199 (Uttaranchal)	Capsicum
30.	Taluka Seed Farm, Maharashtra Agriculture University, Ambajojai – 431 517 (Maharashtra)	Chilli, okra, cauliflower and ridge gourd
31.	Defence Research and Development Organization, Defence Agriculture Research Laboratory, Ministry of Defence, P.B. No. 6, Pithoragarh – 262 501 (Uttaranchal)	Capsicum and cucumber

Contd...

Table 8–Contd...

Sl.No.	Name of Organisations	Crops
32.	Ankur Seed Pvt. Ltd., 27, New Cotton Market Layout, Nagpur – 440 018 (Maharashtra).	Brinjal, tomato, chilli, okra, cucumber, bottle gourd, and bitter gourd.
33.	Sheetal Hybrid Seeds Co. Ltd., P.B. No. 72, A-2, Mama Chowk, Jalna – 431 203	Brinjal, chilli and bitter gourd
34.	Syngenta India Ltd., Seed Sector, WAOGHOLI, Gate No. 2347, Nagar Road Taluk Haveli, Pune	Brinjal, tomato, Chilli, Okra, Cauliflower and Watermelon
35.	Division of Horticulture, ICAR Research Complex for NEH Region, Umiam – 793 1039 (Meghalaya)	Cabbage

Identification of Hybrids for Release

Outstanding hybrids are identified for specific zone at the annual workshops of All India Coordinated Vegetable Improvement Projects. The proposal for identification of hybrids is in an open forum by constituting a committee in the workshop. The respective breeder who has developed the concerned hybrid prepare proposal for the identification of entries. They are prepared according to a prescribed proforma that include performance of hybrid in AHT for at least two years, data from pathological, entomological, quality tests etc. and those from agronomic trials. The criteria for identification of entries vary considerably from one crop to the other. An entry may be identified if it falls into one of the following two categories:

1. An entry significantly superior to the check in yield and comparable to the check in disease resistance.
2. An entry comparable to the check in yield but constantly and markedly superior to it in disease resistance.

These criteria are fixed in the workshop of the concerned coordinated projects and reflect the experiences of the workers with the concerned crops. As a result, they are not fixed entities, and are likely to vary as per the needs of the nation as well as the experiences of the concerned workers. Tremendous progress has been made by public and private sectors in the development of hybrids in several vegetables including tomato, brinjal, chilli, bell pepper, cabbage, cauliflower, cucumber, muskmelon, watermelon, carrot, onion etc. and till date 64 vegetable hybrid varieties have been recommended for cultivation in various parts of country based on multilocational trials conducted under All India Coordinated Vegetable Improvement Project (AICVIP). Name of these hybrids along with parental organization and recommended zones are given in Table 9.

Table 9: List of Some Public and Private Sector Breed Hybrids Identified/Released for Cultivation

Crop/Hybrids	*Developing Centre*	*Recommended Zone(s)*	*Year of Identification*
	Brinjal		
Arka Kusumkar	IIHR Banglore	VIII	1981
Kt-4	IARI, Katrain	VIII	1987
Pusa Hybrid-6	IARI, New Delhi	IV	1990
Arka Navneet	IIHR, Banglore	-	1991
Pusa Hybrid-5	IARI, New Delhi	IV, VII, VIII	1992
ARBH-201	Ankur Seeds, Jalana	IV, V, VI, VII	1993
NDBH-1	NDU and T, Faizabad	IV, VII, VII	1993
MHB-10	MAHYCO Seeds, Jalana	IV, VI, VII	1993
NDBH-6	NDUA and T, Faizabad	IV	1995
Pusa Hybrid-9	IARI, New Delhi	VI	1997
ARBH-541	Ankur Seeds, Jalana	All	2001
PBH-6	Pandey Seeds, Faizabad	All	2001
JBH-1	GAU, Junagarh	All	2001
BH-1	PAU, Ludhiana	IV	2002
BH-2	PAU, Ludhiana	IV, V,	2003
VRBHR-1	IIVR, Varanasi	IV, VI	2003
NDBHL-54	IIVR, Varanasi	IV	2004
ARBH-786	Ankur Seeds, Jalana	IV	2004

Contd...

Table 9–Contd...

Crop/Hybrids	Developing Centre	Recommended Zone(s)	Year of Identification
	Tomato		
ARTH-4	Ankur Seeds, Jalana	IV, VII	1992
MTH-6	MAHYCO Seeds, Jalana	VII, VIII	1992
Pusa Hybrid-2	IARI, New Delhi	I, IV, VI, VII	1993
NA-501	Nath Seeds, Aurnagabad	VI, VII	1995
DTH-4	IARI, New Delhi	VII	1995
Avinash-2	Novartis Seeds, Pune	VI	1998
HOE-303	Novartis Seeds, Pune	IV	1998
Sun-496	Sungrow Seeds, New Delhi	II, IV, VIII	1999
BSS-20	Beejo Sheetal Seeds, Jalana	All	2001
DTH-8	IARI, New Delhi	IV	2002
CHTH-1	HARP, Ranchi	IV	2002
ARTH-128	Ankur seeds, Jalana	VIII	2002
KTH 2	CSAUT, Kanpur	IV, V	2003
JKTH-3055	J K Seeds, Secunderabad	I, IV	2004
KTH-1	CSAUAT, Kanpur	IV	2004
Nun-7730	Nunhems Seeds, Hyderabad	I, IV	2004
	Chilli		
HOE-888	Novartis Seeds, Pune	IV, VIII	1997
ARCH-236	Ankur Seeds, Jalana	IV	1997
Sungro-86-235	Sungro Seeds, New Delhi	IV and VIII	2002
ARCH-228	Ankur Seeds, Jalana	IV, V, VI	2003
	Bell pepper		
KT-1	IARI (Katrain)	I	1990
Lario	Syngenta Seeds, Pune	I	2002
DARL-202	DARI, Almora	I, IV	2003
	Cabbage		
Pusa Synthetic	IARI (Katrain)	I, II, IV	1992
Sri Ganesh Gol	MAHYCO Seeds, Jalana	V	1992
Nath-401	Nath Seeds, Aurangabad	I, IV, V, VI, VII	1993
BSS-32	Bejo Sheetal, Jalana	VII	1995

Contd...

Table 9–Contd...

Crop/Hybrids	Developing Centre	Recommended Zone(s)	Year of Identification
Quisto	Novartis Seeds, Pune	IV	1998
SYCFH-202	Syngenta Seeds Pune	IV, VIII	2004
Summer King	Sungrow Seeds, New Delhi	I, IV	2004
	Cauliflower		
Pusa Hybrid-1	IARI, New Delhi	II, IV	1992
DCH-541	IARI, Ned Delhi	II, IV	2003
	Muskmelon		
M-3 Hybrid	IARI, New Delhi	IV	1990
MHY-5	RAU,Durgapura	VII	2002
	Watermelon		
Arka Jyoti	IIHR, Bangalore		1981
MHW-6	MAHYCO Seeds, Jalana	IV, V, VIII	1999
	Bottle gourd		
PBOG-2	GBPU and T, Pantnagar	VII	1998
PBOG-1	GBPU and T, Pantnagar	IV, VI	1999
NDBH-4	NDUAT, Faizabad	All	2001
	Okra		
DVR-1	IIVR, Varanasi	IV, VII	1998
DVR-2	IIVR, Varanasi	VI	1998
DVR-3	IIVR, Varanasi	All	2001
DVR-4	IIVR, Varanasi	IV, V and VII	2002
	Bitter gourd		
Pusa Hybrid-2	IARI, New Delhi	IV, V, VI	2003
	Cucumber		
NBGH-167	Nirmal Seeds, Maharashtra	IV	2004
Hybrid No.1	Century Seeds, New Delhi	IV, VII	2004

Besides identified hybrids, several vegetable hybrids are popular among the farmers. Seeds of these hybrids are also being marketed by private organizations (Table 10). Realizing potentials of vegetable hybrid technology in India, Indian Council of Agricultural Research (ICAR) initiated a network project entitled "Promotion of Hybrid Research in Vegetable Crops" during 1995-96. This project was initiated in nine vegetables, *viz.*, tomato, brinjal, chilli, capsicum (bell pepper), okra, onion, cabbage, cucumber and bitter gourd with the following objectives: (i) to promote hybrid research in order to increase productivity *per se* the country, (ii) to incorporate biotic stress resistance in the hybrids, (iii) to strengthen hybrid seed research and hybrid seed production technology (iv) to test the proven hybrids for their potentiality and (v) to establish their production technology. This project has completed with considerable achievements. The hybrids promising at various locations are given in Table 11.

Table 10: List of Some Popular Hybrids Developed and Marketed by Private Sector

Crop	*Hybrid Variety*
Tomato	Rupali, Naveen, Rashmi, Tolstoi, Minakshi, Vaishali, Karnataka Hybrid, Lerica NS-815,NS-2530, TH-2312 and TH-802.
Brinjal	Suphal, Manju, Kalptaru and HOE-404.
Chilli	Agni, NS-1101, CH-1, CH-3 NS-1101, Tejashwani and HOE-808.
Capsicum	Indira, Bharat, Lerio, Early Bonty, NS-33 and Hira.
Cabbage	Cabbage No.8, Questo, NS-22, Hari Rani Gol, Krishna and Vishesh.
Bottle gourd	Barad and Kaveri.
Watermelon	Madhur, Milan, Nath-101 and MHW-4.
Muskmelon	Abhjeet, MCH-5 and MHW-4.
Okra	Varsha, Vijay and Vishal.

Release of Hybrids by the Central Subcommittee on Crop Standard and Release of Varieties/Hybrids

After identification, a hybrid is tested at least one year on adoptive research trials. During this period, disease tests and quality tests are also conducted. Based on data from adoptive trials and disease and quality tests, the breeder who has developed the concerned strain submits a proposal for its release as new hybrid for consideration by the Central Subcommittee on Crop Standards and Release of Varieties. The proposal is prepared according to a prescribed proforma as given below. The breeder usually gives a name or he may give it a number, which is included in the proposal. After a hybrid has been released for a zone by the Central Subcommittee on Crop Standard and Release of Varieties, the Director, High Yielding Varieties / Hybrids, Ministry of Agriculture and Irrigation, Government of India, notifies the concerned authorities of the states within that zone for seed of development and distribution of the hybrid / variety. This is known as notification of the hybrids / varieties.

Table 11: Performance of Promising Hybrids at Various Locations

Crop	*Location*	*Promising Hybrid*
Tomato	Varanasi	Avinash-2, Hemlata, TH-2312 and Ratna.
	Bangalore	TH-2312, ARTH-13, Avinash-2 and DTH-6.
	Hisar	ARTH-3
	Coimbatore	Avinash-2
Brinjal	Varanasi	NDBH-8, ARBH-201 and PDBH-13.
	Bangalore	IIHR-500 A x Arka Keshav and IIHR-104 x IIHR-491.
	Coimbatore	190 x Annamalai, 190 x Co-2 and Aruna x SM-8.
Chilli	Varanasi	BSS-188, BSS-141 and HOE-818.
	Bangalore	HOE-888, MHP-59 and ARCH-236.
	Coimbatore	Kiran
	Anand	Kiran, Tejaswini and BSS-138.
Cabbage	Katrain	Kt CBH-127, Kt CBH-95 and Kt CBH-139.
Cucumber	Varanasi	CH-13,
	Bangalore	PCUCH-1
	Rahuri	DH-9, 18 and 17 (developed form gynoecious line).
Onion	Bangalore	MS-65 x PBR-243 and MS-65 x PBR-245.
	New Delhi	Hybrid-63 and Hybrid-35.

Besides public sectors developing high yielding hybrids in vegetable crops, some private seed sectors have also been play pivotal role in developing disease resistant hybrids (Table 12).

Table 12: Disease Resistant Hybrids by Private Sector Seed Companies

	Mahyco, Jalna	
Cabbage	Harirani Gol and Kranti	Resistant to yellows and black rot.
	SriGanesh Gol	Resistant to yellows.
Ridge gourd	Surekha	Tolerance to mildews.
Watermelon	Amrut, MHW-6	Field tolerance to fusarium wilt.
Capsicum	Green Gold and MCPH-11	Resistant to TMV.
Okra	No.7, No.8, No.10	Tolerant to yellow vein mosaic virus.
	Century Seeds, New Delhi	
Tomato	Rishi, Maitri and Manmohan	Marked resistant to fusarium wilt and verticellium wilt.
Cabbage	Geetanjali	Tolerance to black rot and fusarium wilt.
Cauliflower	Early Himlata	Tolerance to black rot.
Muskmelon	Madhumati	Tolerance to downey mildew.
Chilli	Hybrid No.1	Tolerance to mosaic, leaf curl, wet rot and dieback
Okra	Panchalic and Adhunik	Tolerant to yellow vein mosaic virus.
	Beejo-Sheetal Seeds, Jalna	
Cabbage	Swati, Sarita, Bajrang, Shanti, BSS-44, Swarna and Sonali	Resistant to fusarium.
Tomato	Meenakshi and Menka	Resistant to fusarium.
Okra	Tara	Resistant to yellow vein mosaic virus.
	Sungro Seeds, New Delhi	
Tomato	Sungro 501	Tolerant to leaf curl.
	SC-1, SC-3	Tolerant to wilts.
	Sungro 830	Tolerant to alternaria.
Okra	Supriya	Tolerant to yellow vein mosaic virus.

Contd...

Table 12–Contd...

	Namdhari Seeds, Bangalore	
Chilli	NS 2302	Tolerant to virus.
	NS 1820	Tolerant to virus.
Tomato	Suraksha	Tolerant to bacterial wilt.
	Pro-Agro Seeds, New Delhi	
Tomato	Abhiman	Tolerant to early blight.
Cabbage	Bahar	Tolerant to black rot.
	Pragati	Tolerant to black rot.
Chilli	Vardan	Tolerant to anthracnose.
	Kranti and Krishna	Tolerant to TMV.
Okra	Uphar	Tolerant to yellow vein mosaic virus
	Sandoz Seeds, Pune	
Tomato	Rasika and Ratna	Tolerant to fusarium and verticillium wilt.
	Lerica	
	Avinash II	Tolerant to tomato leaf curl virus.
Chilli	Bijli and Agni	Tolerant to TMV.
Capsicum	Indra	Tolerant to TMV and PVY.
Cabbage	Quisto and Vishwas	Tolerant to Fusarium yellows
Cauliflower	Pawas	Tolerant to *Xanthomonas*, *Pernospora*, *Pseudomonas*
	Indo-American Hybrid Seeds (I) Pvt. Ltd., Bangalore	
Okra	Varsha, Vijay and Vishal	Highly tolerant to yellow vein mosaic virus.
Tomato	9501, 9502	Highly tolerant to leaf curl virus.
	Akash, Utpan and Ruchi	Tolerant to foliar diseases.
Chilli	No.5, 6 and 9	Tolerant to leaf curl and dieback.

Source: Seshadri (1998).

Proforma for Submission of Proposal of Release and Notification of Vegetable Varieties/Hybrids to Central Sub-committee on Crop Standards

Notification and Release of Varieties/Hybrids of Vegetable Crops

1. Name of the crop with latin name :
2. (*a*) Name/code/designation under which tested :
   (*b*) Proposed name of the hybrid :
3. Sponsored by
   (*a*) Institution or agency responsible for developing hybrid (with address) :
   (*b*) Name of Person (s)/Scientist responsible for the development of the variety :
4. (*a*) Breeding Objectives :
   (*b*) Breeding Method :
   (*c*) Parentage with details of its pedigree :
5. State the hybrid which most closely resemble the proposed hybrid in general characteristics :
6. (*a*) Whether recommended by workshop/ group meeting :
   (*b*) If so, its recommendation with specific justification for the release of proposed hybrid :
   (*c*) Specific areas/region for which recommended :
7. Recommended ecology including suitability under controlled/field conditions :
8. Description of Hybrid
   (*a*) Plant height :
   (*b*) Distinguishing morphological characteristics :
   (*c*) Characteristics different from those of the parents :
   (*d*) Maturity (range in number of days) seeding/ transplanting to flowering/fruiting) :
   (*e*) Maturity on ground (early medium and late) wherever such classification exists :
   (*f*) Reaction to major diseases under field and controlled conditions :
   (*g*) Reaction to major pests (Under field and controlled including storage pests) :
   (*h*) Agronomic features (*e.g.* resistance to lodging, shattering, fertilizer) :
   (*i*) Quality attributes purpose for which recommended :
      (i) Table/proceeding/report for any other :
      (ii) Table quality attributes/nutritional quality :

Contd...

Contd...

(*j*) Reaction to environmental stress	:
(*k*) Kind of root stock to be used, if applicable	:
9. Description of parents	:
(*a*) Plant height	:
(*b*) Distinguishing morphological characteristics	:
(*c*) Days to 50 per cent harvest	:
(*d*) Maturity/ripening (Range in number of days seed to be used)	:
(*e*) Is there any problem of synchronization? If yes, methods of over it	:
(*f*) Reaction to major disease (under field and controlled conditions (reactions to physiological strains/races bio types to be indicated wherever possible)	:
(*g*) Reaction to major pets (under field and controlled conditions including store pests)	:
(*h*) Agronomic features (*e.g.* fertilizer responsiveness, seed rate etc.)	:
(*i*) Reaction to environmental stresses:	
10. Yield data in AICRP trial/regional trial on farm trials (levels of fertilizer application, density of plant population and superiority over local control standard variety to be indicated	:
11. Premium attributes based on Table 1:	
12. (*a*) Agency responsible for maintaining breeder seed	:
(*b*) Quantity of breeder seed for 5 ha for low volume seed rate and 10 ha for high volume seed rate	:
13. Information on the acceptability of the variety by farmers/consumers/industry attach data generated on these aspects	:
14. Specific recommendations, if any, for seed production/seed technology	:
15. Any other pertinent information	:

Release of a Hybrid by a State Variety Release Committee

In addition to the Central Sub–committee on Crop Standard and Release of Varieties, each state has its own variety release committee. The State Variety Release Committees usually require a separate set of trials conducted at the Regional Research Station of the state, in addition to the data from coordinated multiplication trials. A hybrid released by a State Variety Release Committee must be notified by Ministry of Agriculture, Government of India, before its certified seed can be produced. Further, only the ministry notifies those varieties/ hybrids, which have been identified by the workshop of All India Coordinated Vegetable Improvement Project. The State Variety Release Committee consists of the following (1) Director of Agriculture of the state (Chairman), Director State Seeds Corporation; (3) Director, State Seed Certification agency; (4) Additional Directors of Agriculture of the state; (5) Joint Directors of Agriculture of the state; and (6) Directors of Research of all the agricultural universities of the state. In addition, (7) Deputy Directors of Agriculture of the state; (8) Officers–in–Charge of all Regional Agricultural Testing and Demonstration Centers; (9) Breeders, agronomists, pathologists and entomologists of all agriculture universities and agricultural institutes in the state are also invited. It may be noted that the persons listed under items 7, 8 and 9, infact, not the members of the committee, but they are invited to assist the committee in its deliberations. The prescribed proforma for the proposal of releasing hybrid by State Variety Release Committee is similar to proforma as given above in Central Subcommittee on Crop Standard and Release of Varieties/ Hybrids.

Chapter 9

Seed Production of Vegetables

Family–Solanaceae

Brinjal (*Solanum melongena* L.) 2n = 2x = 24

Land Requirements

The land should be free of volunteer plants. The soil of selected fields should be fertile, rich in organic matter, sandy loam and well drained.

Field Inspection

The field inspection should be done at:

1. Before flowering
2. Flowering and fruiting stage and
3. Mature fruit stage and prior to harvesting.

Figure 44: Brinjal (*Solanum melongena* L.)

Field Standards

General Requirements

Isolation

Contaminants	*Minimum Distance (Meters)*	
	Foundation	*Certified*
1	*2*	*3*
Fields of other varieties	200	100
Fields of the same varieties	200	100

Specific Requirements

Factor	*Maximum Permitted (%)**	
	Foundation	*Certified*
Off types	0.10	0.20
**Plants affected by seed borne diseases	0.10	0.50

* Standards for off types shall be met at and after flowering and for seed borne diseases at final inspection.

** Seed borne diseases shall be: phomopsis blight.

Seed Standards

Factors	Standards for Each Class	
	Foundation	Certified
Pure seed (minimum)	98.0 per cent	98.0 per cent
Inert matter (maximum)	2.0 per cent	2.0 per cent
Other crop seeds (maximum)	None	None
Weed seeds (maximum)	None	None
Germination (minimum)	70 per cent	70 per cent
Moisture (maximum)	8.0 per cent	8.0 per cent
For vapour-proof containers (maximum)	6.0 per cent	6.0 per cent

Brief Cultural Practices

Time of Sowing

Plains–February to March, June to July, October to November; Hills–March to April

Preparation of Land

Prepare the field to a good tilth by ploughing, three to four harrowings followed by levelling.

Source of Seeds

Obtain nucleus/breeder/foundation seed from source approved by a seed certification agency.

Seed Rate

200 to 500 g/ha.

Sowing of Nursery

Like tomato crop.

Fertilizers

Apply 250 to 300 quintal farmyard manure per hectare at the time of land preparation Apply 60 kg/ha nitrogen, 80 kg/ha phosphorus, and 60 kg/ha potash at the time of transplanting. Top dress 60 kg/ha nitrogen in two splits. The first top dressing should be done 15 days after transplanting and the second at flowering time. Irrigate immediately after top dressing.

Spacing

Non–spreading type: 60 x 60 cm between rows and plants.

Spreading type: 75 to 90 x 60 to 70 cm between rows and plants.

Transplanting

Transplant the seedlings when 12 to 15 cm tall, preferably at evening time. Irrigate immediately afterwards.

Irrigation

Irrigate at two weeks intervals during the summer.

Interculture

Two–three hoeings and weedings, from time to time, are necessary to keep the seed field clean of weeds.

Insects and Diseases

Insects/Diseases	*Control Measures*
	Insects
Shoot and fruit borer (*Leucinodes orbonalis*)	Spray cypermethrin @ 30 g a.i./ha or deltamethrin @ 10 g a.i./ha or endosulfan 700g a.i./ha at 14 days intervals.
Leaf roller (*Eublemma olivacea*)	Spraying of 0.1 per cent carbaryl or 0.05 per cent malathion.
Aphids (*Aphis gossypii*) and Jassids (*Amrasca biguttula*)	Application of granules of carbofuran @ 1500 g a.i. during transplanting followed by three sprays of cypermenthrin @ 30 g a.i./ha at 10 days interval.
Root knot nematodes (*Meloidogyne* spp.)	DD mixture at 280.82 l/ha, DBCP (dibromo-chloropropane) at 26.92 l/ha or phorate at 4.92 kg/ha
	Diseases
Phomopsis blight (*Phomopsis vexans*)	Seed treatment by 0.1 per cent mercuric chloride.Hot water treatments of seeds at 50 °C for 30 minutes. Spraying with 1 per cent Bordeaux mixture (4:4:50).
Damping off (*Pythium spp.*, *Phytophthora* spp. and *Rhizoctonia* spp.)	Nursery soil is sterilized before sowing and seed are treated with bavistin prior to sowing. Hot water treatment at 51.7 °C for 30 minutes of seeds.
Fruit rot (*Phytophthora nicotianae* var. *nicotianae*)	Good drainage should be maintained in the field. Staking plants, removing foliage and fruits up to a height of 15-30 cm from ground level. Used a combination of mulching and difoltan sprays (0.3 per cent).

Contd...

Insects/Diseases	*Control Measures*
Fusarium wilt (*Fusarium oxysporum* f sp. *lycopersici*)	Application of bavistin (0.1 per cent), benlate (0.2 per cent) as soil drench.
Blight (*Sclerotium rolfsii*)	Application of two sprays of bavistin @ 0.05 per cent, 30 and 50 days after transplanting.
Bacterial wilt (*Pseudomonas solanacearum*)	Disease control is same as in case of bacterial wilt of tomato.
Brinjal little leaf (Vectors: leafhopper)	Infected plant must be rouged. Seedling dip in tetracycline + soil application of phorate followed by seedling dip in monocrotophos + soil application of phorate.

Roguing

An early roguing for off type plants are possible, since off types may often be detected much before blossoming has occurred. When each plant has several more or less mature fruits, roguing can be based on fruit size, shape and colour, as well as the plants overall performance. In addition to off types, plants affected by diseases such as phomopsis blight and little leaf, etc. should be removed time to time from the field.

Figure 45: Fruits at Harvest

Harvesting and Thresing

Fruits are picked when fully ripe. The fruits are picked and collected. The outer covering is peeled off and the flesh with the seeds is cut into thin slices. Soaking till the seeds are separeted from the pulp then softens these. If the material is allowed in this condition to stand overnight, the separation of seeds from the pulp becomes easier. After separation, the seeds are dipped into water. Those, which float, should be rejected. The seeds should then be dried in partial shade to a moisture content of 8 per cent or below, before storing.

Seed Yield

Average seed yield varies from 100 to 200 kg/ha.

Varieties/Hybrids Characteristics

Varieties/Hybrids Recommended by IARI, New Delhi

DBR-8

This variety has been developed at IARI, New Delhi. Plants are semi-upright (semi-erect stem) and medium-tall (70 cm). Fruits are around (9 cm diameter) and dark purple. It gives an average yield of 650 q/ha.

Pusa Hybrid-5

It is a long fruited hybrid developed from IARI, New Delhi. The plants are vigorous, non-spiny with semi-erect branches. Fruits are long, glossy, attractive and dark purple. It is an early. It gives an average yield of 525 q/ha.

Pusa Hybrid-6

It is a round-fruited hybrid developed from IARI New Delhi. The plants are non-spiny and semi-erect. Fruit are round, glossy, attractive and purple. It is an early-bearing hybrid, which yields 460 q/ha.

Pusa Hybrid-9

It is a round-fruited hybrid. It combines highly desirable quality attributes. The plants are non-spiny with strong upright branches. Fruits are big, oval-round, glossy, highly attractive and dark purple.

Pusa Kranti

The variety has been developed from the cross of PPL x Hyderpur x Wynd Giant. Plants are medium tall, upright erect in growth habit. Fruits are oblong (15-20 cm), dark purple. The average yield is 277.0 q/ha.

Pusa Purple Long

It is an old variety developed and released from IARI, New Delhi. This is a selection from the mixed Batia cultivar grown in Punjab, Delhi and Western Uttar Pradesh. Fruits are long, purple and tender. It is suitable for summer and autumn seasons. Average yield is 281.0 q/ha.

Pusa Purple Round

This cultivar has been evolved at IARI, New Delhi. Plants are very tall with a thick stem of greenish purple colour. This cultivar was found to be resistant to shoot and fruit borer and little leaf.

Varieties/Hybrids Recommended by Tamil Nadu State

Annamalai

It is selection from AC-49A, released from TNAU, Coimbatore. Fruits are somewhat club-shaped and slightly curved in the middle. This variety is resistant to aphids. Yield potential is 300-350 q/ha.

Co. 1

It is pure line selection and released from TNAU, Coimbatore. The plants are erect, compact and bushy. The fruits are oblong and medium size with pale green shade under white background. It is moderately resistant to root knot and reniform nematodes. It yields on an average of 240 q/ha in crop.

Co. 2

It is a pure line selection from the local variety 'Varikkathiri' and released from TNAU, Coimbatore. The plants are compact and erect. The fruits have dark purple strips intermingled with light green colour. They are oblong in shape. The crop yield is around 350 q/ha.

MDU-1

It is a selection from Kallampatti Local and released from TNAU, Coimbatore. The plants are compact and medium spreading. Fruits are bright purple in colour and round. The yield per hectare is 340 q/ha.

Varieties/Hybrids Recommended by Karnataka State

Arka Keshav

It is development from IIHR, Banglore. Plants are 80-100 cm in height. Fruits are purple in colour measuring 18-20 cm in length and 5-6 cm in diameter. An average yield of this variety is 300-400 q/ha.

Arka Kusumkar

It is an improvement over a local collection (IIHR '93'). The fruits are small, finger shaped and light green in colour, borne in cluster. It has good flesh texture and cooking quality. On an average, it produces about 397 q/ha fruits.

Arka Navneet

It is a cross between IIHR 22-1 x Supreme from IIHR, Bangalore. This is a round fruited hybrid with shining deep purple colour Yield potential of this hybrid is 650-700 q/ha.

Arka Nidhi

It is developed from IIHR, Bangalore. Fruits are medium in length and bearing occur in cluster. Fruit colour is black purple. It is resistant to bacterial wilt. Average yield is 485 q/ha.

Arka Sheel

This is improved over Purple Long cultivar has been developed from Coorg and released from IIHR, Bangalore. The plant is well branched and erect. The fruits are medium long, tender and possess an attractive deep purple skin colour. Average yield of 394 q/ha.

Arka Shirish

It is an improvement over the Irangeri brinjal and released from IIHR, Bangalore. The fruits are tender, extra long with attractive light green skin colour. The flesh texture is good with better cooking quality. The average yield potential is 384 q/ha.

Varieties/Hybrids Recommended by Uttranchal State

Pant Rituraj

It is derivative of Kalyanpur T-3 x Pusa Purple Cluster. It is suitable for *kharif* and summer seasons. Plants are 60-80 cm tall and having semi-erect growth habit. Fruits are almost round with slight tapering towards fruit bottom and dark purple in colour. This variety is resistance to bacterial wilt. Average fruit yield is 403 q/ha.

Pant Samrat

It is developed through pure line selection. Plant is 80-100 cm tall and robust. Fruits are medium long, borne in small clusters, attractive purple colour. This variety is resistant to phomopsis blight and bacterial wilt and less affected by shoot and fruit borer and jassids.

Varieties/Hybrids Recommended by Maharashtra State

ARBH-201

This hybrid has been developed at Ankur Seeds Pvt. Ltd., Nagpur. Plants are semi–spreading to spreading. Fruits are purple and oblong with milky-white flesh. Good for long distance transportation. It is tolerant to sucking borer. an average yield of 700-750 q/ha.

MBH-10

It has been developed at Maharashtra Hybrid Seeds Co. Ltd., Jalna. Plants are erect, bushy and spiny. Fruits are round to slightly oval, shining, reddish purple with white strips (bicolour). It gives an average yield of 400-500 q/ha.

MHB-39

It has been developed at Maharashtra Hybrid Seeds Co. Ltd., Jalna. Plants are erect, compact and non-spiny. Fruits are oval, shining, reddish purple. The yield potential of the hybrid is 350 q/ha.

RHRBH-1

It is a cross of RHRB-1 and JB-16 developed at MPKV, Rahuri. Fruits are purple with non-spiny, purple having white stripes. It gives an average yield of 576 q/ha.

RHRBH-2

It is a cross of RHRB-1 and RB-25 developed at MPKV, Rahuri. Plants are short and spreading type. Fruits are purple with white stripes. It gives an average yield of 407 q/ha.

Varieties/Hybrids Recommended by Uttar Pradesh State

KS-224

This variety has been developed at Vegetable Research Station, CSAUA and T, Kanpur; derived from the cross Pusa Purple Round x Arka Sheel. Plants are medium-tall (90 cm). Fruits are round, purple. This variety gives an average yield of 300 q/ha.

NDBH-1

It has been developed at NDUAT, Faizabad. Plants are semi-erect, non-spiny. Fruits are oblong, attractive, purple, soft, fleshy. Plant is moderately resistant to shoot and fruit borer, phomopsis blight and little leaf. Average yield is about 600 q/ha.

NDBH-6

It has been developed at NDUAT, Faizabad. Plants are non-spiny, semi-erect. Fruits are oblong, attractive and purple. It is moderately resistant to shoot and fruit borer, phomopsis blight and little leaf. Average yield is 600 q/ha.

Ram Nagar Giant

This variety is very much popular in the Varanasi region and it's surrounding areas. Plants are tall (100-150 cm in height), erect, branches grow upright. Fruits are light green in colour, very big size (1.5–2.5 kg), soft, very less seeds and used mostly for Bharta preparation. The yield potential of this variety is 300 q/ha.

VRBHR-1

It is a cross of Pant Rituraj and BR-SPS-14 developed at IIVR, Varanasi. Plants are of 70.8 cm height. Fruits are round, light purple. The average yield is 789 q/ha.

Varieties/Hybrids Recommended by Punjab State

Barahmasi

This is long duration fruiting variety. Plants are 80-100 cm in height. Friuts are soft, less seeded and 9-10cm in length and 8-9 cm in diameter. An average yield is 250-300 q/ha.

Jamuni Gola

A variety developed at Ludhiana. Fruits are plump and shining purple coloured. It is an early in maturing.

Punjab Bahar

It is a selection from local material from Punjab. Plants are 93 cm tall and non-spiny. Fruits are round, dark purple, shining. Fruits are suitable for bhartha making.

Punjab Barsati

It is released from PAU, Ludhiana. Fruits are medium long, shining purple and suitable for rainy season. This variety is tolerant to shoot and fruit borer. Average yield is 350 q/ha.

Punjab Chamkila

It is released from PAU, Ludhiana. Fruits are long, thin and dark purple. This variety is good for summer and autumn seasons. The yield potential of this variety is 400 q/ha.

Punjab Sada Bahar

It is developed from PAU, Ludhiana. Fruits are long, thin and deep purple in colour. It is tolerant to shoot and fruit borer and suitable for rainy season. Average yield of the variety is 350 q/ha.

Varieties/Hybrids Recommended by Bihar State

Green Long

This variety has been bred at RAU, Sabour. Plants are upright (erect stem) and medium-tall (73 cm). Fruits are long and green. It gives an average yield of 40 t/ha.

Rajendra Baingan

Plants are tall and erect with strong frame. Fruits are long, light green colour with slight curve. This variety is suitable for growing in rainy and winter seasons. It is less affected by wilting and phomopsis blight. The average yield is 250-325 q/ha.

Varieties/Hybrids Recommended by Haryana State

Hisar Jamuni

It is a cross of Aushey (a round brinjal) and R 34 (a long brinjal). Plants are indeterminate, semi-spreding, medium-tall. Fruits are oblong, bulging at the blossom end, 10-15 cm long, deep bright purple blackish in colour. It is suitable for autumn and spring seasons as well as ratoon crop in early spring. Yield potential of this variety is 320 q/ha.

Hisar Pragati

This variety was developed at CCSHAU, Hisar, derived through pedigree method from the cross R-34 x Sel-26. Plants are compact, dwarf, bushy, indeterminate and erect. Fruits are long, dark bright purple. This variety gives an average yield of 334 q/ha. It is tolerant to little leaf and high temperature.

Varieties/Hybrids Recommended by Orissa State

Utkal Madhuri

This bacterial wilt resistant variety has been developed at OUA and T, Bhubaneshwar, derived through pedigree method from the cross PBR 125-5 x Pipli-4 (local collection). Fruits are green (basal portion white), long and medium. It is resistant to *fusarium* wilt and drought. It gives an average yield of 316 q/ha.

Utkal Tarini

This is a bacterial wilt resistant variety, developed at OUA and T, Bhubaneshwar, derived through pedigree method from the cross Pusa Kranti x Gopa Local. Fruits are oblong, medium sized, deep purple. It gives an average yield of 340 q/ha.

Utkal Jyoti

This variety has been developed at OUA and T, Bhubaneshwar, derived through pedigree method from cross KT-4 x BB-11. Plants are medium-tall and. Fruits are long, small medium, purple. It is tolerant to bacterial wilt. It gives an average yield of 348 q/ha.

Varieties/Hybrids Recommended by Himachal Pradesh State

Pusa Anupam

This variety was developed at Regional Station of the IARI, Katrain through hybridization of 'Pusa Kranti' and 'Pusa Purple Cluster'. The plants are medium in size and bushy. The fruits are cylindrical, medium-long, tender, firm and purple in colour. The fruit is also suitable for pickle making. It gives an average yield of 300-400 q/ha.

Pusa Purple Cluster

It is a early variety and developed at IARI, Regional Research Station, Katrain. Fruits borne in clusters (4-9), 10-12 cm long and deep purple in colour. It is moderately resistant to bacterial wilt.

Other Varieties/Hybrids

CHBR-1

It has been developed at CHES, Ranchi. Plants are prostrate (spreading stem) and medium-tall (89.0 cm). Fruits are round (9.6 cm) and dark violet. It gives an average yield of 756 q/ha.

Junagarh Long

A derivative of the cross Doli-5 x Inbred Panch Mahal. Plant height is 120 cm and semi-spreading. Fruits are medium in size, tender, pinkish purple in colour and long (about 17 cm). Fruits are tapering at the distal end. This variety is free from collar rot disease.

SEL-4

It was developed at APAU, Hyderabad. Plants are erect, 5-6 primary branches. Fruits are medium-long, light purple. Thie variety is suitable for long distance transportation.

Vijay Hybrid

It is developed from Indo-American Hybrid Seeds Ltd., Bangalore.The fruits are purple and oval to oblong shaped. It gives average yield is 400 q/ha.

Tomato (*Lycopersicon esculentum* Mill.) 2n = 2x = 24

Land Requirement

The land should be free of volunteer plants. A well-drained fairly fertile loam with a fair moisture holding capacity is ideal for growing of good crop of tomato.

Field Inspection

A minimum of three inspections should be done, the first before flowering, the second during flowering and fruiting stage, and the third at the mature fruit stage and prior to harvesting.

Figure 46: Tomato

Field Standards

General Requirements

Isolation: Seed field should be isolated from the contaminants as in given in Table 4

Contaminants	*Minimum Distance (Meters)*	
	Foundation	*Certified*
1	*2*	*3*
Field of other varieties	50	25
Field of the same variety not conforming to varietal purity requirement for certification	50	25

Specific Requirement

Factor	*Maximum Permitted (%)**	
	Foundation	*Certified*
Off–types	0.10	0.20
**Plants affected by seed borne diseases	0.10	0.50

*Maximum permitted at final inspection

**Seed borne disease shall be: Early blight, leaf spot and tobacco mosaic virus

Seed Standards

Following seed standard should be maintained.

Factors	Standards for Each Class	
	Foundation	Certified
Pure seed (Minimum)	98.0 per cent	98.0 per cent
Inert matter (Maximum)	2.0 per cent	2 per cent
Other crop seeds (Maximum)	5 kg	10 kg
Weed seeds (Maximum)	None	None
Germination (Minimum)	70 per cent	70 per cent
Moisture (Maximum)	8 per cent	8 per cent
For vapour proof containers (Maximum)	6 per cent	6.0 per cent

Brief Cultural Practices

Time of Sowing

Northen plains: June–August, November to December; Southern states: September to October; In hilly areas: March–April.

Preparation of Land

Prepare the field to a good tilth by ploughing, two to three harrowings followed by levelling.

Source of Seeds

Obtain nucleus/breeders/foundation seed from source approved by a seed certification agency.

Seed Rate

400–500 g/ha

Nursery Raising

Nursery bed should be prepared at the size of 7.5 m long, 1.2 m wide and 10 cm high. Well-decomposed farmyard manure is properly mixed into top soil of the bed at the rate of about 3 kg/m^2.A fertilizer mixture of 0.5 kg of NPK per bed is also mixed in the soil. Seedbeds are also treated with steam or formalin. A half litre of 40 per cent formalin/m^2 area is sufficient.Fumigation may be done with 10 per cent formaldehyde in properly dug the soil.If seeds are sown in rows, a distance of 7.5 cm is kept between the rows. The beds are covered with straw or polythene sheets until the seeds germinate. Every week if required a fungicide such as dithane M-45 or difolaton at 0.25 per cent should be sprayed to reduce the post emergence "damping off".

Fertilizers

Apply 550 quintal of farmyard manure per hectare in the soil at the time of field preparation. Apply 75 kg nitrogen, 60 kg phosphorus, and 60 kg potash/ha at the time of transplanting. Top dress nitrogen @ 75 kg /hain two doses, one after 15-20 days of transplanting and theother at flowering stage (40-45 days after transplanting).

Transplanting

Single healthy seedlings (7.5–10 cm) are transplanted at proper spacing and water is given immediately for better establishment in the soil.

Spacing

Indeterminate Varieties/Hybrids 60-120 cm x 45-75 cm.

Determinate varieties/ hybrids 45-60 cm x 30-45 cm

Irrigation

Irrigate fortnightly during winter and weekly during summer.

Interculture

Shallow cultivation is frequently required especially during the first four weeks. Two-three hoeings and weedings, from time to time, are necessary to keep the seed field clean of weeds. Premergence application of oxyflurofen @ 0.25 kg a.i./ha, combined with one hand weeding at 45 DAT (days after transplanting) has been recommended for controlling weeds.

Staking

Staking is very essential operation in indeterminate group of open pollinated Varieties/Hybrids for good quality fruits. Staking may be done either by using bamboo stick or trained on wire.

Insects and Diseases

Insects/Diseases	*Control Measures*
	Insects
Tomato fruit worm (*Heliothis armigera*)	Fortnightly sprays of monocrotophos or carbaryl or endosulfan or quinalphos.
White fly	Sprays of phosphamidon or monocrotophos.
Leaf minor	Cyromaizin (250 g/ha) at 15 days interval
Root knot nematode (*Meloidogyne incognita, M. javanica* and *M. arenaria)*	Application of aldicarb and carbofuran each at 1kg a.i. /ha with neem cake and urea, each at 100 kg N/ ha at transplanting.
	Diseases
Damping off and Root rot (*Pythium spp, Rizoctonia solani,* and *Phytophthora parasitica)*	Soil treatment with formaldehyde and 1 per cent mercuric chloride. Spraying of seedlings with fytolan or captan.
Early blight (*Alternaria solani*), Late blight (*Phytophthora infestans*)	Spray Difolatan (0.2 per cent), dithane M–45 (0.2 per cent)
Buckeye rot (Fruit rot) (*Phytophthora parasitica*)	Staking plants and and difolatan spray (0.3 per cent).
Fusarium wilt (*Fusarium oxysporum* sp. *lycopersici*)	Application of bavistin (0.1 per cent), benlate (0.2 per cent) as soil drench.
Bacterial wilt (*Pseudomonas solanacearum*)	Growing resistant Varieties/Hybrids. Crop rotation, such as cowpea–maize–cabbage and okra–cowpea–maize. Seedling treatment with streptocycline (1g/40 liter).
Tomato leaf curl virus: (Vector: *Bemisia tabaci)*	Treat the nursery bed furadan @ 1kg a.i./ha. Furadan 2–3 foliar spray of dimethoate (0.05 per cent).
Tomato mosaic virus (TMV) (Spread by contact with clothes, hand of working labourers, touching of infected plants)	Select seed from healthy plant. Soak the seeds for 15 minutes in a solution of trisodium phosphate at 90 g/l.

Roguing

Off type plants should be removed as soon as they are observed. Lants with off type foliage should be removed before they blossom, to reduce any possibility of cross polliantion. As the first fruits are approaching edible maturity, plants with undesirable fruit type must be rogued out. When the fruits begin to show their final colour of red, occasional plants with off colour fruits have to be removed. In addition to off type plants, remove plants affected by early blight, leaf spot and mosaic (TMV) from the seed field.

Harvesting

Fruits on turning red, ripe red and over ripe stages are found to be good for extracting good quality seed. Picking at mature green stage and fruits of last picking should be avoided for seed extraction. Fruits could be harvested at the breaker stage for early seed production without loss of seed viability.

Seed Extraction

There are following methods of seed extractions.

Fermentation

The ripe fruits are crushed in a non-metallic box by hand or try any mechanical method. The entire mass is kept in it till it ferments.

Figure 47: Tomatoes Ready for Harvesting

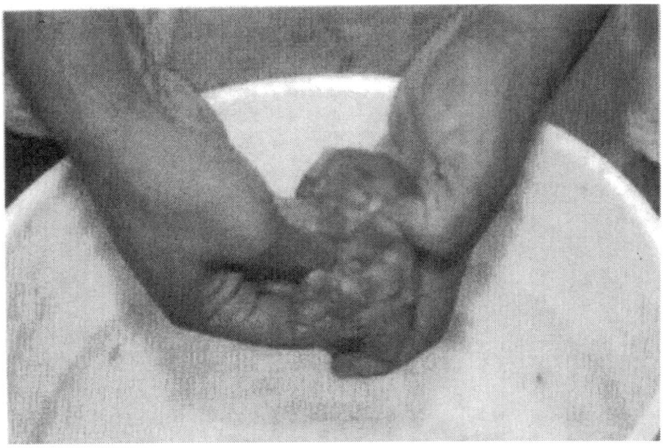

Figure 48: Extraction Seeds by Hand

The duration depends upon the temperature. Generally at 15-20 ^{0}C, 4-5 days are required for complete fermentation. Adequate foam formation on the top and an adhering of seed to tomato flesh indicate that fermentation has been completed. When fermentation is not complete, some chemical present in the parenchymatous tissues suppress the germination of seeds. Long period fermentation is not advisable because it reduces the quality of seeds. The fermentation temperature affects germination and it depends upon the maximum air temperature during the fermentation occurring before seed extraction. The fermented mass and debris on the top are removed and then all the liquid is decanted. The seed settle down at the bottom of the container and they are collected and washed thoroughly with clear water. The seeds are dried in the sun or with a drier and stored in a cool dry place.

Acid Treatment Method

Natural fermentation gives a poor breakdown of the gelatinous coating of tomato seeds. Therefore, HCl is used for a quick break down of the gelatinous coating and extraction of seed. Use of HCl, however, may have an adverse effect on seed germination especially at high temperature. About 250 ml of commercial hydrochloric acid (HCL) is throughly mixed with 10 liter of slurry and left for 30 minute. The pulp is then again stirred and washed thoroughly with clear water and allowed to dry in the sun or with a drier. The pulp from ripe tomato fruit treated with concentrated hydrochloric acid and at the rate of 20 ml/kg fruit for 30 minutes has been effective for seed extraction (Patil and Patil, 1974). The seed extracted by this method have cleaned and germinated well. Herrington (1981) found that with the increase of HCl concentration from 1 to 16 per cent seed germination has decreased considerably in different cultivars.

Alkali Treatment Method

For quick extraction for seeds, the seed slurry is mixed with 10 per cent washing soda in equal volumes and kept overnight in an earthen pot. Next day, all the seeds will settle down at the bottom of the container. Now, the clean liquid is decanted. Seed thus obtained are thoroughly washed with clear water and allowed to dry in the sun or with a drier. This is the safest method of seed extraction. The extracted seeds are treated with 2.4 per cent trisodium orthophosphate for 15-30 min to inhibit tomato mosaic virus.

Mechanical Seed Extraction

Manual extraction of seeds is time consuming, labour intensive and unhygenic. An axial flow vegetable seed-extracting machine has been designated and developed at Punjab Agricultural University, Ludhiana.

Figure 49: Rinsing of Seeds

Figure 50: Drying Seed in Partial Shade

Figure 51: Preparing Seeds for Drying in an Oven

Washing

After extration the seeds are washed with water to remove the pulp and seed. It is stirred thoroughly and is drained out along with pulp and other mucilaginous substance. This process is repeated until the seed are clean.

Seed Drying and Storage

Seed should be dried as rapidly as possible by spreading in thin lying under partial shade is desirable. In artificial seed drying, the seed should be first be dried at low temperature and then the temperature could be gradually increased to 40 ^{0}C. The seed should be dried to moisture of 6-8 per cent. The seed should be packed in cloth bags or moisture proof/resistant containers and kept in a cool and dry place. Carefully and properly stored seed can return its viability for two to three years even under ambient conditions.

Seed Yield

100-120 kg/ha.

Varieties/Hybrids Characteristics

Varieties/Hybrids Recommended by IARI, New Delhi

Pusa Early Dwarf

It is a derivative of the cross Meeruti x Red Cloud. It is suitable for rainy season in plains and for spring summer sowing in hills. Plants are determinate. Fruits are roundish, slightly flattish, medium large, uniform red, ribbed. Yield potential is 250-300 q/ha.

Pusa Gaurav

It has been developed from the exotic segregating generation of a cross between Galmour and 'Watch' and released by the IARI, New Delhi. The fruits are smooth, elliptical (egg shaped). Whole fruit is suitable for processing and canning.

Pusa Ruby

It is a derivative of Improved Meeruti x Sioux. It is early maturing and indeterminate type. Fruits are flattened, grooved, firm and medium size. It is suitable for rainy season.

Pusa Sadabahar

It has been released by IARI, New Delhi.It plants are determinate dwarf. This variety can be grown through out the year except rainy season.

Pusa Sheetal

It has been developed by combining the desired attributes of two cold set lines Balkan from Bulgaria and Temnorrosnij from USSR by hybridization and selection of plants to set fruit at very low temperature. Fruits are flatish round, smooth, attractive and developed uniformly red colour.

Pusa-120

It is released from IARI, New Delhi. Plants are semi-indeterminate. Fruits are attractive, round to flatish round, medium to large in size, smooth, uniform red in colour. It is resistant to nematodes.

SL-120

It has been developed at IARI, New Delhi. Plants are determinate. Fruits are flatish-round, medium-large, uniform red. It is resistant to nematode. It gives an average yield of 305 q/ha.

Varieties/Hybrids Recommended by Karnataka State

Arka Abha

It is developed from IIHR Bangalore. Plants are indeterminate. Fruits are flat, round. An average yield is 350 q/ha. This variety is resistant to bacterial wilt

Arka Abhijit

This hybrid was developed at the IIHR, Bangalore. Plants are semi determinate. Fruits are round. This hybrid yielded 650 q/ha in 140 days. It is resistant to bacterial wilt and suitable for fresh marketing.

Arka Alok

It is developed from IIHR, Bangalore. Fruits are determinate. Fruits are large in size (100 g), round in shape. Average yield is 350 q/ha. It is resistant to bacterial wilt.

Arka Saurbh

It is an improvement over breeding line V-685 from Canada and released from IIHR, Bangalore. Plants are semi–determinate. Fruits are firm, fleshy, deep red. This is a dual-purpose variety suitable for both fresh market and processing purposes. This variety yielded 350 q/ha. It is resistant to fruit cracking.

Arka Shreshtha

It is developed and released from IIHR, Bangalore. The plant is semi–determinate. Fruits are round. It gives 830 q/ha yield. It is resistance to bacterial wilt and suitable for both fresh marketing and processing purposes.

Arka Vardan

It has been developed at IIHR, Bangalore. Plants are indeterminate. Fruits are flat round, firm deep red. It is resistant to nematodes.

Arka Vikas

It is a selection from American variety TipTop and released from IIHR, Bangalore. Plants are semi-determinate. Fruit is oblate, deep red in colour. It is suitable for fresh market purposes. Yield potential is 350-400 q/ha.

Varieties/Hybrids Recommended by Maharashtra State

ARTH-3

It has been developed at Ankur Seeds Pvt. Ltd., Nagpur. Plants are determinate. Fruits are squarish round, deep red. It is moderately tolerant to drought. This variety gives an average yield of 880-950 q/ha.

ARTH-4

It has been developed by Ankur Seeds Pvt. Ltd., Nagpur. Plants are semi-spreading and indeterminate. Fruits are deep red and flat round. It is moderately tolerant to drought. It gives an average yield of 1000-1100 q/ha.

Meenakshi

This hybrid has been developed at Beejo Sheetal Seeds Pvt. Ltd., Jalna. Plants are indeterminate. Fruits are dark red, round global. It is resistant to fusarium wilt. It gives an average yield of 600-1000 q/ha.

Madhuri

It has been developed at Beejo Sheetal Seed Pvt. Ltd., Jalna. Plants are determinate. Fruits are dark red, flat round, crack resistant and can with stand long transportation. It gives an average yield of 500-700 q/ha.

Varieties/Hybrids Recommended by Uttar Pradesh State

Azad T-2

It is a derivative of the cross between HB-5 x Kuber, released from CSAUA&T, Kanpur. Plants are determinate. It is resistant to root knot nematode and moderately resistant to leaf curl virus. It is suitable for processing purposes. Fruit yield is 400-500 q/ha.

DVRT-2

This variety has been developed at IIVR, Varanasi; derived from the cross between *L. esculentum* cv. Sel-7 and *L. hirsutum* f. *glabratum* B 6013, following backcross-pedigree selection. Plants are determinate. Fruits are large, spherical, attractive red. Gives an average yield of 600 q/ha.

VRTH-1

This is an early maturing hybrid having determinate growth habit developed at PDVR, Varanasi. The average yield is 1100 q / ha.. It posses field tolerance to tomato leaf curl virus and early blight.

Varieties/Hybrids Recommended by Tamil Nadu State

CO-2

It is a pure line isolated from USSR type. Fruits are ovate, medium in size, smooth, green when unripe and bright red on ripening. Fruits are devoid of cracking. It gives 410 q/ha fruit in 6-8 harvesting.

CO-3

It is a mutant of Co.1 variety and released from TNAU, Coimbtore. Plants are determinate in growth habit. Fruits are medium in size, roundish, globular in shape, parrot green when unripe and attractive red on ripening. It gives 500 q/ha fruits and 3 q/ ha seed.

Varieties/Hybrids Recommended by Haryana State

H-24

It is a determinate variety resistant to TLCV, derived through backcross-pedigree method from an inter-specific cross *L. esculentum* cv. Sel-7 x *L. hirsutum* f. *glabratum*. Fruits are flatish-round, fleshy, medium and red. It gives an average yield of 400 q/ha.

HS-101

It is suitable for winter season. Plants are determinate type. Fruits are round, small to medium in size. Yields is 250 q/ha.

HS-110

It is a selection from an exotic line. Plants are determinate having potato leaf type. Fruits are large, smooth, fleshy and jointless. It is suitable for table purposes.

Hisar Arun

It is derivative of Pusa Early Dwarf x K-1. It is extremely early and high yielding. Fruits are medium to large, round deep red. It has ability to fruit set at low temperature.

Hisar Lalit

It is a derivative of HS-101 x Resistant Bangalore. Plant is semi–determinate and early fruiting. Fruits are round and medium too large. It is resistant to root knot nematode.

Varieties/Hybrids Recommended by Manipur State

Manikhamnu

It is dwarf, early maturing, determinate. Fruits are medium size, round, smooth, uniform red. It is suitable for processing as well as fresh consumption purposes. It is resistance to fruit cracking and tolerant to leaf curl virus. The yield potential is 533 q/ha.

Manileima

It is an early maturing, determinate. Fruits are medium size, oval, nipple shaped and bilocular. It is suitable for processing and long distance transportation. It is resistance to fruit cracking and tolerance to leaf curl virus. It is also tolerant to moisture stress and suitable for spring summer season. The yield potential of the variety is 630 q/ha.

Varieties/Hybrids Recommended by Uttaranchal State

Pant Bahar

It is a selection from germplasm lines maintained at Pantnagar and released. Plant is indeterminate in growth habit. Fruits are flatish round, medium in size, slightly ridged and red on maturity. It is resistant to verticillum wilt and fusarium wilt. Fruits are good for processing purposes.

Pant-T–1

This variety has been developed from GBPUA&T, Pantnagar. Fruits are round with nipple. This variety is tolerant to tomato leaf curl virus.

Varieties/Hybrids Recommended by Punjab State

Punjab Chhuhara

It is derivative of the cross Punjab Tropic x EC55055, released from PAU, Ludhiana. It is suitable for summer season. Plants are determinate type. Fruits are pear shaped, thick pericarp, small to medium in size. It is good for processing purposes. It is tolerant to root knot nematode.

Punjab Tropic

It is a selection from the introduction of USA and released from PAU, Ludhiana. Plant is indeterminate. Fruits are large, round, red on maturity. Fruit yield is 225 q/ha.

Varieties/Hybrids Recommended by Orissa State

BT-10

This is a bacterial wilt resistant and indeterminate variety, developed at OUA& T, Bhubaneshwar.It is derived from the cross Selection 22 x BT-2. Fruits are round, red, medium-large. It gives an average yield of 404 q/ha.

BT-12

This variety has been developed at OUA&T, Bhubaneshwar. It derived from the cross Punjab Chhuhara x LE-79. Plants are indeterminate. Fruits are pear shaped, red. This variety is resistant to bacterial wilt and tolerant to early blight. It gives yield of 300-400 q/ha.

Other Varieties/Hybrids

Best of All

It is indeterminate and mid season variety. Fruits are medium in size, scarlet red in colour, round, juicy.

Floradade

This variety was brought from Florida. Plants are determinate. Fruits are big in size. This variety is early blight. Yield is 400-500 q/ ha.

Marglobe

It is a mid season variety. Plant is indeterminate type. Fruits are nearly round, deep scarlet colour on ripening, 3-5 locules, juicy, mildly acidic, thick walled, smooth, solid and less seeded.

Ox Heart

Fruit is heart shaped, pink in colour and each weighing 300-400 g. Interior of the fruit extremely solid with thick firm walls and small seed cavities with few seeds.

Roma

It is an introduction form USA. Plants are semi-dwarf, bushy growth habit with luxuriant foliage cover. Fruits are borne in 4-5 clusters, pear shaped with thick pericarp, 7 cm long and 5 cm in girth, red in colour on ripening, free from green shoulder, bilocular, small seed cavity and less seeded. This variety possesses good keeping quality and good for processing.

Shakati

It was developed through pure line selection at College of Horticulture, Vellanikkara, Kerala. Plants are determinate. Fruits are round, white and of medium size. It is resistant to bacterial wilt and fruit cracking. It gives an average yield of 435 q/ha.

Sioux

It is an exotic variety introduced from USA. It is indeterminate type and produces yellowish green, medium large fruit, which becomes uniform red at ripening.

Sweet-72

It is cross of Pusa Red Plum x Sioux and developed at the Regional Agricultural Research Institute, Gwalior. The fruits are round oval and somewhat flattened in shape. It is very sweet in taste with sugar per centage of 1.62. An average yield of this variety is 200 q/ha.

Yashwant-2

This variety has been developed from YS Parmar University of Forestry and Horticulture, Solan. Plants are indeterminate. Fruits are uniform red in colour. Yield is 500 q/ha. This variety is resistant to buckeye rot.

Chilli and Bell Pepper *Capsicum fruitescens* L. (Chilli) *Capsicum annuum* L. (Bell pepper) 2n = 2x = 24

Land Requirement

The land should be free of volunteer plants. The soil of selected fields should be well drained and aerated.

Field Inspection

A minimum of three inspections should be done, the first before flowering, the second during flowering and fruiting stage, and the third at the mature fruit stage and prior to harvesting.

Field Standards

General Requirements

Isolation

Seed fields should be isolated from the contaminants shown in column 1 of the Table below by the distances specified in columns 2 and 3 of the said Table:

Figure 52: Chilli

Figure 53: Bell Pepper

Contaminants	*Minimum Distance (Meters)*	
	Foundation	*Certified*
Fields of other varieties	400	200
Fields of the same variety not conforming to varietal purity requirements for certification.	400	200
Fields of chilli from capsicum and *vice versa*	400	400

Specific Requirements

Factor	*Maximum Permitted (%)**	
	Foundation	*Certified*
Off types	0.10	0.20
** Objectionable weed plants	0.10	0.50

* Maximum permitted at and after flowering in the case of off types and the final inspection in the case of seed borne diseases.

** Seed borne diseases shall be: Leaf blight and Anthracnose (Ripe rot, Die back).

Seed Standards

Factors	Standards for Each Class	
	Foundation	Certified
Pure seed (minimum)	98.0 per cent	98.0 per cent
Inert matter (maximum)	2.0 per cent	2.0 per cent
Other crop seeds (maximum)	5/kg	None
Weed seeds (maximum)	5/kg	None
Germination (minimum)	60 per cent	60 per cent
Moisture (maximum)	8.0 per cent	8.0 per cent
For vapour-proof containers (maximum)	6.0 per cent	6.0 per cent

Brief Cultural Practices

Time of Sowing

Chilli

Northen Plains: *Kharif* crop–June to July, Spring crop–February to March; Eastern and Southern zones-Throughout the year.

Capsicum

Hills–April-May, North–July, November to February, South-September to October

Preparation of Land

Prepare the field to a good tilth by ploughing, two to three harrowings followed by levelling.

Source of Seeds

Obtain nucleus/breeders/foundation seed from source approved by a seed certification agency.

Seed Rate

One to two kg/ha.

Sowing of Seed in Nursery

Seed should be sown by broadcasting in nursery beds, which should be raised 18 to 20 cm high from the ground. Each bed may be 2 to 2.5 meters long and 1 to 1.25 meters wide. Seedling in the nursery should be thinned to avoid "damping off" disease. Apply a band of 50 per cent sevin dust around beds to protect from ants

Fertilization

Apply 250 to 300 quintal and 500 quintal of farmyard manure per hectare for rainfed crop and irrigated crop respectively in the soil at the time of field preparation. Apply 50 kg nitrogen, 80 kg phosphorus, and 60 kg potash/ha at the time of transplanting. Top dress nitrogen @ 50 kg /ha before flowering (40-45 days after transplanting).

Transplanting

Single healthy seedlings (7.5–10 cm) are transplanted at proper spacing and water is given immediately for better establishment in the soil.

Spacing

Chilli–60 x 30 to 45 cm.

Capsicum–45 x 45 cm.

Irrigation

The maintenance of uniform soil moisture is essential to blossom and prevent fruit drop. Generally 8 to 9 irrigations are required, depending upon rainfall, soil type, humidity and prevailing temperature.

Interculture

Two-three hoeings and weedings, from time to time, are necessary to keep the seed field clean of weeds.

Insects and Diseases

Insects/Diseases	*Control Measures*
	Insects
Thrips *(Scritothrips dosalis)*	Spraying dimethoate at 0.25 per cent at fortnightly intervals. Soil application of carbofuran @ 1 kg a.i./ha.
Aphids *(Aphis gossypii, Myzus persiae)*	Foliar application of ethiopencarb or methamidophos 60 EC at 0.05 per cent.
Mites *(Hermitasonemus latus)*	Application of sulphur @ 2.5 kg or dicofol @ 0.1 to 0.3 kg/ha.

Contd..

Insects/Diseases	Control Measures
	Diseases
Damping off (*Pythium spp.* and *Rhizoctonia solani)*	Sterilization of soil by formaldehyde. Seed should be treated with ceresan or agrosan GN or bavistin.
Anthracnose (*Colletotrichum capsici)*	Seed treatment with bavistin at 0.1 per cent against anthracnose.
Phytophthora blight and Fusarium wilt	Soil and foliar application of blitox (0.3 per cent) and Bordeaux mixture (1.0 per cent).
Bacterial wilt (*Pseudomonas solanacearum*)	Seedling treatment with streptocycline (1g/40 liter) for 30 minute and penicillin and argrimycin–100 in Hogland's solution before planting.
Leaf curl (Vector–*Bemesia tabaci)*	Soil application of carbofuran or thimet at 0.1 per cent applied after the seedlings become established.
Tobacco mosaic virus (TMV)	One or two days before sowing, soak the seeds for 15 minutes in a solution of trisodium phosphate at 90 g/l.

Roguing

Off type plants should be removed as soon as they are observed. The small leafed plants can usually be detected among plants with

Figure 54: Chilli Pepper Prepared for Dry Seed Extraction

Figure 55: Sweet Pepper Cut for Fresh Seed Extraction by Hand

large leaf and *vice versa*. As the first fruits are approaching edible maturity, plants with undesirable fruit type must be rogued out. When the fruits begin to show their final colour of red or yellow, occasional plants with off colour fruits have to be removed. In addition to off type plants, remove plants affected by leaf blight, anthracnose and virus diseases from the seed field.

Harvesting and Threshing

Fruits are picked when red ripe. Picking of immature fruits may result in serious germination difficulty. Fruits are crushed, cut or macerated. Seeds are washed free of pulp and skins. After washing, seeds are dried in the sun to below 8 per cent moisture content before storage.

Seed Yield

50 to 80 kg/ha.

Varieties/Hybrids Characteristics

Chilli

Varieties/Hybrids Recommended by IARI, New Delhi

Pusa Jwala

This is derived form a cross between NP46 A and Puri Red. Plant is dwarf, bushy. Fruit is 9 to 10 cm long and thin, light green turning light red at maturity. Dry fruits are wrinkled and liable to break during packing and transport. The fruit contains capsaicin of 0.48 mg/g of fruit. Fruit yield is (fresh) 75 q/ha.

Pusa Sadabahar

This is evolved from a cross between Pusa Jwala x IC31339 (*C. fruitescens*). Plant is erect and tall (60-80 cm). Fruit is erect 6 to 8-cm length in cluster (6 to 14/cluster). Ripe fruit is bright red. Fruit yield (dry) is 15.20 q/ha. It is resistant to CMV, TMV and leaf curl viruses.

Varieties/Hybrids Recommended by Gujarat State

ACH-15

It is a cross of S-49 and CDPS-120 developed at GAU, Anand. Plants are 72 cm tall. Fruits are of 10.4 x 3.4 cm size and 4.85 g average weight. It gives an average yield of 100 q/ha.

ACH-30

It is a cross of S5-5 and Punjab Guchcedar and developed at GAU, Anand. Plants are 56.04 cm tall. Fruits are of 10.2 x 3.5 cm size and average weight of 4.8 g. It gives an average yield of 85 q/ha.

Gujrat Chillies

It is a selection from germplasm material of Reshampatta group of chillies. Plants are medium and bushy in growth habit. The yield potential of this variety is higher than Pusa Jawala variety.

Varieties/Hybrids Recommended by Andhra Pradesh State

Andhra Jyoti

This is evolved from a cross between G-2 and a Bihar variety. Released from Agricultural Research Station, Lam, Guntur. Plants are tall and dense. The fruits are bright red measuring 5.1 cm in length and 6.3 cm in girth. The capsiacin content is 0.65 mg/g of fruit.

Bhaskar

It is an early maturing variety developed at Agricultural Research Station (RARS), Lam Farm, Guntur, derived from the cross G-4 x Yellow Anther Mutant. Fruits are thin, medium-long, thick red with red-flesh and suitable for export. This variety possesses good shelf life and transport quality. It gives an average green fruit yield of 150 q/ha.

Prakash

It has been developed at RARS, Lam Farm, Guntur; derived from the cross G-3 x Huntaka. Fruits are long, thin, bright red at maturity with red flesh. This variety possessing good shelf life and transport quality. This variety is suitable for rainfed condition. It gives an average green fruit yield of 80 q/ha.

Varieties/Hybrids Recommended by Maharashtra State

Musalabadi

This is local selection, released from Mahatma Phule Agricultural University, Rahuri. Plant is tall (90-120 cm). Fruit is small, dark green with black patches. It has been found tolerant to die back and powdery mildew.

Sankeshwar-32

It is selection from Sankeshwar type and released from MPAU, Rahuri. Plant is tall (90-120 cm). Fruit is longest (25-30 cm), thin wirely and highly wrinkled. It retains red colour in storage.

Varieties/Hybrids Recommended by Punjab State

CH-1

It is a cross of MS-12 and Ludhiana Long Selection developed at PAU, Ludhiana. Plants are tall (100 cm), spreading type. Fruits are light green, 6.6 cm long and characterized by 23.05 per cent dry matter, 0.82 per cent capsaicin and 98.37 mg/100 g vitamin C. Gives an average yield of 95 q/ha (red ripe fruits).

Punjab Jal

This variety is derived from a cross between Perennial x Long Red and released from PAU, Ludhiana. Plant is dwarf, bushy. Fruit is erect medium size (4.25 x 0.79 cm) dark green and turning dark red at maturity. Fruits are rich in capsaicin (0.7 per cent), oleoresin and dry matter. This variety is resistant to CMV and leaf curl virus. Average fruit yield as fresh is 100-120 q/ha.

Varieties/Hybrids Recommended by Uttar Pradesh State

Kalyanpur Mohini

This variety has been developed from Vegetable Research station Kalyanpur of CSAUAT, Kanpur. In this variety fruits bears upright.

Chanchal

It is selection from local germplasm. Plants are medium tall and perennial habit. Fruits are 25-30 cm long, 6-7 cm thick, erect, highly pungent, dark green when fresh and shining red on maturity. This variety is suitable for kitchen gardening. It can grow leaf curl disease infested area.

Varieties/Hybrids Recommended by Tamil Nadu State

CO. 1

It is a reselection from Sattur Samba (CA (P) 247) and released from TNAU, Coimbatore. The plants are erect and medium tall. The fruits are 6 to 6.5 cm long with sharp tip. The variety has a high capsaicin content (0.72 mg/g). The yield 21.10 q/ha of dry fruits.

CO. 2

It is a selection form Nambiyar local Gundu type of Periyar district of Tamil Nadu [CA (P) 63] and released from TNAU, Coimbatore. The stem is angular, semi-dwarf and less spreading. The fruits are oblong, thick and bright reds in colour. The green fruit yield is about 110 q /ha.

K-1

It is a pure line selection from local Sattur Samba developed at Agricultural Research Station, Kovilpatti and released from TNAU, Coimbatore. The plants are tall and spreading. The ripe fruits are red, long and contain high capsaicin. The yield potential is 18 q/ha dry pods.

K-2

It is a hybrid derivative of the cross between B-TOA (Assam type) x Sattur Samba and released from TNAU, Coimbatore. The plants are erect with a height of 90.2 cm. The length and girth of pod is 7.0 and 4.0 cm respectively. This variety recorded an increased yield of 29 per cent over K-1.

PKM-1

This is a hybrid derivative of the cross between AC.No.1797 x Co.1 and released from TNAU, Coimbatore. Plants are dwarf (70 cm). It is suitable for cultivation under irrigated conditions. It has very bold fruits, which are dark red in colour. It yield potential is on an average is 30.8 q/ha.

Varieties/Hybrids Recommended by Madhya Pradesh State

Jawahar-218

It is selection from a cross between Kalipeeth and Pusa Jwala and released from JNKVV, Jabalpur. Plant is dwarf. Fruit is green turned rosy red at maturity. The average fruit yield is (dry) 18-22 q/ ha. It is early and tolerant to leaf curl and fruit rot.

Jawahar-283

It has been developed by selection made from local genotype collected from the adjoining areas of Jabalpur. Plants are dwarf (40-45 cm). The fruits are straight, firm and medium in length. The fruits are very attractive with high consumer preference. The red ripe fruits are bright red and pungent. The average yield of green chilli is 79.96 q/ha.

Varieties/Hybrids Recommended by Uttranchal State

Pant C-1

It is an advanced generation selection from a cross between NP46A and a Local cultivar and released from GBPUAT, Pantnagar. Plant is short stature. Fruit is erect, small size, highly pungent, 5.5 cm long, light green and turning light red at maturity. The average yield potential as dry is 12.00 q/ha.

Pant C-2

It is also selection from a cross involving the same parents as that of Pant C-1 and released from GBPUAT, Pant Nagar. Fruit is pendent. The yield potential as (dry) is 14.00 q/ha.

Varieties/Hybrids Recommended by Other States

Arka Lohit

This is a pure line selection and released from IIHR, Bangalore. Fruits are dark green colour in immature and dark red colour in mature. Highly pungent variety contains 0.21 per cent capsiacin. The yield potential is 250 q/ha green fruits and 35 q/ha as dry fuit.

Solan Yellow

This variety has been developed from Y.S. Rao Parmer University of Horticulture and forestry by local germplasm. The fruits of this variety are four to five cm in length having good pungency. An average yield of this variety is 75 to 100 q/ha green fruit.

Capsicum

Varieties/Hybrids Recommended by Karnataka State

Arka Basant

This variety has been developed from IIHR, Bangalore. It is an improvement over a Hungarian variety Soroksari. It is creamish white, conical, erect fruited variety with thick flesh. It is having good aroma and crisp texture. The yield potential of the variety is 150-200 q/ha. It has got good export value of the fruits.

Arka Gaurav

It is a pure line selection from Golden Calwonder from USA and released from IIHR, Bangalore. Plant is indeterminate with dark green fruiting. Fruits are blocky, 3-4 lobed, thick fleshed and medium sized (70-80 g). Fruits turn to orange yellow on maturity. The yield potential of this variety is 180-250 q/ha.

Arka Mohini

It is a selection from variety Titan from USA and released from IIHR, Bangalore. Plant is determinate with medium to large blocky fruits. Fruit is dark green surface, thick fleshed, 3-4 lobed and becomes red on ripening. Yield potential is 200-250 q/ha.

Varieties/Hybrids Recommended by Himachal Pradesh State

KTCPH-3

It is a cross of Yolo Wonder and HL-201, developed at IARI, Regional Station, Katrain. Fruits are blocky, smooth, green, 6 cm x

5.8 cm size. Flesh is very thick, sweet and fines flavoured. Gives an average fruit yield of 27.6 t/ha.

KTCPH-5

It is a cross between Yolo Wonder and EC-143570 and developed at IARI Regional Station, Katrain. Plants are of medium maturity group and upright habit. Fruits are blocky, smooth, dark green, square shaped (6.0 cm x 6.0 cm) with hard skin and thick flesh. Gives an average yield of 258 q/ha.

Pusa Deepti

This variety has been released at IARI Regional Station Katrain. Fruits are smooth, erect, conical, light green with thick flesh, 9-11 cm long and 3-5 cm in diameter. It is early to medium in maturity. It is more suitable for commercial production. It gives an average yield of 350-400 q/ha.

Varieties/Hybrids Recommended by Other States

Punjab-27

This variety is developed and released from PAU, Ludhiana. Fruits are light yellows, sweet and elongated. This variety gives 190 q/ha total yield.

Nishant-1

This variety has been developed at Shalimar Campus, Jammu and Kashimir. Plants are indeterminate and medium (48 cm). Fruits are dark green, turn yellow at maturity, 3-4 lobed, non-pungent.

Potato (*Solanum tuberosum* L.) 2n = 2x = 48

Land Requirement

Well-drained sandy loam and medium loam soils, are most suitable for potato cultivation. A crop of seed potato shall not be eligible for certification if grown on land infested with:

1. Wart (Syn*chytrium endobioticum)* and or cyst forming nematodes
2. Brown rot (*Pseudomonas solanacearum*) or non-cyst forming nematodes within the previous three years
3. Common scab (*Streptomyces scabies*)

Figure 56: Potato

Field Inspection

A minimum of four inspections shall be made as follows:

1. The first inspection should be done about 45 days after planting in the hills and about 35 days after planting in the plains to verify isolation, off types and the extent of disease infection with specific reference to mild and severe mosaics, leaf roll, yellows, brown rot and other relevant factors.
2. The second inspection should be done about 60-65 days after planting for early varieties or at appropriate growth stage depending on the crop duration of the variety concerned to check isolation, off types and extent of disease infection with specific reference to mild and severe mosaic, leaf roll, yellows, brown rot and other relevant factors.
3. The third inspection should be done immediately after haulms cutting/destruction in order to verify that haulms have been cut /destroyed by the prescribed date and in proper manner.
4. The fourth inspection should be done about 10 days after haulms cutting/destruction and before harvesting in order to verify that no re-growth of haulms has taken place.

Field Standards

General Requirements

Isolation

The fields of potato should be isolated from the contaminants shown in column 1 of the Table below by the distances specified in columns 2, 3 and 4 of the said Table:

Contaminants	*Minimum Distance (Meters)*		
	Foundation		*Certified*
	Stage-I	*Stage-II*	
Fields of other varieties	5	5	5
Fields of the same variety not conforming to varietal purity requirements for certification	5	5	5

Specific Requirements

Factors	*Stage*	*Maximum Permissible Limits*		
		Foundation		*Certified*
		Stage-I	*Stage-II*	
Off types	I & II Inspection	0.05%	0.05%	0.10%
Plants showing symptoms of:				
- Mild mosaic	I & II Inspection	1.00%	2.00%	3.00%
- Severe mosaic, leaf roll and yellows	I & II Inspection	0.50%	0.75%	1.00%
* Total virus	-	1.00%	2.00%	3.00%
* Plants infected by brown rot (Syn. Bacterial wilt) (*Pseudomonas solanacearum*)	I & II Inspection	None	None	3 plants per hectare
*** Re-growth of plants after destruction of haulms	IV Inspection	0.50%	0.50%	0.50%

* Of the two inspections, the higher virus percentage will be considered for the purpose of the specified limits of tolerance.

Contd...

** The presence of brown rot infected plants within the specified limits of tolerance shall be permitted in the areas known to be infected with this disease. In case of plants suspected to be infected with brown rot, the neighbouring plants, one on either side should also be rouged along with tubers.

*** Standards for re-growth after destruction of haulms shall be met at fourth inspection to be conducted about 10 days after haulms cutting.

Note:

1. All off types and diseased plants should be rouged out along with the tubers and destroyed.
2. Gaps in the seed plot should not be more than 10.0 per cent
3. Haulms must be destroyed as close to the ground as possible before the date specified by the certified Agency. Failure to destroy haulms in time shall render the crop liable for rejection.

Seed Standards

Specification in respect of size and weight of seed material for Foundation Stage-I, Foundation Stage-II and Certified class shall be as under:

Size	*Mean Length and Two Widths at the Middle of Tuber*	*Corresponding Weight*
(a) Hill Seed (HS)		
Seed size	30 mm-60 mm	25-150 gm
Large size	Above 60 mm	Above 150 gm
(b) Plains Seed (PS)		
Seed size	30 mm-55 mm	25-125 gm
Large size	Above 55 mm	Above 125 gm

Note:

The size of tuber will be decided either on the basis of mean of two widths of a tuber at the middle and that of length or on the basis of corresponding weight of tuber.

In a seed lot, tubers not conforming to specific size of seed shall not exceed more than 5.0 per cent (by number).

(*a*) The seed material should be reasonably clean, healthy, firm and shall conform to the characteristics of the variety. The tubers not conforming to the varietal characteristics shall not exceed 0.05 per cent and 0.10 per cent (by number) for foundation and certified seed classes respectively.

(*b*) Cut, bruised, unshapy, cracked tubers or those damaged by insects, slugs or worms shall not exceed more than 1.0 per cent (by weight).

(*c*) Greenish pigmentation on tubers will not be a disqualification for certification.

Maximum tolerance limit of tubers showing visible symptoms caused by the diseases mentioned below will be as follows;

Diseases	Maximum Permissible Limits		
	Foundation		Certified
	Stage-I	Stage-II	
Late blight (*Phytophthora infestans*), dry rot (*Fusarium caeruleum*) or charcoal rot (*Macrophomina phaseoli*)	1.0 per cent (by number)	1.0 per cent (by number)	1.0 per cent (by number)
Wet rot (*Sclerotium rolfsii*)	None	None	None
* Common scab (*Streptomyces scabies*)	3.0 per cent (by number)	3.0 per cent (by number)	5.0 per cent (by number)
** Black scurf (*Rhizoctonia solani*)	5.0 per cent (by number)	5.0 per cent (by number)	5.0 per cent (by number)
*** Total diseases (by number)	5.0 per cent (by number)	5.0 per cent (by number)	5.0 per cent

* Even, if a single tuber infected with common scab is detected in a seed lot, the entire seed lot shall be treated with approved fungicide before seed lot is declared fit for certification. Seed lots having infected tubers more than the prescribed limits will not be certified even after treatment.

** (a) A tuber carrying 10.0 per cent or above scurfed surface will be considered as one infected unit.

(b) Seed lots having black scurf infection more than the prescribed limits could be certified after treatment with approved chemical/fungicide.

*** For all diseases, the higher disease percentage will be considered for the purpose of the specified limits of tolerance.

Brief Cultural Practices

Time of Planting

Western Indo-Gangetic plains: First week of October; Central Indo-Gangetic plains: Second to third week of October; Eastern Indo-Gangetic plains: Last week of October to first week of November

Preparation of Land

Prepare the field to a good tilth by one deep ploughing, two to three harrowings followed by levelling. The ideal plot/bed size is 3 × 5 m, 4.4 × 8 m and 4.8 × 10 m in manual, bullock and tractor cultivation, respectively. Rows with in the plot/bed should remain open from both the ends for free and uniform movement of water.

Source of Seeds

Obtain nucleus/breeders/foundation seed from source approved by a seed certification agency.

Seed Rate

Traditionally, seed of relatively small size (30 g or less) is because of its better keeping quality, delayed sprouting in ordinary stores and lower seed rate. The seed requirements for a hectare on the basis of seed size are given below:

Large size, 60-100g (3.5-5.0 cm)–25-30 q/ha; Medium size, 25-60 g (2.5-3.5 cm),–15-20 q/ha; Small size below 25 g (2.5 cm)–10-15 q/ha

Pre Sprouting of Seed Tubers

Seed should be taken out from the cold store 10-15 days before planting in the plains. Tubers should be kept in baskets, wooden or plastic trays in the shade having diffused light or may be spread in a thin layer on semi-pucca or pucca floor. These arrangements help in developing multiple, thick stout, green and healthy sprouts.

Methods of Sowing

Ridge and Furrow Method

The ridges are prepared. The length of the ridges depends on slope of the plot. Too long ridges and furrows are not supplied with irrigation water conveniently. The potato tubers are planted on ridges and irrigation water is let into the furrows.

Flat Bed Method

The whole plot is divided into beds of convenient length and width. The shallow furrows are opened and potato tubers are planted at recommended distance. The tubers are covered with the soil of furrows.

Spacing

Diameter of Tuber from Longer Axis	*Planting Distance (Row x Seed)*
2.5-3.5 cm	50 × 20 cm or 60 × 15 cm
3.5-5.0 cm	60 × 25 cm
5.0-6.0 cm	60 × 40 cm

Fertilization

200-250 quintals of farmyard manure/ha should be applied. It needs 100 to 150 kg nitrogen, 80 to 100 kg phosphorus and 80 to 100 kg potassium per hectare. Two-third to three fourth quantity of nitrogen along with whole quantity of phosphorus and potassium should be applied at the time of planting. Remaining one fourth to one third nitrogen should be applied 30 to 35 days after planting *i.e.* at the time of first earthing up or when plants become 25 to 30 cm in hight either in the form of top dressing or as a foliar feeding.

Irrigation

Before coming to the planting operation, it should be kept in mind that the sufficient soil moisture is available for satisfactory sprouting. Irrigation should be provided as and when needed. Stop irrigation 10 days before dehaulming in light and 15 days in heavy textured soils to avoid regrowth and for curing of skin of seed tubers.

Interculture

Earthing should be done when the plants are 15 to 22 cm in height. Generally, earthing is done at the time of top dressing of nitrogenous fertilizers. It necessary, a second earthing may be done after two week of the first one. Pre–emergence application of nitrofen @ 1.0 kg a.i./ha or alachlor @ 2.0 kg a.i./ha may be used.

Insects and Diseases

Insects/Diseases	*Control Measures*
	Insects
Cut worms (*Agrotis spp, Euxoa spp*)	Spray the crop with dursban 20 EC at 2.5 ml per litre. Apply phorate 10 G granules at 10 kg per hectare.
Epilachna beetle	Spray the crop with carbaryl at 0.1 per cent.
Leaf eating caterpillars	Spray with thiodon 35 EC or endocel at 0.15 per cent.
Aphids (*Myzus persicae*) and Jassids	Spray the crop with rogor or metasystox or nuvacron at 1 ml per liter of water.
Potato tuber moth (*Phthorimaea operculella*)	Spray the crop with carbaryl or nvacron or monocil at 0.1 per cent. Disinfect the godown with malathion at 0.05 per cent before storing potatoes.
Root knot nematodes (*Meloidogne spp.*)	Combination of deep ploughing to a depth of 20 cm and application of aldicarb at 1kg a.i./ha.

Contd...

Insects/Diseases	*Control Measures*
	Diseases
Early blight and late blight	Spray the crop with diathane M-45 (2.0 kg/ha) or diathane Z-78 (2.5 kg/ha) or difolatan (2.5 kg/ha)
Black scurf	Treat the seed tubers with 6 per cent agallol or aretan.
Common scab (*Streptomyces scabies*)	Disinfect the tubers by dipping in emisan-6 or agallol-6 at 0.25 per cent concentration for 5 minutes.
Bacterial wilt	Treat the seed tubers in 0.02 per cent streptocycine.
Wart disease	Soil treatment with 5 per cent formalin.
Leaf roll and mosaic	Uproot affected plants. Spray with metasystox at 0.1 per cent.

Roguing

Off type plants should be removed as soon as they are observed. In addition to off type plants, remove plants affected by blight, leaf roll and mosaic diseases from the seed field.

Haulm Cutting

For the seed crop, the haulms should be cut about 10 days before harvesting so that the tuber skim matures and prevents the tubers from getting bruised while harvesting and grading. Halums may be cut with the sickle as close to the ground as possible or killed by spraying with gramaxone (paraquat) @ 2.5 l/ha.

Harvesting and Curing

Harvesting is done either manually or mechanically. For manual harvest, small hand tools like *khurpa* are used to break the ridges and collect the tubers. Animal drawn plough can also be used to expose the ridges and the tubers are picked up manually. For mechanical harvesting, tractor drawn diggers are used in large fields. After harvesting, potato tubers are required to be kept in heaps for about 15 days for curing of skin and shedding of soil

Storage

Treated tubers should be dried in shade. Seed potatoes are packed in gunny bags of 50-80 kg. The treated seed bags should be properly sealed and should be labelled 'poisonous' so that they are not used for human consumption. The seed material in the plains should be sent to cold store by first week of March and latest by the end of March. The temperature in the cold store should be maintained at 2-4°C during storage.

Seed Yield

200 to 250 q/ha

Varieties/Hybrids Characteristics

Recommend for North Indian Plains

Kufri Alankar

It is a derivatives of the cross (Kennebee x O.N.2090) x (Majestic x Ekishiraju). Tubers are oblong, tapering towards end, flesh dull white and eyes medium deep to fleet. It matures 75 days in plains and 140 days in hills. The yield potential is 300 q/ha in plains and 100 q/ha in hills.

Kufri Anand

A derivative of PJ376 x PH/F 1430. Tubers are white, large oval long, flattened, smooth skin, fleet eyes, white flesh. It is a medium maturing variety (100-110 days) having a yield potential of 350-400 q/ha. This variety is resistant to late blight.

Kufri Ashoka

It is a derivative of (EM/C-1020 x Allerfruii Heste Gelbe). Tubers are white, large, oval long, smooth skin and fleet eyes with white flesh. It is an early maturing (70-80 days) having yield potential of 400q/ha.

Kufri Chamatkar

A derivative of the cross Ekishiraju x Phulwa. Tubers are round and white, eyes medium deep, flesh yellow. It is late maturing variety, which matures 110-120 days in plains and 150 days in hills. The yield potential of this variety is 250 q/ha in plains and 75.00 q/ha in hills. It is resistant to early blight.

Kufri Chipsona-1

It is a cross of MEX.750826 x MS/78-79. Tubers are white medium to large oval, smooth skin, fleet eyes, dull white flesh. It is medium maturing variety (90-110 days) having 400q/ha. It is resistant to late blight. The variety is highly suitable for making chips and French fries.

Kufri Chipsona-2

It is a derivative of F-6 x QB/B-92-4. Tubers are white, medium, round oval, smooth skin, fleet eyes, and yellow flesh. It is medium maturing variety yielded 350 q/ha. This variety is resistant to late blight. Due to high dry matter content, low reducing sugars and low phenols, the variety is highly suitable for making chips and French Fries.

Kufri Sindhuri

It is derivative of the cross Kufri Kundan x Kufri Red. Tubers are round, light red, with medium deep eyes. Flesh is dull white. It is late variety and matures in 120 days in plains and 165 days in hills. In plains, it yields 300 q/ha.

Kufri Sutlej

It is a derivative of Kufri Bahar x Kufri Alankar. Tubers are white, large, oval, smooth skin, fleet eyes, and white flesh. It is medium maturing (90-100 days) variety yielded 400 q/ha.

Recommended for North Indian Plains and Plateau Region

Kufri Badshah

It is a cross of Kufri Jyoti and Kufri Alankar. Tubers are mostly large to medium, oval, shining white, smooth, regular with shallow eyes, flesh dull white. It matures in plains 90-100 days.

Kufri Chandramukhi

It is a derivative of the cross S.4485 x Kufri Kuber. Tubers are oval, white, flesh dull white, eye fleet. It possesses good keeping quality. It is mid season variety, matures 80-90 days in plains and 120 days in hills. An average yield is 200 q/ha in plains and 75 q/ha in hills.

Kufri Jawahar

It is a derivative of Kufri Neelamani x Kufri Jyoti. Tubers are creamy white, medium sized, round-oval smooth skin, eyes fleet, pale yellow fleshes. It is an early maturing variety (80-90 days) yielded 400 q/ha. This variety is resistant to late blight.

Kufri Pukhraj

It is a cross of Craig's Defiance x JEX/B-687. Tubers are white, large, oval, slightly tapered, smooth skin, fleet eyes, yellow flesh. It is an early maturing variety (70-90 days). The yield potential is 400 q/ha. It is resistant to early blight.

Recommended for North Indian Hills

Kufri Giriraj

It is a cross of SLB/J-132 x EX/A 680-16. Tubers are white, medium to large, oval smooth skin, fleet eyes, and white flesh. The yield potential is 250 q/ha. This variety is resistant to late blight.

Kufri Jeevan

It is a derivative of the cross M-109-3 x D 698. Tubers are medium, white skin, oval shaped, eyes fleet. It matures 150-160 days in hills and 120 days in plains. It is resistant to wart, highly resistant to late blight.

Recommended for South Indian Hills

Kufri Himalini

It is a derivative of cross SLB/H-140 x SLB/Z-389 (b). Tubers are medium to large, oval, slightly flattened, skin white, eyes shallow to medium deep, flesh dull white. It is highly resistant to late blight. It matures 130-140 days after planting.

Kufri Swarna

It is a cross of Kufri Jyoti x (VIn)^2 (62.33.3). Tubers are white, medium, round oval, smooth skin, fleet eyes, and white flesh. It matures 130-135 days in summer while 100-110 days in autumn. This variety is highly resistant to cyst nematodes (*Globodera rostochiensis* and *G. palida*) and resistant to early and late blight.

Recommended for North and South Indian Hills and North Indian Plains

Kufri Jyoti

It is a derivative of the cross 3069d (4) x 2814 Q (1). Tubers are oval, white, eye fleet. Flesh is light white and waxy. It is resistant to early blight and wart. It matures 100 days in plains and 120 days in Hills. The yield potential is 200 and 150 q/ha in plains and hills respectively.

Kufri Naveen

It is a derivative of the cross D-692 x 3070d (4). Tubers are medium, white and oval with fleet eyes. Flesh is waxy and pale yellow. It is resistant to wart.

Recommended for North Eastern Hills

Kufri Khashigaro

It is a derivative of the cross Taborky x SD 698 D. Tubers are medium, attractive white, round with deep eyes and waxy, pale yellow flesh. It gave 160 q/ha tubers in 120 days crop duration.

Kufri Megha

Tubers are white, medium, round-oval with fleet eyes. This variety is resistant to late blight.

Recommended for Plateau Region

Kufri Lauvkar

It is a derivative of cross Serkoy x Adina. Tuber is round, white, flesh white. It is an early variety, which yielded 250 q/ha tubers in 75-95 days crop duration.

Recommended for North Western Plains

Kufri Dheetman

It is a derivative of the cross Craig Defiance x Phulwa. Tubers are oval and white with fleet eyes. It is resistant to frost and drought.

Recommended for North Bengal Hills and Sikkim

Kufri Sherpa

It is a derivative of the cross Ultimus x Adina. Tubers are medium in size, round flattened, skin smooth, medium thick, white, eyes medium deep and flesh of pale yellow colour. It matures 135 and 100 days in hill and plains respectively and yield is 225-250 q/ha. It is resistant to late blight and phoma leaf spot.

Family–Malvaceae

Okra [*Abelmoschus esculentus* (L.) Moench] 2n = 2x = 72-144

Land Requirements

Land to be used for seed production should be free of volunteer plants. In addition, the soil of selected fields should be fertile, well drained and free from soil borne diseases.

Field Inspection

A minimum of three inspections should be done, the first before flowering, the second during peak flowering and fruiting stage, and the third at the mature fruit stage and prior to harvesting.

Field Standards

General Requirements

Isolation

Seed fields should be isolated from the contaminants shown in

Figure 57: Okra

column 1 of the Table below by the distances specified in columns 2 and 3 of the said Table:

Contaminants	*Minimum Distance (Meters)*	
	Foundation	*Certified*
1	*2*	*3*
Fields of other varieties	400	200
Fields of the same variety not conforming to varietal purity requirements for certification of wild okra (*A. ficuleneus, A. manihot* etc.)	400	200

Specific Requirements:

Factor	*Maximum Permitted (%)**	
	Foundation	*Certified*
Off types	0.10	0.20
** Objectionable weed plants	None	None

* Maximum permitted at and after flowering.

** Objectionable weeds shall be: (*A. ficuleneus, A. manihot* and *A. moschatus*)

Seed Standards

Factors	*Standards for Each Class*	
	Foundation	*Certified*
Pure seed (minimum)	99.0 per cent	99.0 per cent
Inert matter (maximum)	1.0 per cent	1.0 per cent
Other crop seeds (maximum)	None	5/kg
Total weed seeds (maximum)	None	None
Objectionable weed seeds (maximum)	None	None
Other distinguishable varieties (maximum)	10/kg	20/kg
Germination (minimum)	65.0 per cent	65.0 per cent
Moisture (maximum)	10.0 per cent	8.0 per cent
For vapour-proof containers (maximum)	8.0 per cent	8.0 per cent

Brief Cultural Practices

Time of Sowing

In the northen plains, okra can be sown from February to July and in east and south India throughout the year. The best time for sowing the seed crop in the northen plains is 20th June.

Preparation of Land

Prepare the land to a fine tilth by one deep ploughing and three to four harrowings followed by levelling.

Source of Seeds

Obtain nucleus/breeders/foundation seed from source approved by a seed certification agency.

Seed Rate

Kharif 8-10 kg/ha.

Spring crop 10-15 kg/ha.

Method of Sowing

The sowing should be done in rows. Soaked the seeds overnight in water. Sow in full soil moisture behind the plough in shallow furrows. Seed should not be sown more than three cm deep.

Spacing

Kharif crop 60 x 30 to 45 cm and for spring crop 45 x 30 cm.

Fertilizers

Apply 250 to 300 quintal of farmyard manure per hectare two weeks before sowing and incorporate it in to the soil. Apply 50 kg nitrogen, 80 kg phosphorus, 60 kg potash per hectare in the rows before sowing. Top dress 50 kg nitrogen after about a month.

Irrigation

The summer crop needs irrigatrion twice a week. The rainy season crop needs irrigation only, if there is a prolonged drought.

Interculture

Two–three hoeings and weedings, from time to time, are necessary to keep the seed field clean weeds.

Insects and Diseases

Insects/Diseases	*Control Measures*
	Insects
Shoot and fruit borer (*Earias vittella* and *E. insulanta*)	Application of cypermethrin (30 g) or fenvalerate 50 g or deltamethrin (10 g a.i./ha).
Leaf hopper (*Ameasca biguttula ishida*)	Application of carbofuran (1kg a.i./ha) and phorate (1.5 kg a.i./ha).
Mites (*Tetranychus* spp.)	Application of either tetradifon (0.025 per cent) or monocrotophos (0.05 per cent).
	Diseases
Powdery mildew (*Erysiphe cichoracearum*)	Spraying of bavistin (0.1 per cent) or wettable sulphur (0.3 per cent).
Yellow vein mosaic (Vector: white fly)	Soil application of furadan @ 1.5 kg a.i./ha followed by spray of dimethoate (0.05 per cent).

Figure 58: Pods Maturing from the Base of the Plant

Roguing

Roguing of the seed crop should begin with uprooting and destroying of yellow vein mosaic affected plants soon after they are noticed. This should continued up to fruiting stage. Subsequent roguing for off types and wild *Abelmoschus species* should be done prior to flowering. This should continue during the flowering stage also. The off type plants are easily distinguishable on the basis of plant height, leaf and stem characteristics, pigmentation, flower size and shape, fruit shape etc.

Harvesting and Thresing

Pods should be harvested when they have dried (about 25 days old). Varieties with angular pods, which open along sutures, should be harvested promptly to avoid shattering. The pods are usually picked by hand and threshed it. Seed should be dried to at least 10 per cent moisture content before storage.

Seed Yield

12 q/hectare.

Varieties/Hybrids Characteristics

Varieties/Hybrids Recommended by IARI, New Delhi

Pusa Makhmali

It was selected from the material collected from West Bengal. It is an early maturing variety. Pods are smooth, straight, 5-edged, attractive light green, slender, tapered and 15-20 cm long when fully developed. It yields 80-100 q/ha.

Pusa Sawani

It was derived from an intervarietal hybridization between IC-1542, and Pusa Mukhmali. The fruits are smooth (slightly hairy on the edges), 5-edged, dark green and 18-20 cm long. It was reported to free from the yellow vein mosaic virus. But at present, it behaves as susceptible. It yields 125-175 q/ha.

Varieties/Hybrids Recommended by Karnataka State

Arka Abhaya

It is released from IIHR Bangalore. Fruits are 18 cm in length having 5-6 ridges, dark green in colour. It is resistant to yellow vein mosaic virus. The yield potential is 170-200 q/ha.

Arka Anamika

It is yellow vein mosaic virus resistant variety evolved at IIHR, Bangalore. It is of inter specific origin between *Abelmoschus esculentus* and a wild species *A. manihot* spp tetraphyllus. Fruits are medium, green, rough, 5-ridged. It gives 115 q/ha of green fruit.

Varieties/Hybrids Recommended by Tamil Nadu State

Co. 1

It is a pure line selection from the 'Red Wonder' introduced from Hyderabad and released from TNAU, Coimbatore. The pods are long, slender, 5 ridged, glossy, smooth and scarlet red. It has field tolerance to yellow vein mosaic virus. It yields 120 q/ha.

CO. 3

It is a F_1 hybrid between Parbhani Kranti x MDU-1 (Hy–8) and released from TNAU, Coimbatore. It is suitable for growing during *kharif* as well as summer. The fruits are dark green, medium in size. It exhibits moderate resistance to YVM. This variety has on yield potential of 161 q/ha.

Varieties/Hybrids Recommended by Haryana State

Varsha Uphar

It is a cross between 'Lam Selection 1' and 'Parbhani Kranti'and developed at CCSHAU, Hissar. The fruits are smooth, dark green with 5 ridges. It gives 98 q/ha as vegetable and 15–20 q/ha as seed. It is resistant to yellow vein mosaic virus.

Hisar Unnat

It is developed from CCSHAU, Hissar. This variety is moderately tolerant to leaf hopper. It gave fruit yield of 102 q/ha and was free from yellow vein-mosaic virus infection.

Varieties/Hybrids Recommended by Uttar Pradesh State

VRO-3

It is a YVMV resistant variety developed at IIVR, Varanasi, by pedigree method of selection. Fruits are green with five ridges. Gives an average yield of 112 q/ha.

VRO-4

It is a YVMV resistant variety developed at IIVR, Varanasi by pure line selection. Fruits are green with five ridges. It is suitable for cultivation under summer as well as for rainy season. Gives an average yield of 160 q/ha.

Varieties/Hybrids Recommended by Other States

Perking Long Green

It is a mid season variety. Fruit is long, tapered, bright green, spineless and mostly eight edged. It is especially suitable of hilly areas.

Parbhani Kranti

It is a yellow vein mosaic resistant variety evolved at MPAU, Parbhani, from interspecific cross between *A. esculentus* cv. Pusa Sawani and *A. manihot*, a Japanese species. The fruits are dark green, smooth, tender, slender, 5-ridged with long and narrow top. It gives green fruit yield of 85-90 q/ha during summer and 115 q/ha during rainy season. It gives seed yield 10 q/ha from rainy season crop and 5-6 q/ha from the summer crop.

Punjab Padmini

It evolved from a cross between *Abelmoschus esculentus* and *A. manihot* ssp. *manihot*. Fruits are dark green, shining smooth, thin, 15-20 cm long, 5-ridged and remain tender for a long period (3-4 days). It possesses field resistance to yellow vein mosaic virus and tolerance to jassids and cotton boll worm. it gives yields 100-125 q/ha of green pods. As a seed crop it gives around 12.5 q/ha of seed.

Family–Cruciferae

Radish (*Raphanus sativus* L.) 2n = 2x = 18

Land Requirements

Land to be used for seed production shall be free of volunteer plants. The soils of selected fields should be deep, light and friable and the fields well drained.

Figure 59: Radish

Field Inspection

Mother Root Production Stage

A minimum of two inspections shall be made as follows:

1. The first inspection should be done after 20-30 days of the sowing in order to determine isolation, volunteer plants, out crosses, off types and other relevant factors;
2. The second inspections should be done after the mother roots have been lifted to verify the true characteristics of roots.

Seed Production Stage

A minimum of one inspection should be done during flowering to check isolation, off types and other relevant factors.

Fields Standards

General Requirements

Isolation

Seed fields should be isolated from the contaminants shown in column 1 of the Table below by the distances specified in columns 2 and 3 of the said Table:

Contaminants	Minimum Distance (Meters)			
	Mother Root Production Stage		Seed Production Stage	
	Foundation	Certified	Foundation	Certified
1	2	3	4	5
Fields of the other varieties of the same species	5	5	1600	1000
Fields of the same variety not confirming to varietal purity requirement for certification and rat tail radish (*Raphanus caudatus*)	5	5	1600	1000

Specific Requirements

Factor	Maximum Permitted (%)*	
	Foundation	Certified
*Roots not confirming to the varietal characteristics including forked roots	0.10 (by number)	0.20 (by number)
**Off types (plants)	0.10	0.20
***Plant affected by seed borne diseases at final inspection	0.10	0.50

*Maximum permitted at second inspection at mother root production stage.

** Maximum permitted at and after flowering at seed production stage.

*** Seed borne diseases shall be: Black rot and Black leg.

Seed Standards

Factors	Standards for Each Class	
	Foundation	Certified
Pure seed (minimum)	98.0 per cent	98.0 per cent
Inert matter (maximum)	2.0 per cent	2.0 per cent
Other crop seeds (maximum)	5/kg	10/kg
Total weed seeds (maximum)	10/kg	20/kg
Germination (minimum)	70 per cent	70 per cent
Moisture (maximum)	6.0 per cent	6.0 per cent
For vapour-proof containers (maximum)	5.0 per cent	5.0 per cent

Method of Seed Production

Both seed to seed and root to seed methods can be employed for raising seed of radish. Seed to seed method for varieties, which do not stand transplanting well, is preferred for raising certified seed. The nucleus seed, however, is invariably produced by root to seed method. In this method, the planting of whole roots should be done before these get pithy. In India, the radish seed is largely produced by root to seed method.

Root to Seed Method

First Season-raising of Stecklings

Time of Sowing

For hills: The autumn sowings preferably may be done in early October. Spring sowing can also be done in March, as soon as the land can be prepared. However, the autumn sowings gives higher seed yields.

Preparation of Field

Prepare the field to fine tilth by ploughing three to four harrowings, followed by levelling.

Source of Seeds

Obtain nucleus/breeders/foundation seed from source approved by a seed certification agency.

Seed Rate

(i) Asiatic varieties-10 kg/ha. (ii) Temperate varieties-12 kg/ha.

Method of Sowing

The seed should be sown on ridges. Row to row spacing for Asiatic types are 45 cm and for temperate types 30 cm.

Fertilizers

Apply 200-250 quintals per hectare of well-rotted farmyard manure, at the time of preparation of land. At the time of final levelling mix in the soil 35 kg nitrogen, 50 kg phosphorus and 50 kg potash per hectare as basal application. Another dose of 35 kg nitrogen is top-dressed when the roots just start growing.

Thinning

Thin the crop after 20-25 days to maintain a plant to plant distance of 10-15 cm.

Irrigation

Irrigate the crop at 8-10 days interval depending on the weather conditions and requirements. During early warm weather, it may be necessary to irrigate at 4 to 5 days interval.

Interculture

One weeding and one earthing up during the early stages of growth are necessary for proper development of roots.

Insects and Diseases

Insects/Diseases	*Control Measures*
	Insects
Aphids (*Myzus persicae, Brevicory brassicae)*	Spray with malathion (0.01 per cent) or monocrotophos (0.05 per cent)
Flea beetle (*Phylloreta sp.)*	Spray malathion (0.15 per cent)
Mustard sawfly *(Athalia proxima)*	Spray carbaryl (0.2 per cent) or malathion (0.05 per cent).
	Diseases
White rust *(Albugo candida*) and Alternaria blight *(Alternaria brassicae*)	Difoltan (0.3 per cent) and dithane M-45 (0.2 per cent)
Root rot of radish (*Erwinia rphapontici*)	Spray Agrimycin-100 (100 ppm)
Radish mosaic	Soil application of furadon @ 1.5 kg a.i. followed by 2-3 foliar sprays of either dimecron (0.05 per cent).

Harvesting of Roots (Stecklings)

Uproot the plants (stecklings) when they have attained edible maturity (30-70 days depending upon the variety).

Selection of Roots for Transplanting

After harvesting, each steckling is critically examined for foliage and root characteristics namely, size, shape, colour, texture, sponginess etc. Stecklings having off type foliage or roots not conforming to variety characteristics must be rejected. Also, the diseased, malformed, forked or any other undesirable types are rejected.

Seed Yield

Average seed yield varies from 2-3 q/ha.

Second Season

Preparation of Selected Roots for Transplanting

After selection the top are cut off in such a way as to leave the small undeveloped leaves together with about 2 cm of the petioles of older leaves. It is the usual practice to cut $1/3^{rd}$ of the lower root portion in Asiatic types, but the whole roots are set in European varieties. These are then reset in a well-prepared field.

Preparation of Field

Same as described for the previous season.

Method of Transplanting

The stecklings selected and prepared in the manner described earlier are reset in the field at a distance of 60 x 45 cm. The soil is then firmed and tapped around the roots. Soon after transplanting the field is irrigated. The field is then checked after 10-15 days for mortality and the gaps are filled well in time to have uniform crop stand.

Fertilizers

Apply 200-250 quintals per hectare of well-rotted farmyard manure should be given at the time of preparation of land. At the time of final leveling, mix 40 kg nitrogen, 50 kg phosphorus and 100 kg potash per hectare in the soil as basal application. Another dose of 40 kg nitrogen is top-dressed at the time of bolting and pre flowering stages.

Rouging

The plant with off type foliage and the direct bolters are eliminated from the prior to flowering and the crop is left to set seeds. When the roots are lifted they critically examined for trueness to type. Besides, small sized or oversized stecklings, diseased, forked

Figure 60: Radish Roots and Ripening Seed Pods

and off types should be discarded. Normally, three rouging should be done, namely before maturity of roots for off type foliage, at the time of replanting for verifying root characteristics, such as size, shape, colour, texture etc. and the last at flowering stage for early and late plants affected by designated, diseases (black-leg and black rot) and eliminate them.

Harvesting and Threshing

The crop is cut when plants are fully mature. The crop is cut by sickle and brought to threshing floor for threshing. There is more often considerable difficulty in threshing the seed from the pod. Frequently, the pod will not break and let the seed escape. It is, therefore, important to thoroughly dry them before commencing the threshing. Threshing can be done by beating with sticks. The seed after sifting should be dried to 6 per cent moisture, before storage.

Seed Yield

Average seed yield varies from 600-800 kg/ha.

Varieties/Hybrids Characteristics

Varieties/Hybrids Recommended by IARI, New Delhi

Japanese White

It was originally named as Shiroaguri-Kyo and released by IARI, New Delhi. Roots are 25-30 cm long, 5 cm in diameter, cylindrical and blunt at the tip. Skin is pure white and smooth. Flesh is snow white, crisp, solid and mild pungent.

Pusa Chetki

It is a selection made from a material collected from Denmark. Roots are medium large and stumpy, pure white, tender and smooth and mildly pungent. it can tolerate high temperature, therefore, it has been found suitable for sowing from the middle of March to the middle of August.

Pusa Desi

It is suitable for sowing from middle of August to early October in the northern plains. It is selection from local material. Roots are pure white, 30-35 cm long, tapering with green stem end, pungent and heavy yielder, tolerant to slightly higher temperature.

Varieties/Hybrids Recommended by Uttar Pradesh State

Kalianpur No. 1

It is a selection from local material. Roots are 22-23 cm long, smooth, crispy, and white with green shoulder, thick and tapering. It is free from mustard sawfly, aphids and white rust.

IIVR 1

It is developed through selection from Japanese materials suitable for growing in spring, summer, autumn and winter seasons.

Roots are 25-30 cm long, 3.5-4 cm in diameter, straight, tapering with blunt tip. This variety yielded about 500q/ha.

IIVR-2

It is suitable for planting from September to February. Roots are straight and tapering types, 35-40 cm long and 3.5-4.2 cm diameter. This variety yielded 450q/ha.

Varieties/Hybrids Recommended by Punjab State

Punjab Ageti

It is suitable for April to August sowing. Roots are long, thin with lower half-white and upper red. This variety gives an average yield 315 q/ha.

Punjab Safed

It is a derivative of cross White5 x Japanese White. Roots are white, tapering smooth, mild in taste, free of forking, 30-40 cm in length and 3.5 cm thick.

Varieties/Hybrids Recommended by Himachal Pradesh State

Chinese Pink

It is an introduction identified by Dr. YS Parmar University of Horticulture and Forestry, Nauni (Solan). Roots are 12-15 cm long, semi stumpy to stumpy roots, pink with white colour towards the tip.

Nadauni

This is developed from a local material. The roots are long, tapering, smooth and light pink in colour.

Pusa Himani

It is developed through hybridization between a temperate type (Black) and a popular Asiatic Type (Japanese White) and released from IARI, Regional Station Katrain. It is the only variety, which can be grown throughout the year in the hills farming. The roots are 30-35 cm in length and 10-12 cm in girth and they are semi-stump to tapering with short tops.

Pusa Reshmi

It is suitable for mid September to early October sowing. Roots are 30-35 cm long, tapering white with green shoulder, mildly pungent, tolerant to slightly higher temperatures medium light green top with cut leaves.

Varieties/Hybrids Recommended by Karnataka State

Arka Nishant

It is an improvement over a collection (IIHR-72) from Singapore, developed through mass selection and released by IIHR, Bangalore. Roots are medium-sized (25 cm x 3-4 cm), marble white, crisp texture, pleasant aroma, free from early bolting, pithiness, splitting and forking. It yields 200-300 q/ha.

IIHR-1-1

Roots are thin long (30 cm), each root weighing about 300 g and free from pre-mature bolting, pithiness, splitting and forking and mildly pungent. Root surface is smooth and shining white. It is highly resistant to white rust.

Other Varieties/Hybrids

CO-1

It is a selection from germplasm type (RS-44-2-1) and released from TNAU, Coimbatore. The roots are milky white, long (22 cm), thick (12.5 cm girth). This variety is suitable for growing round the year. It yields 200-250q/ha.

Rapid Red White Tipped

It is an extra early European variety. Roots are small, globular, bright red with white tip. Flesh is pure white, crisp and snappy

Scarlet Long

It is an early variety having 15-20 cm length. Roots are long and tapering to a point. Flesh is white and crisp.

White ICICLE

It is a medium-short variety. The skin is pure white, thin and tender whereas the flesh is white icy, crisp, juicy, mild and sweet flavour with juicy enough pungency to appeal the appetite. The root is solid Icile–shaped straight and tapered.

Cauliflower (*Brassica oleracea* var. *botrytis*) 2n = 2x =18

Land Requirements

Land to be used for seed production should be free of volunteer plants. The soil of selected field should be deep, fertile, well supplied with organic matter and a pH value 5.5. The field should be well drained.

Field Inspection

A minimum of three inspections shall be made, the first before the marketable stage, the second at the marketable stage and the third at flowering stage.

Field Standards

General Requirements

Isolation

Seed fields should be isolated from the contaminants shown in column 2 and column 3 of the said Table:

Figure 61: Cauliflower

Contaminants	*Minimum Distance (Meters)*	
	Foundation	*Certified*
1	*2*	*3*
Fields of other varieties	1600	1000
Fields of the same variety not conforming to varietal purity	1600	1000

Specific Requirements

Factor	*Maximum Permitted (%)**	
	Foundation	*Certified*
Off types	0.10	0.20
**Plants affected by seed borne diseases	0.10	0.50

* Standards for off types shall be met at and after flowering and for seed borne diseases at final inspection.

** Seed borne diseases shall be: Black leg, black rot, soft rot

Seed Standards

Factors	*Standards for Each Class*	
	Foundation	*Certified*
Pure seed (minimum)	98.0 per cent	98.0 per cent
Inert matter (maximum)	2.0 per cent	2.0 per cent
Other crop seeds (maximum)	5/kg	10/kg
Weed seeds (maximum)	5/kgt	10/kg
Germination (minimum)	65 per cent	65 per cent
Moisture (maximum)	7.0 per cent	7.0 per cent
For vapour-proof containers (maximum)	5.0 per cent	5.0 per cent

Method of Seed Production

There are two methods of seed production:

1. *In Situ* Method (Seed to Seed Method)
2. Transplanting Method (Head to Seed Method)

For seed production, seed to seed method is recommended since the head to seed method in India has not been very successful. In seed to seed method (in *situ* method) the crop is allowed to over winter and produce seed in the original position, where they are first planted in the seedlings stage.

Brief Cultural Practices for in *situ* Method of Seed Production

Main Season and Late Varieties (Seed Production in Hills)

Time of Sowing and Transplanting

The last week of August is the optimum sowing time for the crop. The seed is sown in a nursery. Transplanting of seedlings should be completed by the end of September.

Method of Sowing Nursery

Seeds may be sown on raised nursery beds (15 to 20 cm high from the ground), in rows 5 cm apart. Cover the seeds with fine leaf mould and water with a sprinkler. Twenty five nursery beds of 2 to 2.5 meters long and 1 to 1.25 meters wide will raise enough seedlings to plant one hectare.

Source of Seed

Obtain nucleus/breeder's/foundation seed from source approved by a seed certification agency.

Seed Rate

Main season and late varieties: 300–500 g per hectare.

Preparation of Land for Transplanting

The field should be prepared to fine tilth by deep ploughing, 3 to 4 harrowings followed by leveling.

Fertilizers

The cauliflower seed crop requires heavy manuring as it removes large quantities of major nutrients from the soil. For good seed yield 200 to 300 quintals of farmyard manure is applied per hectare. Apply 50 kg nitrogen, 60 kg phosphorous and 60 kg potassium before transplanting. Top dress in two doses, 50 kg nitrogen during growing period (one application during October to November, another in February to March is essential), still higher doses of nitrogen may be applied if it is necessary.

A cauliflower crop often shows boron and molybdenum deficiency symptoms when grown either on an alkaline or highly acidic soil. In such cases 10 to 15 kg of borax per hectare may be applied to the soil, or two sprays with 0.3 per cent borax applied on the seedlings may correct the boron deficiency. Molybdenum deficiency symptoms occur in highly acidic soils and can be corrected by liming, or application of about 1 to 1.5 kg per hectare of sodium molybdate.

Transplanting

Transplant the seedlings when 12 to 15 cm long, preferably at evening time, and irrigate immediately afterwards.

Spacing

Row to row 60 to 90 cm, Plant to plant: 45 to 60 cm.

Irrigation

Irrigate the field according to the soil requirements and climatic conditions. A crop after transplanting may need irrigation twice a week and later once a week. At later stages, irrigation may be given if there is a long gap between rains. Adequate moisture supplies during flowering and seed maturation are necessary to obtain high yields.

Interculture

Frequent shallow cultivation should be given to the soil to kill weeds and provide soil mulch. Earthing up of plants four to five weeks after transplanting is highly desirable.

Insect Pests and Diseases

Like cabbage.

Early Varieties (Seed Production in Plains)

Most of the cultural practices for seed production of early varieties in the plains are same, except for the time of sowing (June to August), it varies with the varieties and a little closer spacing (60 x 45 cm). In addition, scooping is normally not required.

Rouging

Selection of curds is done when the curds are well developed. Off type plants and those forming poor curds, should be removed at this stage. Subsequent roughing for off types and diseased plants infected by black leg, black rot, leaf spot and phyllody should be done from time to time as required.

Harvesting and Threshing

Harvesting can be when pods brown. Too ripe pods dehisce. Seed should not crush or split when rubbed between the hands. The harvesting may be done in two lots. Generally the early plants are harvested first, when about 60-70 per cent of the pods turns brown and the rest of the crop changes to yellowish brown. After harvesting it is piled up for curing. After 4-5 days, it is turned upside down and allowed to cure for another 4-5 days in the same way. It is then threshed with sticks and sifted with hand sifters. After through drying of seed in the sun (7 per cent moisture content) it is cleaned and stored.

Seed Yield

Average seed yield varies from 500 to 600 kg per hectare.

Varieties/Hybrids Characteristics

Varieties/Hybrids Recommended by IARI, New Delhi

Pusa Aghani

This variety has been developed from IARI, New Delhi. Curds are ready for harvesting in the month of November-December. Curds are big in size, solid and white in colour. An average yield of this variety is 150-160 q/ha.

Pusa Deepali

This has been developed through inbreeding from the local material at IARI, New Delhi. Curds are compact, self-blanching, white and medium in size.

Pusa Early Synthetic

It is released by IARI, New Delhi. Curd is small to medium in size, flat, creamy white and compact. It is suitable for early. The average yield is 117 q/ha. It is resistant to riceyness.

Pusa Him Jyoti

It is released from IARI, New Delhi and suitable for transplanting in hill tract. Curd is quietly white, solid and round. Yield of this variety is 160 q/ha.

Pusa Hybrid-2

It is released from IARI, New Delhi. Curd is creamy-white and highly compact. The yield potential of this variety is 230-250 q/ha. It is resistant to downy mildew.

Pusa Katki

This is one of the earliest variety released from IARI, New Delhi. It is suitable for sowing the middle of May.

Pusa Sharad

Curds are white, knobby, very compact, semi-dome shaped and about 750-1000 g in weight. Average yield is 260q /ha.

Pusa Shubra

Curds are compact and white having average curd weight is 700-800 g. It is highly tolerant to riceyness. The yield potential of this variety is 250-300 q/ha. This variety is resistant to black rot.

Pusa Synthetic

It is a mid-season variety, synthesized from 7 inbred lines with good combining ability released from IARI, New Delhi. Curds are somewhat creamy white to white, compact becomes ready from mid-December to mid-January (12-15°C).

Varieties/Hybrids Recommended by Himachal Pradesh State

Pusa Snowball-1

It is a derivative of the cross between EC 12013 and EC 12012 and released from IARI, Regional Station, Katrain (Kullu Valley). It is a late variety and suitable for cool season. Curds are very compact, medium in size and snow white.

Pusa Snowball-2

It is suitable for late sown conditions. The Curd remains white even on exposure. This variety could not become popular because of its poor seeding ability.

Pusa Snowball K-1

This is also developed at IARI Regional Station, Katrain (Kullu Valley) and is tolerant to black rot. Amongst the snowball types, it has best quality curds which are snow white in colour and retain it even if the harvesting is delayed.

Varieties/Hybrids Recommended by Uttaranchal State

Pant Gobhi-2

It is composite variety developed from GBPUA&T, Pantnagar. Curds are ready for harvesting from October onwards. An average yield of this variety is 100 q/ha.

Pant Gobhi-3

It is a synthetic variety combining eight inbred lines of katki group and released from GBPUA&T, Pantnagar. Curds are hemispherical, creamy white, medium compact, non-ricey. Average yield potential is 120 q/ha.

Pant Gobhi-4

It is released from GBPUA&T, Pantnagar by simple recurrent selection of Aghani group. Curds are round, creamy white and solid. This variety is free from receness.

Pant Shubra

It is developed through simple recurrent selection and released from GBPUA&T, Pantnagar. Curds are compact, slightly conical, creamish white in colour, non-ricey and non-leafy. It is suitable for rainy season.

Varieties/Hybrids Recommended by Punjab State

Early Kunwari

It is an early variety selected by PAU, Ludhiana. Curds are semi-spherical with even surface and ready for harvesting from mid September to mid-October.

Punjab Giant-26

It is a main season variety developed by PAU, Ludhiana. Curds are solid, white and medium in size.

Punjab Giant-35

It is suitable for late season cultivation. Curds are compact, medium sized and snow white. The average yield of this variety is 225 q/ha.

Varieties/Hybrids Recommended by Other States

Improved Japanese

It is an introduction from Israel. It can not tolerate hot season. Curds are compact, white and maturing in late November to mid-December at 16-20°C.

Hisar-1

It is suitable for mid late season. It bears medium to large size white heads. It yields 250 q/ha.

Cabbage (*Brassica oleracea* var. *capitata* Linn.) 2n = 2x = 18

Land Requirements

Land to be used for seed production of cabbage should be free from volunteer plants. Early cultivars grow well in light soils whereas late maturing ones perform better on heavy soils.

Field Inspection

A minimum of three inspections should be done, the first before

Figure 62: Cabbage

the marketable stage, the second at the marketable stage and the third at flowering stage.

Field Standards

General Requirements

Isolation

Seed fields should be isolated from the contaminants shown in column 2 and column 3 of the said Table:

Contaminants	*Minimum Distance (Meters)*	
	Foundation	*Certified*
1	*2*	*3*
Fields of other varieties	1600	1000
Fields of the same variety not conforming varietal purity	1600	1000

Specific Requirements

Factor	*Maximum Permitted (%)**	
	Foundation	*Certified*
Off types	0.10	0.20
**Plants affected by seed borne diseases	0.10	0.50

* Standards for off types should be met at and after flowering and for seed borne diseases at final inspection.

** Seed borne diseases shall be: Black leg, black rot and soft rot.

Seed Standards

Factors	*Standards for Each Class*	
	Foundation	*Certified*
Pure seed (minimum)	98.0 per cent	98.0 per cent
Inert matter (maximum)	2.0 per cent	2.0 per cent
Other crop seeds (maximum)	5/kg	10/kg
Weed seeds (maximum)	5 /kg	10/kg
Germination (minimum)	65 per cent	65 per cent
Moisture (maximum)	7.0 per cent	7.0 per cent
For vapour-proof containers (maximum)	5.0 per cent	5.0 per cent

Methods of Seed Production

Being a biennial, cabbage requires two seasons to produce seed. In the first season the heads are produced and in the following season seed production follows. The seed crop can be left in situ or transplanted during autumn. In situ method is usually followed for certified seed production and the latter for nucleus seed production.

In situ method, the crop is allowed to over winter and produce seed in their original position, *i.e.* where they are first planted in the seedling stage. In the transplanting method, the mature plants are uprooted. After removing whorls, the plants are immediately reset in a well prepared new field, in such a way that the whole stem below the head goes underground with the head resting just above their surface.

There are three methods to produce seed of cabbage.

Stump Method

When the crop in the first season is fully mature, the heads are examined for true to type. The plants with off type heads are removed. Then heads are cut just below the base by means of a sharp knife, keeping the stem with outer whorl of leaves intact. The beheaded portion of the plant is called "stump". The heads are marketed and the stumps either are leaf *in situ*, or replanted in the second season *i.e.*, during autumn. The following spring, after the dormancy is broken, the bud sprout forms the axils of all the leaves and leaf scars.

Advantages

1. Gives extra income by way of scale of heads.
2. The crop matures 12-15 days earlier than the head intact method.
3. Seed yield is slightly increased.

Disadvantages

In this method, flowering shoots are decumbent and require very heavy staking, otherwise they breakdown very easily while interculturing or spraying.

Stump with Central Core Method

When the crop is fully mature in the first season, the heads are examined for true to type. Plants with off type heads are removed and rejected. Then the heads are chopped on all sides with downward perpendicular cuts in such a way that the central core is not damaged. This is an improvement over stump method in that the shoots arising from the main stem are not decumbent. During the last week of February and until 15th March when the heads start bursting, two vertical cross-cuts are given to the head. Taking care that the central growing point is not injured. In the absence of such cuts, the heads burst out irregularly and sometimes the growing tip is broken. The operation is completed by going around the field twice or thrice during this period.

Advantages

1. Shoots arising from the main stem are not decumbent, hence very heavy staking is not required.
2. Seed yield is increased.

Disadvantage

The chopped heads cannot be marketed.

Head Intact Method

When the crop is fully mature in the first season the heads are examined for true to type. The plants with off type heads are removed from the field. The head is kept intact and only a cross cut is given to facilitate the emergence of a stalk.

Advantages

1. The removal of heads (stump method) or chopping of heads on all sides (central core intact method) is not required. This saves time and labour.
2. Very heavy staking is not required.

Disadvantage

The seed yield is slightly low as compared to stump, or stump with central core intact method.

Brief Cultural Practices (*In situ* Method)

Time of Sowing

Early varieties-First to third week of July; Medium-late varieties: Second fortnight of June; Late varieties–First fortnight of June

Method of Nursery Sowing

The seeds are sown in raised nursery beds in a manner as described earlier in commercial production of cabbage.

Source of Seed

Obtain nucleus/breeder's/foundation seed from source approved by a seed certification agency.

Seed Rate

For main season and late varieties, 375 to 400 g seeds/ha and early varieties, 600-750 g seed/ha.

Preparation of Land for Transplanting

Prepare the land to a fine tilth by repeated ploughing and harrowing followed by leveling.

Fertilizers

Cabbage grows satisfactorily only when the supply of organic matter is liberal. For best results apply 500 to 600 quintal of farmyard manure per hectare at the time of land preparation. Apply 100 kg nitrogen, 60 kg phosphorus and 60 kg of potash by drilling, or by broadcasting sufficiently before transplanting the seedling. Give another dose of 50 kg nitrogen as surface application at the time of seed stalk emergence during March. Extra application of nitrogen may be given as and when there is a need before flowering starts, depending upon the condition of the crop.

Transplanting

Three to four-week-old seedlings are transplanted. Transplanting should preferably be done in the evening and the field irrigated immediately afterwards.

Spacing

Late varieties–60 x 60 cm; Medium varieties 60 x 45 cm; Early varieties 45 x 45 cm.

Irrigation

Cabbage requires a continuous supply of moisture. Irrigate the crop as frequently as required. Heavy irrigation should, however, be avoided when the heads have formed. A sudden heavy irrigation after a dry spell may cause bursting of heads.

Hoeing and Weeding

At least three weedings and hoeing till the end of October are essential. One weeding and earthing up during November and December and the second during March, when seed stalks have emerged, control weeds and also help in proper drainage during winter and thereafter.

Staking

After the flower stalks are sufficiently developed, stalking is necessary to keep the plants in an upright position.

Handling the Mature Head

After the planted crop has fully developed heads at the close of autumn, the next step is the handling of these plants for seed production. Handling of plants can be done by any one of the three methods *i.e.* stump, stump with central core intact methods, described earlier.

Insects and Diseases

Insects/Diseases	*Control Measures*
	Insects
Mustard sawfly (*Althalia proxima*)	Dusting with sevin @ 20-25 kg/ha or spraying the same @ 5 g per litre of water.
Mustard aphids (*Lipaphis erysimi*) and Painted bug (*Bagrada cruciferarum*)	Spraying with dimethoate @ 1 ml/litre of water. Soil application of phorate 10 G @ 15 kg/ha.
Cabbage borer (*Hellula undalis*), and Diamond back moth (*Plutella xylostella*)	Foliar spray of permethrin (0.2 kg a.i.). Use of mustard as trap crop.
Cabbage butter fly (*Pieris brassicae*)	Sprays of 0.15 per cent carbaryl and 0.4 per cent thuricide.
	Diseases
Downy mildew (*Perenospora parasitica*), Black spot (*Alternaria brassicae*) and Black leg (*Phoma lingum*)	Spray the seedling with copper oxychloride (0.3 per cent)

Contd...

Insects/Diseases	*Control Measures*
Wire stem (*Rhizoctonia solani*)	Seed treatment and soil drenching with dithane M-45 (0.2 per cent).
Black rot (*Xanthomonas campestris* pv. *campestris*)	Seeds are treated with plantomycin (100 ppm) or streptocyclin (100 ppm)

Rouging

The first rouging is done at the time of handling the mature heads. All off type plants, diseased, or otherwise undesirable types, are removed at this stage. The second rouging is done before the heads start bursting. The loose leafed, poorly heading plants and those having a long stem with heavy frame must be rouged out at this stage. It is highly undesirable to keep such poor plants in the seed plots. Subsequent rouging for off types, diseased plants affected by phyllody, black leg, soft rot or leaf spot should be done from time to time as required.

Harvesting and Threshing

To avoid shattering of seeds, the whole crop is harvested in two or three lots with sickles. Generally, the early plants are harvested first, and when the pod colour in about 60-70 per cent of the rest of the crop changes to yellowish brown, it is harvested completely and piled up for curing. After 4-5 days, it is then threshed with sticks and sifted with hand sifters. After thoroughly drying, seeds are cleaned and stored.

Seed Yield

450-650 kg per hectare.

Varieties/Hybrids Characteristics

Varieties/Hybrids Recommended by Himachal Pradesh State

Pride of India

An introduction recommended by Dr. Y.S. Parmar University of Horticulture and Forestry, Solan. The plant type of this variety is similar to 'Golden Acre'. But it is about a week later in maturity. It has bigger sized heads weighing about 1.5 to 2.0 kg.

Pusa Drum Head

It is released by IARI, Regional Station, Katrain (Kullu Valley). Heads are solid and flat weighing about 3 to 4 kg. It is probably the earliest variety among the Drumhead Group. It also possesses field resistant to black leg.

Pusa Mukta

This is a new variety developed at IARI Regional Station, Katrain (Kullu Valley) by hybridization between EC 24855 x EC 10109. It has short stalk, medium frame. The heads are compact, slightly flatish round with a loose wrapper leaf at the top. The average head weight is 1.5–2.0 kg. It is resistant to black rot.

Pusa Synthetic

This variety has been developed from IARI, Regional Station, Katrain (Kullu Valley). It is an early synthetic variety. The yield potential of this variety is 350-460 q/ha.

Varieties/Hybrids Recommended by Maharashtra State

BSS-115

The hybrid has been developed at Beejo Sheetal Seeds Pvt. Ltd., Jalna. Heads are highly compact. This hybrid can be grown round the year. The average weight of head is 2-3 kg. This hybrid is resistant to fusarium wilt.

BSS-44

The hybrid has been developed at Beejo Sheetal Seeds Pvt. Ltd, Jalna. The head is roundish, average weight 2.5 kg. This hybrid can grow well both in cold and hot weather conditions. This hybrid is resistant to fusarium wilt.

BSS-50

The hybrid has been developed at Beejo Sheetal Seeds Pvt. Ltd., Jalna. Heads are dark green, smooth, highly compact. It is tolerate high temperature tolerant (up to 36°C) and resistant to fusarium wilt. Average head weight is 1 kg.

Sri Ganesh Gol

The hybrid has been developed at Maharashtra Hybrid Seeds Co. Ltd., Jalna. Heads are round, compact, attractive and bluish green. It is resistant to yellowing. It gives an average yield of 500 to 750 q/ha.

Varieties/Hybrids Recommended by Other States

Early Drumhead

It is an early variety having light green foliage. Leaves are medium to large, curving inward and enclosing the head loosely. Leaves forming the head fold over each other at the centre. The heads are flat.

Golden Acre

This is an earliest variety evolved by selection from the Copenhagen Market and recommended by IARI for cultivation. The heads are solid, short core and weighing of about 1.0-1.5 kg. The heads should be harvested immediately after head formation otherwise cracking may take place. The variety has also been highly suitable for spring and summer growing in the hills.

Red Cabbage

All the red cabbage cultivars are tolerant to diamond back moth insect. It has distinct coat of wax and produces a head of 1-2 kg taking.

September

An introduction from German Democratic Republic. The stalk is long and heads usually tilt on one side after formation. It has solid, flatish round to slightly oblong heads weighing 3-5 kg. It has very good keeping quality.

Turnip (*Brassica campestris* var. *rapa*) 2n = 2x = 20

Land Requirements

Land to be used for seed production should be free of volunteer plants. The soils of selected field should be deep, light, friable and well drained.

Field Inspection

Mother Root Production Stage

A minimum of two inspections shall be made as follows:

1. The first inspection should be done after 20-30 days of sowing in order to determine isolation, off types and other relevant factors.

2. The second inspection should be done after the mother roots have been lifted to verify the true characteristics of roots.

Figure 63: Turnip

Seed Production Stage

A minimum of one inspection should be done during flowering to check isolation, off types, designated diseases and other relevant factors.

Field Standards

General Requirements

Isolation

Seed fields shall be isolated from the contaminants shown in column 1 of the Table below by the distances specified in columns 2, 3, 4 and 5 of the said Table:

Contaminants	*Minimum Distance (Meters)*			
	Mother Root Production Stage		*Seed Production Stage*	
	Foundation	*Certified*	*Foundation*	*Certified*
1	*2*	*3*	*4*	*5*
Fields of other varieties	5	5	1600	1000
Fields of the same variety not conforming to varietal purity requirements for certification and from other species of genus Brassica	5	5	1600	1000

Specific Requirements

Factor	*Maximum Permitted (%)**	
	Foundation	*Certified*
Off types	0.10	0.20
** Plants affected by seed borne diseases	0.10	0.50

* Maximum permitted at and after flowering in the case of off types and at the final inspection in case of seed borne diseases.

** Seed borne diseases shall be: Black rot and Black leg.

Seed Standards

Factors	Standards for Each Class	
	Foundation	Certified
Pure seed (minimum)	98.0 per cent	98.0 per cent
Inert matter (maximum)	2.0 per cent	2.0 per cent
Other crop seeds (maximum)	5/kg	10/kg
Weed seeds (maximum)	5/kg	10/kg
Germination (minimum)	70 per cent	70 per cent
Moisture (maximum)	6.0 per cent	6.0 per cent
For vapour-proof containers (maximum)	5.0 per cent	5.0 per cent

Method of Seed Production

Both seed to seed and root to seed methods can be employed for raising seed of turnip.

Seed to Seed Method

The crop is allowed to over winter in the field and produce seed in the following spring in their original position.

Root to Seed Method

The turnip seed is usually produced by root to seed method. In this method, during autumn the roots, when fully developed, are uprooted and selection of true to type roots is made. Underdeveloped deformed, diseased and off type roots are rejected. After pruning, the taproot and clipping the tops, leaving the crown intact, the selected roots are replanted in prepared soil in such a way that the whole root is covered under the soil, leaving the crown exposed.

Brief Cultural Practices

Root to Seed Method (First Season-Raising of Stecklings)

Time of Sowing

Last week of August or first week of September.

Preparation of Field

Prepare the field to fine tilth by ploughing, three to four harrowing, followed by levelling.

Source of Seeds

Obtain nucleus/breeders/foundation seed from source approved by a seed certification agency.

Seed Rate

Three to four kg/ha. The roots produced on one hectare are sufficient for transplanting of 2.5-3 hectares.

Method of Sowing

The seed should be sown on ridges. Row to row spacing is 45 cm, Seeds should be sown thinly. Keep the ridges moist until germination is completed. Thin out plants to a distance of 7 to 8 cm within the rows when the plants are 10-15 days old.

Fertilizers

Apply 200-250 quintals per hectare of well-rotted farmyard manure at the time of preparation of land. At the time of final levelling mix in the soil 35 kg nitrogen, 50 kg phosphorus and 50 kg potash per hectare as basal application. Another dose of 35 kg nitrogen is top-dressed when the roots just start growing.

Irrigation

Irrigate at 8-10 days interval depending on the weather conditions and requirements. During early warm weather, it is necessary to irrigate at 4 to 5 days interval.

Interculture

One weeding and one earthing up during the early stages of growth are necessary for proper development of roots.

Insects and Diseases

Insects/Diseases	*Control Measures*
	Insects
Aphids (*Myzus persicae, Brevicory brassicae)*	Spray with malathion (0.01 per cent) or monocrotophos (0.05 per cent)
Mustard sawfly (*Athalia proxima*)	Spray carbaryl (0.2 per cent) or malathion (0.05 per cent).
	Diseases
White rust (*Albugo candida*)	Spray with difoltan (0.3 per cent).
Downey mildew	Spray with zineb at 0.2 per cent.

Harvesting of Roots (Stecklings)

Uproot the plants (stecklings) when they have fully developed (70-80 days from sowing).

Second Season

Selection of Roots for Transplanting

The uprooted roots should be laid bare side by side to facilitate selection. Stecklings having off type foliage or roots not conforming to variety characteristics must be rejected. The diseased, malformed, forked or any other undesirable types are also rejected. After selection, the tops of the roots are clipped and their tails are pruned. They are then reset in a well-prepared field.

Preparation of Field

Same as described for the previous season.

Time and Method of Transplanting

Steckling are best planted during the second half of December, as later plantings are prone to serious attacks of aphids. The stecklings selected and prepared in the manners described earlier are reset in the field at a distance of 60 x 30 cm. The soil is then firmed and tapped around the roots. Soon after transplanting the field is irrigated.

Fertilizers

Apply 100-150 quintals per hectare of well-rotted farmyard manure at the time of preparation of land. Supplement it with 50 kg phosphorus per hectare at the time of sowing. Top dress 25 kg nitrogen per hectare in the pre bolting stage and another dose of 25 kg nitrogen before flowering starts.

Interculture

One hoeing and earthing up are necessary during March. Staking is essential as it gives support to the plants, which may otherwise lodge because of heavy winds during the flowering and fruiting season.

Supplementary Pollination

Insects pollinate the turnip. To ensure good seed set, pollination of maximum flowers is necessary. Honeybees are important agents of pollination. In large seed fields, it is advisable to place beehives nearby, to increase pollination.

Rouging

Rogue out undesirable roots, namely, off type, diseased or malformed root, at the time of transplanting. Subsequent rouging for off type and other crop plants may be done at the bloom stage. Rogue out diseased (seed borne diseases) and severely insect damaged plants from the field as required from time to time.

Harvesting and Threshing

Harvesting of the seed crop may be done when the seed pods are reddish brown. Harvesting before this stage gives shrivelled seeds and late harvest results in seed shattering. Harvesting is done by sickle. Cut seed stalks near base of crown, preferably in the early morning. The threshing can be done by beating the seed stalks with sticks, or by thresher, and the seeds can be cleaned by winnowing. Dry the seeds to 6 per cent moisture content before storage.

Seed Yield

600-800 kg/ha.

Varieties/Hybrids Characteristics

Varieties/Hybrids Recommended by IARI, New Delhi

Pusa Kanchan

It has been developed through hybridization between Asiatic type (Local Red Round) and European type (Golden Ball). The skin is red but the flesh is creamy yellow. It has excellent flavour and taste.

Pusa Sweti

It is an Asiatic cultivar suitable for sowing from August to October. The roots are attractive white.

Varieties/Hybrids Recommended by IARI Regional Station, Katrain (Kullu Valley) Himachal Pradesh

Early Milan Top

It is very shy seeder and highly susceptible to turnip malformation. It takes about 65-70 days from sowing to root formation.

Golden Ball

Roots are globe shape, soft with medium size. An average yield of this variety is 200-250 q/ha.

Pusa Chandrima

This has been developed through hybridization between Asiatic and European types. Roots are medium to large in size, 8-9 cm long, smooth, pure white, fine grained with sweet and tender flesh.

Pusa Swarnima

This cultivar has been developed by hybridization between Asiatic type (Japanese White) and European type (Golden Ball) followed by selection. The roots are flattish round, 6-7 cm long and 7-8 cm in diameter with creamy yellow skin. Flesh is amber coloured, fine-textured and mild-flavoured.

Family–Alliaceae

Onion (*Allium cepa* L.) 2n = 2x =16

Land Requirements

Land used for seed production should be free of volunteer plants. The soil of the selected field should be rich in organic matter and have a good water holding capacity.

Field Inspection

Mother Bulb Production Stage

A minimum of two inspections should be done as follows:

1. The first inspection should be done after transplanting of seedlings in order to determine isolation, volunteer plants, off types including bolters and other relevant factors.
2. The second inspection should be done after the bulbs have been lifted to verify the true characteristic of bulb.

Seed Production Stage

A minimum of four inspections shall be made as follows:

1. The first inspection should be done before flowering in order to determine isolation, volunteer plants, off types including bolters and other relevant factors.

Figure 64: Onion

2. The second and third inspections should be done during flowering to check isolation, off types and other relevant factors.
3. The fourth inspection should be done at maturity to verify the true nature of plant and other relevant factors.

Field Standards

General Requirements

Isolation

Onion seed fields shall be isolated from the contaminants shown in column 1 of the Table below by the distances specified in columns 2, 3, 4 and 5 of the said Table:

Contaminants	*Minimum Distance (Meters)*			
	Mother Root Production Stage		*Seed Production Stage*	
	Foundation	*Certified*	*Foundation*	*Certified*
1	*2*	*3*	*4*	*5*
Fields of other varieties	5	5	1000	500
Fields of the same variety not conforming to varietal purity requirements for certification	5	5	1000	500

Specific Requirements

Factor	*Maximum Permitted (%)**	
	Foundation	*Certified*
*Bulb not conforming to the varietal characteristic	0.10 per cent (by number)	0.20 per cent (by number)
**Off types	0.10 per cent	0.20 per cent

* Maximum permitted at second inspection at mother bulb production stage.

** Maximum permitted at and after flowering at seed production stage.

Seed Standards

Factors	*Standards for Each Class*	
	Foundation	*Certified*
Pure seed (minimum)	98.0 per cent	98.0 per cent
Inert matter (maximum)	2.0 per cent	2.0 per cent
Other crop seeds (maximum)	5 /kg	10 /kg
Weed seeds (maximum)	5 /kg	10 /kg
Germination (minimum)	70 per cent	70 per cent
Moisture (maximum)	8.0 per cent	8.0 per cent
For vapour-proof containers (maximum)	6.0 per cent	6.0 per cent

Method of Seed Production

There are two methods of seed production.

Seed to Seed Method

In this method, the first season bulb crop is left to over winter in the field so as to produce seed in the following season.

Bulb to Seed Method

The bulbs produced in the previous season are lifted, selected, stored and replanted to produce seed in the second year.

Mostly the bulb to seed method is used for seed production because of the following advantages over the seed to seed method.

1. It permits selections of true to type and healthy bulbs for seed production.
2. Seed yields are comparatively very high.

The seed to seed method, however, can be practiced for varieties having a poor keeping quality.

Brief Cultural Practices (Bulb to Seed Method)

Production and Storage of Bulbs (First Year)

Sowing Time (Nursery)

Middle of October to middle of November in the plains and April to June in the hills. For raising seedlings for one hectare 1/20 hectare area of nursery is sufficient.

Seed Rate

8-10 kg/ha.

Fertilizers

Add 200 q per hectare of well rotted farmyard manure at the time of land preparation and 60 kg/ha nitrogen, 60 kg/ha phosphorus and 80 kg /ha potassium per hectare at the time of planting. Remaining 60 kg/ha nitrogen may be applied as top dressing in two to three doses during the growing period.

Transplanting

Eight weeks old seedlings are planted in small beds in well-prepared fields. January is the best period.

Spacing

Spacing depends upon variety and bulb size and varies from 10 to 15 cm.

Irrigation

Fortnightly irrigation during winter and weekly irrigation during the hot weather. Irrigate sparingly during maturity.

Interculture

Keep field free from weeds. Frequent intercultural operation is essential for good bulb development. For controlling weeds, post emergence application of tenoran at 2 kg per hectare in 800 litres of water, two to three weeks after transplanting, is recommended. Oxadiazon 1 kg a. i./ ha has also given effective control of weeds.

Insects and Diseases

Insects/Diseases	*Control Measures*
	Insects
Onion thrips (*Thrips tabaci*)	Apply malathion or nuvacron at 0.1 per cent or phorate 10 G at 1 kg per hectare
Borer (*Heliothis armigera*)	Apply endosulfan at 0.1 per cent.
Onion maggot (*Hylemia antiqua*)	Soil application of thimet 10 G is beneficial. Spray malathion at 0.05 per cent.
	Diseases
Downy mildew (*Peronospora destructor*)	Difolatan (0.1 per cent), and ridomil MZ (0.1 per cent)
Onion smut (*Urocystis cepulae*)	Seed treatment with thiram or captan (3 g/kg seed).
Purple blotch (*Alternaria porri*)	Seed treatment with captan (2.5 g/kg seed). Applications of dithane-M-45 (0.25 per cent).
Bulb rot (*Fusarum oxyxporum*)	Spray Benlate (0.025 per cent).

Harvesting and Curing of Bulbs

Well-matured bulbs should be harvested. The tops drooping just above the bulb indicates maturity, while the leaves are still green. After harvesting, the bulbs should be topped leaving a half inch neck. Before storage, at through selection and curing of bulbs should be done. The length of time required for curing depends largely on weather conditions and may take three to four weeks.

Storage

The essentials of successful storage are:

1. The bulbs should be well ventilated, dried and cured before storage.
2. Storage should be done in shallow trays with perforated bottoms.
3. Storage temperatures should range 0 to 4.5°C until three to four weeks prior to planting, when the temperature should be increased to around 10°C.

Planting of Bulbs and Seed Production (Second Year)

Time of Planting Bulbs

The best time for planting of bulbs is the second fortnight of October.

Preparation of Land

Prepare the field to good tilth by one deep ploughing, followed by three to four harrowings and land levelling.

Seed Rate

The seed yield is affected by the size of the bulb. The bigger bulb size caused high seed yield. However, very large sized of 2.5 to 3.0 cm diameter is used for planting. Approximately 15 quintals of bulbs per hectare are required if the bulb diameter is 3 to 4 cm.

Fertilizers

Same as described for first year.

Method of Planting and Spacing

Selected bulbs are planted 8 to 10 cm deep in the soil at a distance of 45 x 30 cm. The size of beds depends upon the source of irrigation. The sprouted bulbs are planted as such. In unsprouted bulbs, the upper half portion should be removed, leaving the disc like stem and roots intact. The removal of the upper tops hastens sprouting.

Intercultural Operations, Disease and Insects

Same as described for first year.

Rouging of Seed Crop

First Year

It is desirable to begin rouging in the field before bulbs are harvested. Any plants having a different foliage colour, or plant type, or late maturing bulbs should be rouged out from the field. After harvesting, the bulbs should be carefully rouged for colour and off types such as thick-necks, doubles, bottlenecks, as well as any other types, which do not conform to varietal type.

Second Year

Plant only selected true to type bulbs and remove plants not conforming to varietal characters before flowering.

Harvesting and Threshing

Seed is ready for harvest when first formed seed in the heads get blackened. Two to three pickings may be necessary to harvest the heads, at just the right stage. Seed heads are cut, or snapped off, keeping a small portion of the stalk attached. Seed heads, after harvest, should be thoroughly dried. Air circulation is important while drying seed heads. Hence, the trays or canvases should be filled only to a depth of 15 cm.

Heads can be threshed when seed separates easily from them. Much of the seed falls from capsules during drying. The seeds can then be cleaned by immersing them in clean water for five to ten minutes and then drying in the sun or artificially. Before storage, the seed must be dried for 6 to 8 per cent moisture.

Seed Yield

850 to 1000 kg per hectare.

Varieties/Hybrids Characteristics

Varieties/Hybrids Recommended by IARI& NBPGR, New Delhi

Pusa Ratnar

It is selected from a hybrid cultivar introduced from the USA under the name of Granex and released from NBPGR, New Delhi. Bulbs are bronze deep red, obovate to flat globular, less pungent. Bulb yield is 500 q/ha.

Pusa Red

This variety was developed at IARI, New Delhi from an indigenous collection. Bulb is red, medium size, globular and less pungent. Bulb yield is 250-300 q/ha. It is grown throughout the country for *rabi* season.

Pusa White Flat

This variety was developed at the IARI, New Delhi. Bulbs are attractive white, flat, medium to large. Bulb yield is 300 q/ha.

Pusa White Round

It was developed from a local collection (Line 106) at IARI, New Delhi. The bulbs are white, roundish flat medium size. Average bulb yield is 300 q/ha.

Varieties/Hybrids Recommended by Maharashtra State

Agrifound Dark Red

The variety has also been developed by NHRDF, Nasik from a local stock of kharif onion. Bulbs are deep red, globular with tight skin. It is suitable for growing in kharif season. Bulb yield is 300–400 q/ha.

Agrifound Light Red

It is also developed by National Horticultural Research and Development Foundation (NHRDF), Nasik and most suited for rabi season. Bulbs are light red and globular. The yield is 300-325 q/ ha.

Agrifound Red

The variety has also been developed by National Horticultural Research and Development Foundation (NHRDF), Nasik. An average weight of bulb is 65-68 g. An average yield is 180-200 q/ha.

N-2-4-1

This variety was developed at MPKV, Rahuri. Bulbs are light red and globe shaped. The bulbs are good keeping quality. Yield is 250-300 q/ha.

N-257-9-1

This variety was developed at MPAU Rahuri. This variety has attractive white bulbs. Bulb yield is 300 q/ha. It is suitable as *rabi* crop.

Varieties/Hybrids Recommended by Karnataka State

Arka Bindu

It is developed from local rose onion collection 'IHR 402' through mass selection and released from IIHR, Bangalore. Its bulbs are deep red, small sized, flatish-globe and uniform in shape, size and maturity. It has high TSS, pungency and dry matter content. It exclusively meets our export requirement.

Arka Kalyan

This variety was developed through mass selection at the IIHR, Bangalore from a local collection (IIHR 145). Bulbs are globe-shaped, pink with fleshy outer and internal scale and suitable for *kharif* season. This variety is having a yield potential of about 450 q/ha.

The variety is moderately resistant to purple blotch disease and performs very well in *kharif* season.

Arka Niketan

It was developed through mass selection from a local collection (IIHR 153) at IIHR, Bangalore. Bulbs are globose with thin neck, red in colour, and highly pungent. The average bulb yield is 330 q/ha and it can be grown both in *rabi* and *kharif* seasons.

Arka Pitamber

This is a short-day yellow onion. Its bulb are uniformly yellow in colour with perfect globe shape, very firm single centered, thin neck, mild pungent and medium sized with good keeping quality. It yields 35 t/ha. It is suitable for the export to the European countries, the USA and Japan.

Arka Pragati

It is developed at IIHR, Bangalore from a local collection (IIHR-149) from Nasik. Bulbs are attractive red, globe shaped, uniform size, thin neck, highly pungent and early maturing. It can be grown successfully during *rabi* and *kharif* seasons and yields 330 q/ha.

Varieties/Hybrids Recommended by Punjab State

Punjab Selection

It was developed at PAU, Ludhiana by mass selection from indigenous material. Bulbs are globular, red, bulb diameter 5-6 cm. Bulb yield is 300 q/ha.

Punjab-48

It is developed at PAU, Ludhiana. Bulbs are flatish round, TSS is 12-13 per cent and had good storage qualities. Bulb yield is 250-300 q/ha.

Varieties/Hybrids Recommended by Rajasthan State

Udaipur 101

This variety was developed at RAU, Rajasthan. Bulbs are deep red and flatish globe. It is good for salad purpose. Yield is 200-300 q/ha.

Udaipur 103

This variety was developed at RAU, Rajasthan. Bulbs are red but lighter than Udaipur101, oblate, globular, sweet but slightly more pungent. Yield is 250-300 q/ha.

Varieties/Hybrids Recommended by Other States

Brown Spanish

This variety has been developed from IARI Regional Station, Katrain (Kullu Valley) Himachal Pradesh. Bulbs are solid, attractive, reddish-brown. It is a medium long day variety. An average yield of this variety is 280 q/ha.

Early Grano

It is an American variety, which produces globular bulb of yellow colour. It is suitable for salad purposes. It has given yield 475 q/ha.

Hisar-2

It is developed at CCSHAU, Hisar by mass selection from indigenous material. Bulbs are bronze red and flatish globular shapes. This variety is tolerant to salinity.

Kalyanpur Red Round

It is a popular variety of Uttar Pradesh, developed at CSAUAT, Kalyanpur. Bulbs are bonze red globular and TSS ranges from 13-14.9 per cent. Yield ranges from 250-300 q/ha.

Patna Red

Bulbs are of bronze red colour, flat to slightly flatish globular and very pungent.

Family–Leguminosae

Pea (*Pisum sativum* L.) 2n = 2x = 14

Land Requirement

The land should be free of volunteer plants. The soil of selected fields should be well drained and aerated. The highly organic soils are unsuitable for pea as their moisture reserve leads to excessive vegetative growth and poor pod formation.

Figure 65: Pea

Field Inspection

A minimum of three inspections should be made, the first before flowering, the second during flowering and fruiting stage, and the third at the edible pod stage.

Field Standards

General Requirements

Isolation

Pea seed fields shall be isolated from the contaminants shown in column 1 of the Table below by the distances specified in columns 2 and 3 of the said Table:

Contaminants	Minimum Distance (Meters)	
	Foundation	Certified
1	2	3
Fields of other varieties	10	5
Fields of the same variety not conforming to varietal purity requirements for certification	10	5

Specific Requirements

Factor	Maximum Permitted (%)*	
	Foundation	Certified
Off types	0.10	0.20

* Maximum permitted at the final inspection.

Seed Standards

Factors	Standards for Each Class	
	Foundation	Certified
Pure seed (minimum)	98.0 per cent	98.0 per cent
Inert matter (maximum)	2.0 per cent	2.0 per cent
Other crop seeds (maximum)	None	5/kg
Weed seeds (maximum)	None	None
Other distinguishable varieties (maximum)	5/kg	10/kg
Germination including hard seeds (minimum)	75 per cent	75 per cent
Moisture (maximum)	9.0 per cent	9.0 per cent
For vapour-proof containers (maximum)	8.0 per cent	8.0 per cent

Brief Cultural Practices

Time of Sowing

Plains: October to middle of November

Hills: Middle of March to end of May

Darjeeling: June to August

North India: March-April, August-September and November-December.

Preparation of Land

Prepare the field to a good tilth by ploughing, two to three harrowings followed by levelling.

Source of Seeds

Obtain nucleus/breeders/foundation seed from source approved by a seed certification agency.

Seed Rate

Early varieties: 100-120 kg/ha

Mid late varieties: 80-90 kg seed per hectare

Methods of Sowing

Seeds can be sown on flat or raised beds either by broadcasting or behind deshi plough in furrows, which are covered by usual planking. Some farmers drill the seed through pora (tube) attached to deshi plough and there is no need of planking. A depth of 5-7.5 cm is optimum.

Spacing

22.5-30 cm x 3.75-5 cm.

Fertilization

Farm yard manure should applied in quantity of 80-100 q/ha at 15 days before sowing the seeds. The nitrogen, phosphorous and potassium should be applied in ratio of 30:50:25 kg/ha respectively. This nutrition should be given at the time of sowing of seeds as basal doses.

Irrigation

Irrigation should be provided as and when needed. Abundant water supply during flowering should be ensured. Late irrigation during warm weather should be avoided which may cause sun-scalding of plants as also plants may tend to lodge and some rotting of vines may occur if the soil is kept too wet.

Interculture

Land should be kept clean by weeding and mulching. Lasso @ 0.75 kg a.i./ha pre emergence spray or tribunil @ 1.5 kg a.i./ha pre emergence or basalin @ 2 kg a.i./ha pre plant incorporation along with one hand weeding at 45 days after sowing (DAS) should be used.

Insects and Diseases

Insects/Diseases	*Control Measures*
	Insects
Pea stem fly	Foliar sprays of endosulfan at 0.7 kg a.i./ha.
Pea aphids	Spray of rogor (0.03 per cent).
	Diseases
Powdery mildew (*Erysiphe polygoni*)	Dust or spray sulphur fungicides (elosal 5 kg/ha, sulfex 2.5 kg/ha) at frequent intervals.
Fusarium wilt and rust	Seed treatment with carbendazim (bavistin 2g/kg seed) and spraying with bavistin (0.1 per cent).
Bacterial blight (*Pseudomonas* spp)	Remove affected plants. Treat the seed with streptomycin sulphate (2.5 g/kg of seed).
Pea mosaic (Vector: *Aphis* spp.)	Control insect vectors carefully. Pull out affected plants and burn them.

Roguing

The off-type and diseased plants affected by pea mosaic, foot-rot and blight should be rogued out from the seed field.

Harvesting and Threshing

Maturity of seed crops take about 130-140 days. To test the maturity a common practice is to squeeze the seed between fingers. If the cotyledons break away from each other and free moisture is not visible, the crop may be considered mature enough for harvest. Vines along with the pods are harvested from the field and dried in the threshing floor under sunshine. Threshing is done by beating with stick when sufficiently dry.

Storage

Threshed seeds are cleaned by winnowing, dried to reduce seed moisture content to 12 per cent for temporary storage. For longer storage pea seed should be stored in sealed containers at 10 per cent moisture content and in air cooled rooms.

Seed Yield

10-20 qt/ha.

Varieties/Hybrids Characteristics

Varieties/Hybrids Recommended by IARI, New Delhi

Arkel

It is an early, wrinkled-seeded, exotic introduced from England. Pods are dark green, sweet and sickle shaped. The seeds are green, wrinkled and bold in size.

Asauji

The selection of this variety has been made from the material collected from Amritsar. It is a dwarf, early, green and smooth seeded cultivar suitable for early sowing. The pods are about 8 cm long, curved, dark green having 7 seeds /pod.

NP-29

This cultivar has been developed through selection. It is medium-tall double-podded cultivar. Pods are green, straight, about 7.5 cm long and 6-7 seeded. It is highly suitable for dehydration.

Varieties/Hybrids Recommended by Uttar Pradesh State

Azad P-1

It is a derivative of the cross 6416 x 6405. Pods are smooth, dark green, 8-10 cm long. Mature seed is wrinkled. This variety escapes powdery mildew and rust. Seed yield is 17 q/ha.

Azad P-2

It is a powdery mildew resistant variety, developed at Vegetable Research Station, CSAU and T, Kanpur, derived from the cross Bonneville x 6587. Pods are medium, nearly straight, light green, smooth and firm. Seeds are wrinkled; suitable for powdery mildew prone areas.

VRP-2

This is an early maturing variety developed at IIVR, Varanasi. The pods are straight, medium sized, light green filled with bold ovules.

Varieties/Hybrids Recommended by Madhya Pradesh State

Jawahar Matar-1

It is a derivative of the cross T-19 x Greater Progress. Pods are straight, green colour, medium thick pod wall and 8-9 seeds/pod. The mature seeds are green in colour, wrinkled.

Jawahar Matar-3

A very early variety developed from a cross of 'T19' and 'Early Badger'.This variety produces light pods, roundish oval, with a length of 5-7 cm. The seeds are green, wrinkled and bold.

Jawahar Matar–4

It has been developed from the cross 'T19' x 'Little Marvel'. Plants are 55-60 cm tall and posses medium size green pods (7.0 cm). The matured seeds are green, wrinkled.

Jawahar Peas 4

It is a cross of 'Local Yellow Batri' x ('6588' x '46c'). The variety possesses medium size pods. The seeds are yellow and medium sizes. This variety is most suitable to hilly areas.

Jawahar Peas 71

It is a progeny of double cross (Arkel x JP 829) x (JP 501 x JM 1) having powdery mildew resistance. The seeds are green and wrinkled.

Jawahar Peas-83

It is a mid-season, powdery mildew resistant from a double cross (Arkel x JP-829) x (46-C x JP-501). The pods are bigger in size.

Varieties/Hybrids Recommended by Uttranchal State

Pant Uphar

It is a medium tall variety (70 cm). Pods are medium size, 7.5 cm long having 7 seeds/pod.

Vivek-6

It is a cross between Pant Upahar and VL Matar 3. The pods are smooth, straight, medium sized (6-7 cm), light green and completely filled. It is tolerant to cold and moisture stress conditions.

VL Ageti Matar-7

It is an early maturing developed from VPKAS, Almora. The green grains are dimpled bold. It is free from the incidence of powdery mildew.

Other Varieties/Hybrids

Bonneville

This is a wrinkled-seeded and introduced from USA. Pods are light green, 8 cm in length. The matured seeds are green and wrinkled bold.

Early Badger

It is a dwarf, early, wrinkled-seeded evolved at Wisconsin. It is suitable for sowing in early October. The yellowish green pods are 7.5 cm long. It is a good canning cultivar. It is resistant to fusarium wilt.

Harbhajan

It is an introduction. Pods are green and long. Seeds are small, round, wrinkled and yellow in colour.

Matar Ageta 6

It is an improvement over 'Arkel' and 'Harabona' and released by PAU, Ludhiana. Pods are long.

Lincoln

It is a dwarf to medium-tall. Pods are dark green, 9.5–10 cm long, 6-7 seeded. It is suitable for late sowing.

Little Marvel

This cultivar has been bred in England from the cross Chelsea Gem x Suttons Alaska. Pods are about 8 cm long. The peas are fine in quality.

Metor

It is a round smooth seeded early variety and introduced from England. Pods are dark green, 8.7 cm long, well filled with 7 seeds.

Sylvia

It is an edible-podded and tall growing variety.

Fenugreek (*Trigonellafoenum graecum* L. and *T. corniculata*) 2n = 2x =16

Land Requirement

Land to be used for seed production shall be free of volunteer plants. Light soil is most suitable for fenugreek cultivation.

Figure 66: Fenugreek

Field Inspection

A minimum of two inspections shall be made, the first before flowering, the second during flowering and fruiting stages.

Field Standards

General Requirements

Isolation

Seed fields shall be isolated from the contaminants shown in column 1 of the Table below by the distances specified in columns 2 and 3.

Contaminants	*Minimum Distance (Meters)*	
	Foundation	*Certified*
1	*2*	*3*
Fields of other varieties	10	5
Fields of the same variety not conforming to varietal purity requirements for certification	10	5

Specific Requirements

Factor	Maximum Permitted (%)*	
	Foundation	Certified
Off types	0.10	0.20
Objectionable weed plants: (*Melilotus* spp.)	0.010	0.020

* Maximum permitted at the final inspection.

Seed Standards

Factors	Standards for Each Class	
	Foundation	Certified
Pure seed (minimum)	98.0 per cent	98.0 per cent
Inert matter (maximum)	2.0 per cent	2.0 per cent
Other crop seeds (maximum)	10/kg	20/kg
Total weed seeds (maximum)	10/kg	20/kg
Objectionable weed seeds (maximum)	2/kg	5/kg
Other distinguishable varieties (maximum)	10/kg	20/kg
Germination including hard seeds (minimum)	70 per cent	70 per cent
Moisture (maximum)	8.0 per cent	8.0 per cent
For vapour-proof containers (maximum)	6.0 per cent	6.0 per cent

Brief Cultural Practices

Time of Sowing

Third week of October to second week of November.

Preparation of Land

Prepare the field to a good tilth by one deep ploughing, two to three harrowings followed by levelling.

Source of Seeds

Obtain nucleus/breeders/foundation seed from source approved by a seed certification agency.

Seed Rate

20-25 kg seeds per hectare

Method of Sowing

The seed is generally broadcast in beds and the surface is raked to cover it. Improved seeds should be sown after inoculation with *Rhizobium* culture at 2 to 3cm depths on well-prepared friable soil.

Fertilizer

150 quintal of farmyard manure is required per hectare land and it should be incorporated during land preparation. 20 to 30 kg per hectare nitrogen as side dressing just after first cutting, which should be followed by a light irrigation.

Spacing

20 x 25 cm or 30 × 30 cm

Irrigation

Frequent watering is necessary to obtain quick and constant growth of crop. Four to six irrigations are required at fortnightly intervals.

Interculture

The crop should be thinned when three to four leaves start sprouting. Two weedings and hoeings are necessary during the early stages of the crop. First weeding and hoeing is done at the time of thinning and the second within 45-50 days after sowing.

Insects and Diseases

Insects/Diseases	*Control Measures*
	Insects
Leaf eating caterpillars and pod borers	Spraying malathion 0.05 per cent or quinalphos 0.04 per cent.
	Diseases
Powdery mildew (*Leveillula taurica* and *Erysiphe polygoni*)	Application of thiovit or karathane or sulphur dust at the rate of 2.5 a.i. kg/ha.
Downy mildew *(Peronospora trigonella)*	Spraying of any one of the fungicides viz; 0.2 per cent of difolatan, cupramar and blitox-50.

Roguing

Rogue out off types and wild *Trigonella spp*. from seed fields prior to flowering and during flowering.

Cutting

Not more than 2 to3 cuts should be taken from the seed crop.

Harvesting and Threshing

The crop should be harvested at maturity. At the time of maturity, all except the upper leaves becomes yellow and fall. After harvesting, plants are allowed to dry under the sun. Seeds are separated by threshing and cleaned by winnowing.

Seed Yield

Common methi–12-15 qt./ha and kasuri methi–6-7 qt/ha.

Varieties/Hybrids Characteristics

Varieties/Hybrids Recommended by IARI, New Delhi

Kasuri Selection

It is a late cultivar of kasuri methi. It takes about 165 days for its seed maturity. It is slow growing with flowers orange yellow. Pods are sickle shaped.

Pusa Early Bunching

A quick growing variety. This variety produces upright branches, and suitable for bunching flowers and white in colour. Pods have bold seed. Average yield is 700-800 (green leaves)/ha.

Varieties/Hybrids Recommended by Rajasthan State

RMT-1 (Rajasthan Methi)

It is developed by RAU, Rajasthan through pure-line selection from a local collection. The plants are semi–erect, tall with bold and yellow-coloured grains. It produces average yield 14.7 q/ha. It is moderately resistant to root rot and tolerant to powdery mildew disease (Edison and Johny, 1991).

RMT-143

It is developed through pure-line selection in a local collection from Jodhpur area. The grains of this cultivar are bold with typical yellow colour. It produces an average yield 16.0 q/ha. It is moderately resistant to powdery mildew.

Varieties/Hybrids Recommended by Haryana State

Hissar Sonali

It is developed by CSSHAU, Hisar through pure-line selection from local germplasm. The plants are bushy, semi-erect with bold, yellow attractive grains (13.15 g/1000 grains). It produces average yield 19.0 q/ha. It is moderately resistant to leaf spot and root rot.

HM-103

It is developed by CCSHAU, Hisar through pure-line selection from local germplasm. The plants are bushy, semi-erect with bold and yellow attractive grains (13-15 g/1000 grains). It produces an average yield 20.10 q/ha. It is moderately resistant to leaf spot disease.

Varieties/Hybrids Recommended by Other States

CO-1

A reselection from TG-2336 from IARI, New Delhi and released from Tamil Nadu. Plants are 50 cm tall vigorous and bushy. Pods are 10-12 cm long with long beak containing 15-20 seeds/pod. It produces an average yield of 6.80 q/ha. It is moderately tolerant to root rot.

French Bean (*Phaseolus vulgaris* Linn.) 2n = 2x = 22

Land Requirement

The land should be free of volunteer plants. The land should also be well levelled and well drained so that water does not stand.

Field Inspection

A minimum of two inspections shall be made, the first before flowering, the second during flowering and fruiting stage.

Field Standards

General Requirements

Isolation

Seed fields shall be isolated from the contaminants shown in column 1 of the Table below by the distances specified in columns 2 and 3 of the said Table:

Figure 67: French Bean

Contaminants	Minimum Distance (Meters)	
	Foundation	Certified
1	2	3
Fields of other varieties	10	5
Fields of the same variety not conforming to varietal purity requirements for certification.	10	5

Specific Requirements

Factor	Maximum Permitted (%)*	
	Foundation	Certified
Off types	0.10	0.20
**Plant affected by seed borne diseases	0.10	0.20

* Maximum permitted at the final inspection.

** Seed borne diseases shall be: bacterial blight, anthracnose, ascochyta blight (for hill areas only), and bean mosaic.

Seed Standards

Factors	Standards for Each Class	
	Foundation	Certified
Pure seed (minimum)	98.0 per cent	98.0 per cent
Inert matter (maximum)	2.0 per cent	2.0 per cent
Other crop seeds (maximum)	None	None
Weed seeds (maximum)	None	10/kg
Other distinguishable varieties (maximum)	5/kg	10/kg
Germination including hard seeds (minimum)	75 per cent	75 per cent
Moisture (maximum)	9.0 per cent	9.0 per cent
For vapour-proof containers (maximum)	7.0 per cent	7.0 per cent

Brief Cultural Practices

Time of Sowing

In the plains: July to September and January to February

In hills: March to June.

Preparation of Land

Land should have a good tilth and be free from clods and lumps of soil.

Source of Seeds

Obtain nucleus/breeders/foundation seed from source approved by a seed certification agency.

Seed Rate

For bush type: 85 to 95 kg per hectare

For pole type: 25 to 30 kg per hectare

Method of Sowing

The seeds are sown 5 cm deep. The two seeds are sown at one place and one week seedlings are removed after germination is completed.

Spacing

Bush types: 45-60 cm x 10-15 cm

Pole types: 60-75 cm x 30 cm.

Fertilization

200 to 300 quintals per hectare well rotten farmyard manure should be incorporated at the time of land preparation. Besides it, apply 60kg nitrogen, 40kg phosphorus and 40kg potassium per hectare. Nitrogen is applied in two splits. First at the time of sowing with whole quantity of phophrous and potash, second at flowering stage.

Irrigation

Irrigation is required in the early phases of crop growth and during bloom and pod development period. Plants are susceptible to water stress, therefore, irrigation at regular interval are necessary.

Interculture

Very shallow cultivation should be done at early stages of crop growth. Usually, dwarf varieties don't require support, but climbing varieties do not grow well without support. The bamboo sticks or other wooden sticks or branches, which are locally available, can be used for the support. Earthing up around the base of the plant, up to the first set of leaves gives additional support to the stem and encourages root growth.

Insects and Diseases

Insects/Diseases	*Control Measures*
	Insects
Stemfly (*Ophiomyia phaseoli*)	Applied phorate at 0.25 kg a.i./ha. Foliar sprays of endosulfan at 0.7 kg a.i./ha.
Bean aphids (*Aphis craccivora*) and Bean thrips	Spray malathion (0.1 per cent) or phosphamidon (0.05 per cent) or methyl demeton (0.025 per cent).
Hairy caterpillers	Spray with endosulfan (0.07 per cent).
	Diseases
Anthracnose	Dithane Z-78 treatment (2g/kg seed).
Powdery mildew	Dust or spray sulfex 2.5 kg/ha) at frequent intervals.
Angular leaf spot (*Isariopsis griseola)*	Application of benlate (0.25 per cent) powder used as seed dressing.
Root rot (*Rhizoctonia solani*)	Seed treatment with benlate (3g/kg seed) or bavistin (3g/kg seed) or vitavax (6g/kg seed)
Root rot (*Phythium spp.*)	Application of Bordeaux mixture (1 per cent).
Common bean mosaic (vector: aphid)	Application of furadan @ 1.5 kg a. i. /ha at the time sowing. Rouging of infect plants from the field.

Roguing

Constant roguing for off type plants and plants affected by diseases like anthracnose, rust and mosaic must be done right from the beginning.

Harvesting and Threshing

Pods are harvested when they have turned yellow but have not been completely dry and seeds in the pods are firm, well developed and have just began to break free from the inside of the pod. Harvesting is best done in the morning to avoid shattering loss.

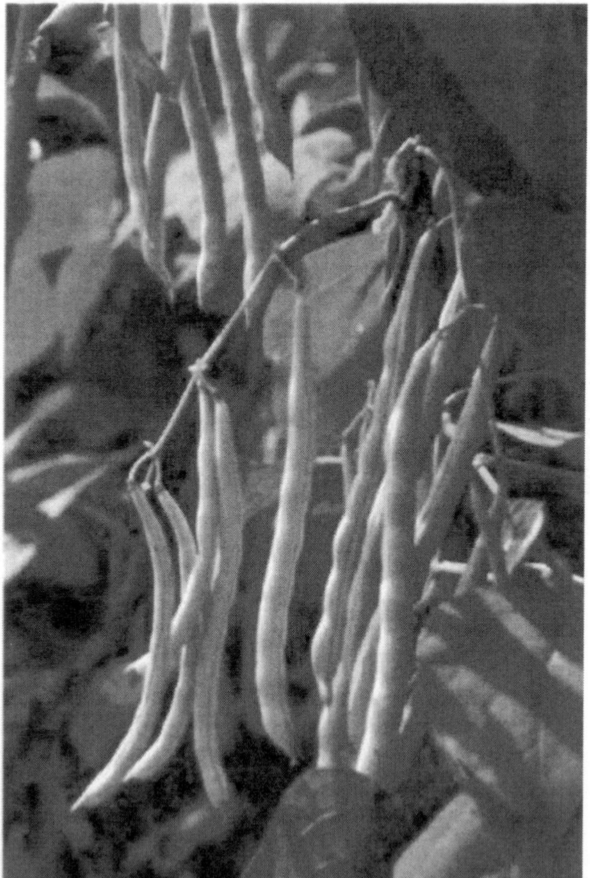

Figure 68: Maturing Bean Pods

Whole plants are harvested either by hands or by machine. The harvested plants are staked for 7-10 days and then they are threshed either by bullocks or by machine.

Storage

Soon after threshing seed should be cleaned and dried. Drying to seed moisture content below 11 per cent will be necessary for temporary storage in ventilated sheds within the RH of 75 per cent.

Seed Yield

12 to 18 quintals per hectare.

Varieties/Hybrids Characteristics

Varieties/Hybrids Recommended by IARI, New Delhi

Pusa Parvati

It is developed through X-ray irradiation of an American variety, 'Wax Pod'. It is an early bush type with attractive round, meaty with light green pod. It gives about 12.5 to 25.0 q/ha of edible pods. It is resistant to mosaic and powdery mildew diseases.

Varieties/Hybrids Recommended by Uttranchal State

Pant Anupama

It is developed through selection at GBPUAT, Pantnagar. Plant is bushy type, early bearing. Pods are round, smooth, tender, stringless and green. The yield potential is 89 q/ha. It is resistant to angular leaf spot and moderately resistant to common mosaic virus.

VLBONI-1

This variety has been developed from VPKAS, Almora. Plants are dwarf (40cm). The pods are long, round bushy and stringless with pole green colour. The average yield is about 105-115 q/ha.

Varieties/Hybrids Recommended by Tamil Nadu State

TKD-1

It is a pole type variety selected from germplasm population at Horticultural Research Station, Thadiyankudisai and release from TNAU, Coimbatore. Pods are less fibrous. The yield potential of this variety is 50-60 q/ha green pods and 27.8 q/ha seeds.

YCD-1

It is a pure lines selection from a local indigenous type collected from Shevroy hill and recommended by TNAU, Coimbatore. The pods are slightly flat, long (15.1 cm) and broad (1.13 cm). Seeds are bold and attractive dark purple in colour. It gives yield of 97.5q/ha.

Varieties/Hybrids Recommended by Karnataka State

Arka Komal

It is an improvement over Australian collection IIHR–60. Plant is bushy. The pods are green, straight, flat and tender. The yield potential is about 90 q/ha. It is very good in transport purposes.

IIHR-909

This variety has been bred at IIHR, Bangalore; derived through pedigree selection from the cross Blue Crop x Contender. Plants are bushy and photosensitive. Pods are straight, oval, light green, fleshy and stringless. Pod yield potential is 190 q/ha.

Other Varieties/Hybrids

Contender

It is an introduction from USA. It is bush type. Pods are round, green, 13-14 cm long, stringless and slightly curved. Seed is light brown. The average pod yield is 80-95 q/ha.

Jampa

It is a Mexican variety having outstanding performance in Maharashtra. The pods are round in shape. Average pod yield is 80-85 q/ha. It is highly resistant to wilt diseases.

Kentucky Wonder

It is an introduction from USA. A pole type variety. It posses 4-5 pods/ cluster, 20 cm long, flatish, appear to be seedy, meaty, stringless. Seed is light brown. Average pod yield is 100-125 q/ha.

Lakshmi

It has been bred from a cross between Contender (bush) x Local (pole) varieties. Plants are indeterminate. Pods are formed in cluster of three, 13-14 cm long, stringless, green, round and are quite attractive. Seed colour is white. Average green pod yield is 120-140 q/ha.

Pusa Himlata

It has been developed through pure line selection of an exotic introduction at the IARI Regional Station, Katrain. It is climbing or pole type. Pods are medium size (14 cm), round, stringless and light green. The seeds are white. The average yield is 26.3 q/ha.

Cowpea (*Vigna unguiculata* L. Walp) 2n = 2x = 22

Land Requirement

Land to be used for seed production shall be free of volunteer plants. It can be grown successfully in different types of soil from loamy to sandy loam. The optimum soil pH should be between 6.0 to 7.0.

Field Inspection

A minimum of two inspections should be done, the first before flowering, the second during flowering and fruiting stage.

Figure 69: Cowpea

Field Standards

General Requirements

Isolation

Seed fields should be isolated from the contaminants shown in column 1 of the Table below by the distances specified in columns 2 and 3 of the said Table:

Contaminants	Minimum Distance (Meters)	
	Foundation	Certified
1	2	3
Fields of other varieties	10	5
Fields of the same variety not conforming to varietal purity requirements for certification	10	5

Specific Requirements

Factor	Maximum Permitted (%)*	
	Foundation	Certified
Off types	0.10	0.20
** Plants affected by seed borne diseases	0.10	0.20

* Maximum permitted at and after flowering.

** Seed borne diseases are: Ashy stem blight, anthracnose, ascochyta blight (for hills areas only) and cowpea mosaic.

Seed Standards

Factors	Standards for Each Class	
	Foundation	Certified
Pure seed (minimum)	98.0 per cent	98.0 per cent
Inert matter (maximum)	2.0 per cent	2.0 per cent
Other crop seeds (maximum)	None	10/kg
weed seeds (maximum)	None	10/kg
Other distinguishable varieties (maximum)	5/kg	10/kg
Germination including hard seeds (minimum)	75 per cent	75 per cent
Moisture (maximum)	9.0 per cent	9.0 per cent
For vapour-proof containers (maximum)	8.0 per cent	8.0 per cent

Brief Cultural Practices

Time of Sowing

Northern plains: End of February to middle of April.

Monsoon crop: Middle of June to the end of July

Southern plains: December-January in.

In the hills: April-May

Preparation of Land

Land should be well prepared by ploughing and harrowing, 15 cm raised beds of 120 cm wide is to be prepared keeping a 40 cm wide drain between beds.

Source of Seeds

Obtain nucleus/breeders/foundation seed from source approved by a seed certification agency.

Seed Rate

Spring season: 20-25 kg seed per hectare

Rainy season: 12-15 kg seed per hectare.

Methods of Sowing

Seeds are sown by hand on the beds. Two lines are drawn on a bed at a recommended spacing from each edge of the bed. Seeds are sown in rows 2-3 cm deep and about 10-15 cms apart. Seeds should be treated with thiram at the rate of 3.5 to 4 g per kg of seed.

Planting Distance

Kharif planting: 40 to 60 cm x 10-15 cm

Summer and spring planting: 25 to 30 cm x 10-15 cm

Fertilizer

Being a leguminous crop, cowpea requires less nitrogen. However, 10-20 kg/ha nitrogen should be incorporated in the soil before sowing. Phosphorus and potassium 50 to 70 kg per hectare each should be drilled in the soil before sowing.

Interculture

Two to three weedings and hoeings would be required to check the weed growth. About 150-175 cm high trellis are to be provided for every two lines of a bed, trellis to be created in such a way that the tops of sticks are bounded together to form a triangular structure with the bed. Sufficient sticks to be provided to help the plants to climb up.

Irrigation

Sufficient irrigation should be provided when flowering starts. Previous irrigations should be light and as per requirements.

Insects and Diseases

Insects/Diseases	*Control Measures*
	Insects
Pod borers (*Heliothis armigera*)	Spray with sevin at 0.2 per cent or spray with endosulfan at 0.05 per cent.
Aphids (*Aphis spp*) and Jassid (*Amrasca kerri*)	Spraying with nuvan 0.05 per cent or malathion 50 EC @ 2 ml per liter water.
Galerucid beetle (*Madhurasia obbscurella*)	Soil application with phorate or aldicarb 10 G @ 10 to 18 kg per hectare.
	Diseases
Cercospora leaf spots (*Cercospora spp.*)	Spray with benomyl (0.2 per cent) or thiram (0.25 per cent) at 15 days interval.
Pod rot (*Choanephora cucurbitarum*)	Spray with blitox 50 at 0.3 per cent
Anthracnose (*Colletotrichum lindemuthianu*)	Treat the seed with 0.125 per cent solution of ceresan or benomyl 0.2 per cent.
Powdery mildew (*Erysiphe polygoni*)	Spray with 0.5 per cent wettable sulphur or, spray with benlate or bavistin at 0.15 per cent
Cowpea mosaic virus: (Vector: Aphids)	Spraying dimethoate (0.05 per cent) or nuvacron (0.05 per cent) at 10 days interval.

Roguing

Constant roguing for off type plants and plants affected by mosaic or other diseases should be done right from the beginning. Affected plants should be removed as a whole immediately on sight.

Figure 70: Cowpea Pods Hunds in Bundles for Drying

Harvesting and Threshing

Pods are harvested when they have turned yellow but have not been completely dry and seed in the pods are firm and well developed and have just begun to break free from the inside of the pod. Two-three harvestings will be necessary. Harvested pods are dried in the sun and threshed by hand with care to avoid loss due to breaking injuries.

Storage

Seeds are cleaned and dried soon after threshing. Drying seed to a moisture level below15 per cent will be necessary for temporary storage in ventilated sheds. For long storage, moisture content should be around 11 per cent.

Seed Yield

10-12 q per hectare.

Varieties/Hybrids Characteristics

Varieties/Hybrids Recommended by IARI and NBPGR, New Delhi

Pusa Barsati

It is an exotic selection from Philippines introduction. It is the earliest variety yielding 2 or 3 flushes of light green pendent pods, 25-28 cm long containing large green seeds. An average yield is 90-95 q/ha.

Pusa Do-fasli

It has been evolved by crossing Pusa Phalguni with a cultivar from the Philippines. It is a bushy cultivar. Pods are light green, pendant and 26-28 cm long. The yield potential is about 75-80 q/ha green pods.

Pusa Komal

It is a cross of P85-2 and PU26 cowpea strains. It is compact, bushy without vines or trailers. The pods are bright green, non-fibrous, cylindrical, soft and meaty. Mature pods shrivel and do not shatter. The seeds have rough surface, whitish buff coloured and prominent brown hilum.

Pusa Phalguni

It has been selected form a Canadian cultivar. It is a bushy type. It gives dark green erect pods of 12.5 cm length with small cylindrical white seeds. Yield potential is 50 to 100 q/ha.

Pusa Rituraj

It is a selection developed by NPBGR, New Delhi. Due to high photo-thermo-insensitive nature, this variety can be grown in summer as well as in rainy seasons. The plant is bushy. The pods are 22 to 24 cm long, thin and palatable. It gives about 80 to 85 q/ha green pods.

Varieties/Hybrids Recommended by Karnataka State

Arka Garima

This popular variety has been bred at IIHR, Bangalore; derived through backcross and pure line selection from the cross TUV-762 x *Vigna unguiculata* spp. *sesquipedalis*. Pods are light green, long thick, round, fleshy and stringless. It is tolerant to heat and drought and gives an average yield of 18 t/ha.

IIHR-16

It is an early maturing, developed at IIHR, Bangalore; derived through pedigree selection from the cross Arka Garima x Pusa Komal. Pods are green, medium thick, medium long (15-18 cm), tender, fleshy without parchment. It gives an average yield of 190 q/ha.

Varieties/Hybrids Recommended by West Bengal State

Bidhan Barbati-1

It is an advance generation collection of cross EC-243954 of culti group, *unguiculata* and EC-305827 of culti group *sesquipedalis* developed at BCKV, Mohanpur. Pods are green, tender, thick, fleshy and 25.2 cm long. Seeds are bold, thick with buff and brown. It shows resistant to cow pea mosaic and cow golden mosaic virus. It gives 134 q/ha green pods.

Bidhan Barbati-2

It is an advance generation collection of cross V-70 of culti group, *biflora* and Sel Tm-3 of culti group *sesquipedalis* and developed at BCKV, Mohanpur. Plants are semi-determinate. Pods are light green, tender, thick, solid, 25.8 cm long. Seeds are flat, buff and dark tan. It shows very low incidence of cowpea mosaic and cow golden mosaic virus. It gives 159 q/ha green pods.

Varieties/Hybrids Recommended by Other States

Birsa Sweta

It is selection from local (cultivar plot) and released from Bihar. It is suitable for rainy season. Plant height is 1.5–2.0 m, indeterminate growth habit hence staking is required. Pods are white, fleshy and soft, 28.5 cm long and 1.10 cm in girth. Seeds are bicolour (dark brown with light yellow colour) and 15.5 g/100 seeds.

Family–Chenopodiaceae

Spinach Beet (Palak) (*Beta vulgaris* var. *bengalensis* Hort.) 2n = 2x =18

Land Requirement

Land to be used for seed production shall be free of volunteer plants. The soil of selected fields should be rich in organic matter.

Figure 71: Spinach Beet

Field Inspection

A minimum of two inspections shall be made, the first before flowering and the second during flowering stage.

Field Standards

General Requirements

Isolation

Seed fields shall be isolated from the contaminants shown in column 1 of the Table below by the distances specified in columns 2 and 3 of the said Table:

Contaminants	*Minimum Distance (Meters)*	
	Foundation	*Certified*
1	*2*	*3*
Fields of other varieties	1600	1000
Fields of the same variety not conforming to varietal purity requirements for certification	1600	1000
Fields of the swiss chard: *Beta vulgaris* L. var. *cicla*, sugar beet (*Beta vulgaris* L.), and garden beet (bet root, mangels or stock beet): *Beta vulgaris* (L.) var. *rubra*.	1600	1000

Specific Requirements

Factor	Maximum Permitted (%)*	
	Foundation	Certified
Off types	0.10	0.20

* Maximum permitted at and after flowering.

Seed Standards

Factors	Standards for Each Class	
	Foundation	Certified
Pure seed (minimum)	96.0 per cent	96.0 per cent
Inert matter (maximum)	4.0 per cent	4.0 per cent
Other crop seeds (maximum)	5/kg	10/kg
Weed seeds (maximum)	5/kg	10/kg
Germination (minimum)	60 per cent	60 per cent
Moisture (maximum)	9.0 per cent	9.0 per cent
For vapour-proof containers (maximum)	8.0 per cent	8.0 per cent

Brief Cultural Practices

Time of Sowing

October to November.

Preparation of Land

Land should be well prepared by deep and good tilth and well levelled.

Source of Seeds

Obtain nucleus/breeders/foundation seed from source approved by a seed certification agency.

Seed Rate

25-30 kg/ha.

Method of Sowing

Sowing of seeds is to be done in lines at a depth of about 2 cm. Seeds are to be soaked over night in water and kept spread between the wet gunny bags in order to keep moderately moist and to accelerate germination. The seeds are ready for planting when visible sprouting spots/cracks of rooting are observed.

Spacing

30-35 cm x 15-20 cm

Fertilization

Well rotten farm yard manure @ 35-40 tonne/ha should be applied at the time of preparation of land. Urea 80-90 kg, SSP 125-130 kg and MOP 70-80 kg/ha may be applied. Entire SSP and MOP to be applied at last ploughing during land preparation. The urea should be applied as top-dressing in two equal instalments, the first 12-15 days after germination and the second before bolting.

Irrigation

Frequent irrigations according to season are required.

Interculture

The field should be kept completely clean of pigweed (bathua), after the crop has been left for seed. Thinning will start as the leaves of seedlings start touching each other and continue till the plant-to-plant distances stands at about 15-20 cm.

Insects and Diseases

Insects/Diseases	*Control Measures*
	Insects
Leaf webber, leafhopper, stem weevil, caterpillar and ants	Spray of malathion at the rate of 1.5 to 2 ml per litre of water.
	Diseases
Leaf spot (*Cercospora* spp)	Spray Bordeaux mixture (5:5:50).
White rust (*Albugo bliti*)	Spray dithane M-45 at 0.2 per cent
Mosaic	Application of 3-4 foliar sprays of monocrotophos (0.05 per cent).

Cutting

Not more than 2 to 3 cuts should be taken from the seed crop.

Roguing

Careful rouging for off types, early bolters, and pigweed plants is necessary. The first rouging may be done at pre flowering stage to remove off types on the basis of foliage characters. Subsequently early bolters, off type plants and pigweed plants may be removed immediately, as and when they are noticed.

Harvesting, Threshing and Storage

When the seed stalks and seeds turn yellow and dry, harvesting is to be done by cutting the branch. Harvested branches to be dried in the sun over drying floor/canvas. After drying is complete, the seeds can be threshed by beating them with sticks. After cleaning the seed must be dried to 9 per cent moisture, before storage.

Seed Yield

10 to 15 qt/ha.

Figure 72: Seed Stalk of Beet

Varieties/Hybrids Characteristic

Varieties/Hybrids Recommended by IARI, New Delhi

Pusa Bharti

It plants flower 175-180 days after sowing. It can be grown in summer and *kharif* season.

Pusa Jyoti

It is evolved through an induced polyploid of cultivar All Green with two per cent colchicine for 24 hrs and selection.

Pusa Palak

It is selection from a cross between Swish Chard and Local Palak. It is a late bolting cultivar.

Varieties/Hybrids Recommended by Other States

Arka Anupama

It has been developed through cross of IIHR-10 and IIHR-8 followed by pedigree method of selection and released from IIHR, Bangalore. It is resistant to cercospora leaf spot.

Banerjee Giant

It is a very popular cultivar evolved by SP Banerjee in West Bengal. This cultivar has been developed from a cross between Local Palak and Beet Root.

Jobner Green

This cultivar was evolved at University of Udaipur, Jobner Campus as a result of spontaneous mutation detected from the population of a local collection Serial No. 5.

Ooty-1

It is released from TNAU, Coimbatore through selection of Local type.

Palak No. 51-16

This cultivar has been developed by the Department of Agriculture, Maharashtra State.

Pusa Harit

This cultivar has been developed by hybridization between sugar beet and local palak at IARI, Regional Station, Katrain.

Family–Umbelliferae

Coriander (*Coriandrum sativum* L.) 2n = 2x = 22, 24

Land Requirement

The land should be free of volunteer plants. The soil of selected fields should be well drained and aerated.

Figure 73: Coriander

Field Inspection

The field inspection should be done before flowering, maturing stage and prior to harvesting.

Field Standards

General Requirements

Isolation Distance

Contaminants	*Minimum Distance (Meters)*	
	Foundation	*Certified*
1	*2*	*3*
Fields of other varieties	200	100
Fields of the same varieties	200	100

Specific Requirements

Factor	*Maximum Permitted (%)**	
	Foundation	*Certified*
Off types	0.10	0.20
Plants affected by seed borne diseases	0.10	0.50

* Standards for off types shall be met at and after flowering and for seed borne diseases at final inspection.

Seed Standards

Factors	Standards for Each Class	
	Foundation	Certified
Pure seed (minimum)	98.0 per cent	98.0 per cent
Inert matter (maximum)	2.0 per cent	2.0 per cent
Other crop seeds (maximum)	None	None
Weed seeds (maximum)	None	None
Germination (minimum)	70 per cent	70 per cent
Moisture (maximum)	8.0 per cent	8.0 per cent
For vapour-proof containers (maximum)	6.0 per cent	6.0 per cent

Brief Cultural Practices

Time of Sowing

Rabi crop–15 October to 15 November

Kharif crop–June-July (Coimbatore conditions of Tamil Nadu)

August-September (Malwa region of MP)

Preparation of Land

Prepare the field to a good tilth by one deep ploughing, two to three harrowings followed by levelling.

Source of Seeds

Obtain nucleus/breeders/foundation seed from source approved by a seed certification agency.

Seed Rate

15-30 kg per hectare

Methods of Sowing

The seeds are split into halves before sowing. Split seeds germinate little earlier than whole seeds. Soaking the seeds in water for 12 to 24 hours and drying them in the shade for 12 hours, which causes quicker germination. Soaking coriander in warm water, 10 ppm IAA, 20 ppm NAA or 50 ppm GA increased germination percentage. Before sowing, seed treatment is given with thiram at the rate of 2.5 g/kg seed. Sowing is done with gorru (seed drill).

Under rainfed conditions, seeds are broadcast and covered with a country plough.

Planting Distance

25 to 30 cm x 5 to 10 cm

Fertilizers

Apply 100 to 120 q/ha well rotten farmyard manure at the time of land preparation. Fertilizer recommendation in the ratio of 30:30:20 kg nitrogen, phosphorous and potash/ha respectively. Full quantity of phosphorus and potassium and half quantity of nitrogen may be applied at the time of sowing and remaining quantity of nitrogen is applied four to five weeks after sowing.

Irrigation

Three to four irrigations are required. First irrigation is given after 30 to 35 days from date of sowing, second after 60-70 days and third after during seed development stage. Subsequent irrigations are given according to requirement of crop.

Interculture

The first hoeing and weeding are done in 30-35 days after sowing. Thinning of plants is also done simultaneously to plant distance in line-sown crops, leaving two plants per hill. Second and third weeding is generally done at an interval of 30 days depending upon the weed growth.

Insects and Diseases

Insects/Diseases	*Control Measures*
	Insects
Thrips *(Thrips tabaci* and *Scirtothrips dorsalis*)	Monocrotophos at the rate of 0.5 kg/ha
Spodoptera litena	Chloropyrifos @ 0.5 kg/ha
Aphids (*Hyalopterus pruni* and *Bervicoryne coriandri*)	Spraying of 0.05 per cent endosulfan
Leaf feeding caterpillar (*Spodoptera* spp.)	Flooding the field, collecting the egg masses and spraying of 4 per cent.

Contd...

Insects/Diseases	*Control Measures*
Bugs (*Nezara viridula* and *Agonoscelis nubilis*)	0.03 per cent dimethoate or quinalphos
	Diseases
Powdery mildew (*Erysiphe polygoni*)	Spraying the plants with 0.2 per cent solution of wet sulfur or karathane.
Stem gall (*Protomyces macrosporus*)	Four foliar sprays of captafol (0.2 per cent) and carboxin (0.1 per cent)

Nipping

Not more than 2 to3 nips should be taken from the seed crop.

Roguing

Careful rouging for off types, early bolters, and weed plants is necessary. The first rouging may be done at pre flowering stage to remove off types on the basis of foliage characters. Subsequently early bolters, off type plants and weed plants may be removed immediately, as and when they are noticed.

Harvesting and Threshing

The crops come to maturity in about 90-150 days. Harvesting is done when the grain changes its colour to straw or light brown colour. The plants are cut or pulled and filled into small stocks in the field to wither for two or three days. Threshing is done by beating with sticks or by rubbing with hands. The produce is winnowed, cleaned and dried in partial shade.

Storage

Seeds should not contain more than six per cent moisture. After drying, produce is stored in gunny bags lined with paper or in polythene bags to avoid entrance of moisture.

Seed Yield

8-12 q/ha

Varieties/Hybrids Characteristics

Varieties/Hybrids Recommended by Kerala State

Sadhna

It is developed by APAU, Kerala through mass selection from Alur collection. The plants are semi-erect with medium-sized (18.0 g/1000 grains), oval, straw-coloured grains. It gives an average yield of 10.3 q/ha. The cultivar is tolerant to white fly and mites.

Sindhu

It is developed by APAU, Kerala through mass selection from Warrangal Local. The plants are dwarf with medium sized (16.5 g/ 1000 grains), oval, straw-coloured grains. It produces an average yield of 10.5 q/ha. The cultivar is resistant to aphids.

Swathi

It is developed by APAU, Kerala through mass selection from Nandyal collection. The plants are semi–erect with medium-sized (16.8 g/100 grains), oval brownish yellow grains. It produces an average yield of 8.8 q/ha.

Varieties/Hybrids Recommended by Tamil Nadu State

CO-1

It is developed by TNAU, Coimbatore through selection. The plants are small. Seeds are globular, small (14.3 q/1000) and dusty brown. It gives an average yield of 4.00 q/ha.

CO-2

It is developed by TNAU, Coimbatore through selection from P_2 cultivar of Gujarat. It can be grown in water logged, drought, saline and alkaline conditions. The plants are erect with oblong, medium-sized grain and dull yellowish brown grains. It gives an average grain yield of 5.2 q/ha.

Varieties/Hybrids Recommended by Andhra Pradesh State

LAM CS-2

It is a medium tall (80-cm) variety with more branching and bushy in nature. Mericarp is straw coloured. The average yield is 1300-1400 kg/ha.

LAM CS-4

A. It is a fairy tall variety with very good branching habit. Mericarp is straw-coloured. The average yield is 1400-1500 kg/ha.

Varieties/Hybrids Recommended by Gujarat State

GC-1

It is developed by GAU, Gujarat through selection in local material. The plants are erect. Seeds are medium size (13.2 g/1000 grains), round and yellow. It produces an average yield of 11.0 q/ha.

GC-2

It is developed by GAU, Gujarat through selection form Co.1. The plants are semi spreading habit and bold grains (14.8 g/1000 grains). It produces an average yield of 14.5 q/ha.

Varieties/Hybrids Recommended by Madhya Pradesh State

Gwalior No. 5365

Seeds are medium in size. The yield potential of this variety is 8-10 q seed /ha. It is resistant to fusarium wilt.

Merrocan

This variety has high oil content. Crop duration is 140-160 days. Seed yield is 10-12 q/ha.

Varieties/Hybrids Recommended by Rajasthan state

Kiran

The yield potential of this variety is 12 q /ha in 130-140 days of crop duration. This variety is resistant to stem gall.

RCR-20

It is developed by RAU, Rajasthan through recurrent half-sib selection from a local collection of Jaipur. The plants are bushy-spreading with medium height. It produces oval grain of large size (18.0 g/1000). The cultivars produces an average yield of 10 q/ha.

RCR-41

It is also developed by RAU, Rajasthan through recurrent half-sib selection from a local collection of Kota. The plants are tall, erect with small-sized (9.3 g/1000) grains. It produces an average yield of 9.2 q/ha. The cultivar is highly resistant to stem gall and wilt.

Varieties/Hybrids Recommended by Other States

Pant Haritama

This variety was released from GBPUA&T, Pantnagar, Uttaranchal. It is a fine seeded variety having late maturity. It is resistant to stem gall. The yield potential of this variety is 15–20 q/ha.

Rajendra Swathi

It is developed by RAU, Bihar through selection from Muzaffarpur Local. The plants are medium-sized with fine round aromatic grains (12.5 g/1000 grains). It produces an average yield of 13 q/ha. It is resistant to stem gall.

Carrot (*Daucus carota* L.) 2n = 2x =18

Land Requirements

There are no requirements as to the previous crop, but land to be used for seed production should be free of volunteer plants. The soils of selected fields should be deep, light and friable and the fields well drained.

Field Inspection

Mother Root Production Stage

A minimum of two inspection shall be made as follows:

1. The first inspection should be done after 20-30 days of the sowing in order to determine isolation, volunteer plants, out crosses, off types and other relevant factors;
2. The second inspection should be after the mother roots have been lifted to verify the true characteristics of roots.

Figure 74: Carrot

Seed Production Stage

A minimum of four inspections shall be made as follows:

1. The first inspection should be done before flowering in order to determine isolation, volunteer plants, out crosses and another relevant factors;
2. The second and third inspections should be done during flowering to check isolation, off types and other relevant factors;
3. The fourth inspection should be done at maturity to verify the true nature of umbels.

Field Standard

General Requirements

Isolation

Carrot seed fields should be isolated from the contaminants shown in column 1 of the Table below by the distances specified in columns 2 and 3 of the said Table.

Contaminants	Minimum Distance (Meters)			
	Mother Root Production Stage		Seed Production Stage	
	Foundation	Certified	Foundation	Certified
1	2	3	4	5
Fields of the other varieties	5	5	1000	800
Fields of the same hybrid not conforming to varietal purity requirements for certification.	5	5	1000	800

Specific Requirements

Factor	Maximum Permitted (%)*	
	Foundation	Certified
*Roots not confirming to the varietal characteristics including forked roots	0.10 (by number)	0.20 (by number)
**Off types (plants)	0.10	0.20

* Maximum permitted at second inspection at mother root production stage.

** Maximum permitted at and after flowering at seed production stage.

Seed Standards

Factors	Standards for Each Class	
	Foundation	Certified
Pure seed (minimum)	95.0 per cent	95.0 per cent
Inert matter (maximum)	5.0 per cent	5.0 per cent
Other crop seeds (maximum)	5/kg	10/kg
Total weed seeds (maximum)	5/kg	10/kg
Other distinguishable varieties (maximum)	5/kg	10/kg
Germination (minimum)	60 per cent	60 per cent
Moisture (maximum)	8.0 per cent	8.0 per cent
For vapour-proof containers (maximum)	7.0 per cent	7.0 per cent

Method of Seed Production

Like other biennials, both roots to seed and seed to seed method can be followed. Generally, the root to seed method is followed because in the seed to seed method, root rot is usually very high as compared to the transplanted roots. The whole roots with the tips cut (to examine the colour) are planted, keeping the crown exposed.

Root to Seed Method

First Season–Mother Root Production

Time of Sowing

The sowing of seed crop should be done from 15th July to 17th August, depending upon duration of variety and climatic conditions.

Preparation of Field

Carrots need deep, loose soil for their best development. Hence pulverization of soil is essential. Prepare the field to a fine tilth by repeated ploughing, harrowing and followed by levelling.

Source of Seeds

Obtain nucleus/breeders/foundation seed from source approved by a seed certification agency.

Seed Rate

2.5 to 3.5 kg seed per hectare. The roots produced on one hectare

are sufficient for transplanting three to four hectares under seed production.

Method of Sowing

For better development of roots, sowing on ridges is preferred to flat sowing. Double row ridges 75 cm apart produce larger number of well-developed roots than single row ridges. The seed takes 8-10 days for germination. For uniform germination, the ridges should remain moist till germination takes place. Hence, the field should be irrigated just after sowing. Afterwards, when the plants are 5 to 6 cm in height, thin out plants to a distance of 6 to 7 cm.

Fertilizers

Apply 150-200 quintals per hectare of well-rotted farmyard manure at the time of preparation of land. At the time of final levelling mix 40-50 kg/ha each of phosphorus and potash as basal application in the soil. Nitrogen @75-100 kg/ha is top dressed after weeding.

Irrigation

It should be done at 8-10 days interval depending on the weather conditions and requirements.

Interculture

Frequent weeding and hoeing is necessary in the early stages. One earthing up by end of September or early October will keep the crop clean till the close of autumn, when it is uprooted for planting.

Insects and Diseases

Insects/Diseases	*Control Measures*
	Insects
Carrot rust fly (*Psila rosae*)	Application of organophosphate or organocarbamate as drench.
Nematodes	Used Nemagon or sannema or DD mixture
	Diseases
Cercospora leaf blight (*Cercospora carotae.*), Alternaria blight (*Alternaria dauci*)	Hot water at 50°C for minutes. Copper oxychloride (0.3 per cent).
Carrot yellows (Vector: leafhopper)	Spray insecticide

Harvesting of Roots

Uproot the plants when they have fully developed roots.

Selection of Roots for Transplanting

Though selection of roots is made on the basis of character of tops, whether short or heavy, colour of skin, shape and size of roots. The colour of flesh, colour and size of the core are the most important characters to be considered. The core should be of the same colour as the flesh and as small as possible.

Second Season–Seed Production

Planting of Selected Roots

The selected roots after their tops are clipped and the tips pruned are reset in a well-prepared field.

Method of Transplanting

The trimmed roots as described earlier are reset in the field at a distance of 75 x 22.5-30 cm. The soil is then firmed and tapped around the roots. Soon after transplanting the field is irrigated.

Fertilizers

Apply 200 quintals per hectare of well rotten farmyard manure, at the time of preparation of land. At the time of final levelling mix 50 kg phosphorus and 50 kg potash per hectare as basal application in the soil. Make surface application of 100 kg /ha nitrogen during April-May after hoeing and earthing.

Interculture

One weeding during March and another hoeing and earthing up during April-May is required.

Insects and Diseases

Same as desribed in first year.

Rouging

Rouging should be done at bloom stage. Early bolters and off types should be removed from time to time as required.

Harvesting and Threshing

The best time for harvesting is when the secondary umbels (heads) are fully ripe and tertiary heads begin to turn brown. The

Figure 75: Cluster of Mature Seeds

crop ripens unevenly. Seed is commonly harvested by hand picking. Two-three pickings are required. After drying, the heads are threshed and cleaned. After cleaning, the seed is rubbed by hand to remove the bristles on the surface and graded by means of sifters and sieves. Before storage, the seed moisture content should be reduced to 8 per cent.

Seed Yield

Average seed yield varies from 500-600 kg/ha.

Varieties/Hybrids Characteristics

Varieties/Hybrids Recommended by IARI, New Delhi

Pusa Kesar

This is a selection from a cross between Local Red and Nantes Half Long. The roots develop with narrow central core, which is also sufficiently red coloured. It contains higher amount of carotene (38mg/100 g edible protein) than Local Red (26-mg/100 g).

Pusa Meghali

It is an advance generation selection from a cross between 'Pusa Kesar' and Nantes'. It has a short top, smooth roots, orange flesh, self-coloured core and stumpy to slightly tapering roots. It is suitable for sowing as early (August-September) as well as late (October-November) crop. It produces 250-280 q/ha carrots. The variety is capable of setting in the plains.

Varieties/Hybrids Recommended by Himachal Pradesh State

Nantes Half Long

It is suitable for sowing during winter and released from IARI Regional Station, Katrain. Roots are cylindrical, stumpy, well shaped with abrupt tail, orange scarlet with self-coloured core. Flesh is sweet, fine grained with good flavour and orange scarlet in colour.

Nantes

The roots are half-long, slim, well shaped, cylindrical with stumped end forming a small thin tail, deep orange-red cortex and core. It ranks first in quality but has a week, brittle top that makes difficult in pulling.

Pusa Yamdagni

It is developed by hybridization between EC-9981 x Nantes and released by IARI Regional Station, Katrain. Roots are 15-16 cm long, orange, self-coloured core and slightly tapering stumpy to semi-stumpy ending. It is high yielder and richer in carotene content.

Varieties/Hybrids Recommended by Punjab State

Sel-233

It is a derivative of a cross of Nantes x No.29 and released by PAU, Ludhiana. Roots are 15.1 cm long, smooth, semi cylindrical orange with light orange core colour, less cracking and forking.

No. 29

It is an Asiatic type developed through selection from Local material and released from PAU, Ludhiana. It is long tapering light red and thin roots.

Varieties/Hybrids Recommended by Tamil Nadu State

Zeno

This is very popular variety for the Nilgiris hills, introduced from German Democratic Republic. The roots are 15-17 cm long, slightly tapering toward the end. The exposed portion of the root becomes pinkish green, medium top and self-coloured.

Ooty-1

This is a selection from half-sib progeny of local type (DC. 3) and released from TNAU, Coimbatore. The roots are long and slightly tapering with attractive deep orange colour. This is a suitable for growing in through out the year.The crop is free from pre-mature bolting and resistant to powdery mildew, leaf spot and drought. It yields 491q /ha roots and 700-1000 kg/ha seeds.

Other Varieties/Hybrids

Chantenay

It is European type and an excellent cultivar for canning and storage. It is having attractive roots with deep reddish orange colour. Roots are long, conical to near conical, thick at the shoulders with gradual tapering with distinct stumpy end. The flesh is beautiful rich orange, tender, sweet and fine textured with indistinct core.

Early Nantes

It has almost cylindrical roots terminating abruptly in small thin tail, 12-15 cm long and fine textured. Flesh is orange with self-coloured core.

Hisar Gairic

It has been developed through mass selection and released from CCSHAU, Hisar. Roots are long, tapering, attractive, light brick red, thin self-core, less fibre and forking. It is good for early sowing.

Imperator

It is a cross between Nantes and Chanteny. Roots are long with short tapered end, deep orange cortex and slightly less pigmented core.

Family–Cucurbitaceae

Cucumber (*Cucumis sativus* L.) 2n = 2x = 14

Land Requirement

The land should be free of volunteer plants. The soil of selected fields should be well drained and aerated.

Figure 76: Cucumber

Field Inspection

A minimum of three inspections should be made, the first before flowering, the second during flowering and fruiting stage and the third at the mature fruit stage and prior to harvesting.

Field Standards

General Requirements

Isolation

Cucumber seed fields shall be isolated from the contaminants shown in column 1 of the Table below by the distances specified in columns 2 and 3 of the said Table:

Contaminants	*Minimum Distance (Meters)*	
	Foundation	*Certified*
1	*2*	*3*
Fields of other varieties	1000	500
Fields of the same variety not conforming to varietal purity requirements for certification and from *Cucumis hardwickii*	1000	500

Specific Requirements

Factor	Maximum Permitted (%)*	
	Foundation	Certified
Off types	0.10	0.20
** Objectionable weed plants	None	None

* Maximum permitted at and after flowering

** Objectionable weed shall be: *Cucumis hardwickii*

Seed Standards

Factors	Standards for Each Class	
	Foundation	Certified
Pure seed (minimum)	98.0 per cent	98.0 per cent
Inert matter (maximum)	2.0 per cent	2.0 per cent
Other crop seeds (maximum)	5/kg	10/kg
Total weed seeds (maximum)	None	None
Objectionable weed seeds (maximum)	None	None
Germination (minimum)	60 per cent	60 per cent
Moisture (maximum)	7.0 per cent	7.0 per cent
For vapour-proof containers (maximum)	6.0 per cent	6.0 per cent

Brief Cultural Practices

Time of Sowing

North-Summer-January to February

Kharif-June to July

South-October to November

Preparation of Land

Prepare the field to a good tilth by one deep ploughing, two to three harrowings followed by levelling.

Source of Seeds

Obtain nucleus/breeders/foundation seed from source approved by a seed certification agency.

Seed Rate

Two to five kg/ha.

Planting

There are two methods, which are generally used for sowing of cucumber.

1. Pits are prepared at proper spacing by adding well rotted farmyard manure and a number of seeds are sown on each hill.
2. Furrows are made at row to row spacings (2 meters) and seeds are sown on the edged of the furrows 60 to 90 cm apart on both sides, above the water level so that capillary movement receives the water. Two–three seeds may be sown at each place.

Fertilizers

Apply 250 to 300 q of farmyard manure per hectare in the soil at the time of preparation. A mixture of 50 kg nitrogen, 50 kg phosphorus, and 50 kg potash per hectare should be prepared and mixed at 500 g/pit. Top dress 50 kg nitrogen at the time of flowering.

Thinning

When the seed is sown on hills, the plants should be thinned so that not more than three plants are left standing on each hill. When sown along the furrows, the plants at each place are thined to one or two.

Irrigation

Generally 8 to 10 irrigations are given, depending upon rainfall, soil type, humidity and prevailing temperature.

Interculture

The beds should be kept free of weeds, especially in the early stages. Later on, rapidly spreading vines suppress the weeds. It is desirable to stake this crop during the rainy season.

Sex Expression and Sex Ratio

The sex expression and sex ratio are varietal characters. They are, however, modified by the environment. Lower fertility, higher temperature, longer light period all induces maleness. Certain gases and chemicals also affect sex. Both auxins and anti-auxins at proper concentrations modify sex. Gibberellic acid at higher concentrations induces maleness, but at the lower concentration of 10 to 25 ppm increases the number of female flowers. Two sprays, one at two leaf stage and again at four leaf stage with 25 to 100 ppm malic hydrazide, 100 ppm of alphanaphthalene acetic acid, 3 ppm of ethrel, 3 ppm of boron or 3 ppm of molybdenum can suppress the number of male flowers and increase the number of female flowers, fruit set and ultimate yield.

Insect and Diseases

Insects/Diseases	*Control Measures*
	Insects
Red pumpkin beetle (*Aulacophora foveicollis*)	Spraying of carbaryl 0.2 per cent or rogor 0.1 per cent
Aphids (*Aphis spp.*)	Spraying metasystox 0.1 to 0.2 per cent
Fruit fly (*Dacus cucurbitae, Dacus dorsalis*)	Using light traps and poison baits. Spray thiodan @ 6 ml per 4.5 litres of water.
	Diseases
Downy mildew (*Pseudoperenospora cubensis*)	Ridomil (0.2 per cent) and blitox (0.2 per cent). Seed treatment with apron (2 g/kg seed) followed by dithane M-45 (0.3 per cent).
Powdery mildew (*Erysiphe cichoracearum* and *Sphareotheca fuliginea*)	Spraying of benlate (0.1 per cent), karathane (0.05 per cent)
Anthracnose (*Colletotrichum*	Seed treatment with diathane M-45 (2 g/kg seed). Spraying of difolatan (0.2 per cent) *lagenarianum*)
Cucumber mosaic (Vector: Aphids)	Spray the crop with nuvacron or dimecron (0.05 per cent). Mineral oil, krishi oil @ 1.5 per cent.
Cucumber green mottle mosaic	Seeds from healthy plants should be sown. Soil fumigation with methyl bromide.

Roguing

The removal of off type plants is an important task and oviously should be carreid out before flowering and natural cross-pollination can occur. However, in practice, it is rather impractical to determine the rogues in cucumber prior to fruit development. Any plant with removing the plant is a practice definitely to be avoided, especially in the early stages of growth off type fruits should be removed from the field as soon as it is noticed, in order to prevent any further contamination by its pollen. Since plants of one variety have the same spine colour white or black, any plant having the wrong colour should be rogued. Later, when the fruits are fully mature, any plants, which are off type in spine colour, are easily distinguished and because of the striking difference in the fruit colour as already described. While the removal of the plant at that time is helpful in that it prevent the mixture of its seed with the rest, considerably cross-pollination may have already occurred and will result in off type plants in succeeding generations. To remove off type fruits without when the plant is still in flower.

Harvesting and Extraction of Seed

The seed crop is ready for harvest when the fruits have turned pale yellow or golden. The seeds are obtained by cutting individual fruits in half, longitudinally and scraping them out with knife. At the time of extraction, the seed will not separate completely from the pulp surrounding it fermentation mechanical means and chemical extraction break its adherence to this material. The seeds can be extracted by following methods:

Fermentation

In this method, the pulp is poured into wooden barrels where it is allowed to ferment for several days. The fermenting material is occasionally stirred to prevent any discolouration or blackening of seed from mold, which may form on the material floating at the

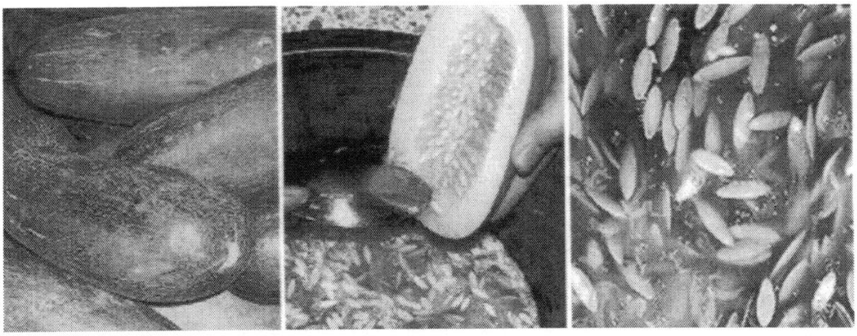

Figure 77: Mature Cucumber, Seed Extraction and Fermenting Seed

surface. As the seed separates from the disintegrating pulp, it tends to sink to the bottom of the container. The seeds are then washed and dried in the sun to moisture contents below 7 per cent, before storage.

Mechanical Means

Seed can be extracted by the use of special machines also. In these machines, the fruits are not only cut and macerated, but also the pulp is squeezed so that the seed is forcibly separated from it. In this operation, considerable water is used for washing towards the end of the operation. The seeds are later washed and dried to moisture content below 7 per cent.

Chemical Extraction

Either acid or an alkali may be used to speed up the extraction of cucumber seed. Hydrochloric acid (100 CC) /11-12 kg of pulp, or 12 parts of 25 per cent technical grade NH_3/1000 parts of material, is thoroughly stirred into the pulp. After about 30 minutes, water is added while stirring. The pulp, other impurities, and empty seeds will float and mature seeds will sink to the bottom. Later, the seeds are washed and dried to moisture content below 7 per cent.

Seed Yield

Average seed yield varies from 110 to 130 kg/ha.

Varieties/Hybrids Characteristics

Varieties/Hybrids Recommended by Himachal Pradesh State

Japanese Long Green

It is a temperate variety, released by IARI, Regional Station, Katrain. This is an extra early maturing (45days) variety. Fruit is 30-40 cm long, flesh light green, crisp, white surface with white spines.

KTCH-11

This is a gynoecious line based prolific bearing hybrid, developed at IARI, Regional Station, Katrain. Fruits are attractive, 25-28 cm long, cylindrical, yellowish light green, smooth, sweet with brown spines and crispy flesh. Gives an average yield of 186 q/ha.

KTCH-8

It is a cross of Poinsett and LC-3.This early maturing hybrid has been developed at IARI, Regional Station, Katrain. Fruits are attractive, dark green, 25-30 cm long, sweet, and cylindrical with crispy flesh and white spines.. It gives average yield of 182 q/ha.

Pusa Sanyog

It is a hybrid between Japanese gynoecious line and Green long Naples and released by IARI, Regional Station, Katrain. It is an early maturing (50days). Fruit is 28-30 cm long, cylindrical, dark green with yellow stripes.

Straight Eight

It is released by IARI Regional Station, Katrain. It is an early maturing.Fruit is 20 to 25 cm long, straight with round end and light green in colour.

Varieties/Hybrids Recommended by Maharashtra State

PH-18

It is a cross of GYC-4 and Poona Khira and developed at MPKV, Rahuri. Fruits are yellow, with average weight of 146 g. It gives an average yield of 586 q/ha.

Phule Priyanka

It is a cross of RHRB5-5 and RHRBG-4-loped at MPKV, Rahuri. Fruits are dark green, 20 cm long. This hybrid is suitable for cultivation under rainy and summer seasons. It is tolerant to downy mildew. It gives an average yield of 282.7 q/ha.

Sheetal

This variety has been developed from Konkan Krishi Vidyapeeth, Dapoli (M.S.). Fruits are light green colour. This variety is suitable for high rainfall area. An average yield of this variety is 421 q/ha.

Varieties/Hybrids Recommended by Other States

CO-1

It is a selection from a local type of Kanyakumari district. Fruits are long (60-65 cm), slightly curved, tapering towards stalk end. The yield potential of this variety is 250-280 q/ha.

Poinsette

It is originally developed at Charleston (South Carolina), USA and multiplied by National Seeds Corporation, New Delhi. Fruit is dark green and 20-25 cm long. It is resistance to downy mildew, powdery mildew, anthracnose and angular leaf spot.

Muskmelon (*Cucumis melo* L.) 2n = 2x = 24

Land Requirement

The land should be free of volunteer plants. The soil of selected fields should be well drained and aerated.

Field Inspection

A minimum of three inspections should bedone, the first before flowering, the second during flowering and fruiting stage and the third at mature fruit stage and prior to harvesting.

Field Standards

General Requirements

Isolation

Seed fields shall be isolated from the contaminants shown in column 1 of the Table below by the distances specified in columns 2 and 3 of the said Table:

Figure 78: Muskmelon

Contaminants	Minimum Distance (Meters)	
	Foundation	Certified
1	2	3
Fields of other varieties	1000	500
Fields of the same not conforming to varietal purity requirements for certification and from longmelon (*Cucumis melo* L. var. *utilissimus*), oriental picking melon (*Cucumis melo* L. var. *conomon*); snap melon (*Cucumis melo* L. var. *momordica*.), *Cucumis prophetarum*; weed melon (*Takmek*) (*Cucumis melo* (L.) var. *agrestis* and other non-dessert forms of *Cucumis melo* (L.), known to cross or suspected to being able to cross	1000	500

Specific Requirements

Factor	Maximum Permitted (%)*	
	Foundation	Certified
Off types	0.10	0.20
** Objectionable weed plants	None	None
*** Plants effected by seed borne disease	0.10	0.20

* Standards for off types and objectionable weeds shall be met at any inspection conducted at and after flowering and for seed borne disease at final inspection.

** Objectionable weeds shall be: *Cucumis prophetarum*; weedmelon (*Takmek): Cucumis melo* (L.) var. *agrestis* and other non-dessert forms of *Cucumis melo* (L.)

*** Seed borne disease shall be: Cucumber Mosaic Virus (CMV).

Seed Standards

Factors	Standards for Each Class	
	Foundation	Certified
Pure seed (minimum)	98.0 per cent	98.0 per cent
Inert matter (maximum)	2.0 per cent	2.0 per cent
Other crop seeds (maximum)	5/kg	10/kg

Contd...

Factors	Standards for Each Class	
	Foundation	Certified
Total weed seeds (maximum)	None	None
Objectionable weed seeds (maximum)	None	None
Germination (minimum)	60 per cent	60 per cent
Moisture (maximum)	7.0 per cent	7.0 per cent
For vapour-proof containers (maximum)	6.0 per cent	6.0 per cent

Brief Cultural Practices

Time of Sowing

February to March.

Preparation of Land

Prepare the field to a good tilth by one deep ploughing, two to three harrowings followed by levelling.

Source of Seeds

Obtain nucleus/breeders/foundation seed from source approved by a seed certification agency.

Seed Rate

3-4 kg/ha.

Method of Sowing

The seeds are sown after soaking them in water overnight, in the manner similar to that described for cucumber.

Fertilizers

Apply 250 to 300 q of farmyard manure per hectare in the soil at the time of preparation. A mixture of 50 kg nitrogen, 50 kg phosphorus, and 50 kg potash per hectare should be prepared and mixed at 500 g/pit. Top dress 50 kg nitrogen at the time of flowering.

Thinning

When the seed is sown on hills, the plants should be thinned so that not more than three plants are left standing on each hill. When sown along the furrows, the plants at each place are thinned to one or two.

Irrigation

The crop requires an abundance of moisture during the early part of vine growth. Once the melons are fully-grown, over watering should be avoided.

Interculture

Regular interculture operations are necessary to keep weeds under control.

Insects and Diseases

Insects/Diseases	*Control Measures*
	Insects
Aphids (*Aphis* spp.)	Spraying malathion or metasystox @ 0.2 per cent.
Red pumpkin beetle (*Aulacophora foveicollis*)	Spraying of carbaryl 0.2 per cent or rogor 0.1 per cent
Fruit fly (*Dacus cucurbitae, Dacus dorsalis)*	Using light traps and poison baits. Spray of thiodan @ 6 ml per 4.5 litres of water.
	Diseases
Downy mildew (*Pseudoperenospora cubensis*)	Ridomil (0.2 per cent) and blitox (0.2 per cent). Seed treatment with dithane M-45 (0.3 per cent).
Powdery mildew (*Erysiphe cichoracearum* and *Sphareotheca fuliginea*)	Spraying of benlate (0.1 per cent), karathane (0.05 per cent).
Anthracnose (*Colletrotrichum lagenarianum*)	Seed treatment with diathane M-45 (2 g/kg seed). Spraying of difolatan (0.2 per cent).
Muskmelon mosaic (Vector: Aphids)	Spray the crop with dimecron (0.05 per cent). Mineral oil, krishi oil @ 1.5 per cent.

Roguing

What has been said for roguing of cucumber applies equally to muskmelons. Any plant bearing off type fruits should be destroyed as soon as observed, so as to prevent its pollen from further fertilizing adjacent plants with good type fruit. The plant and its entire crop of fruit rather than the individual fruit should be the basic unit in the roguing operation. Further, check the interior quality by cutting the fruit at the time of seed extraction.

Harvesting and Extraction of Seed

Fruits are ready to harvest when the point of attachment with the stem shows a wide crack (full ripe stage) and the seed is practically mature. It is advisable to wait until sufficient fruits ripen before harvesting. Fruits, after picking, are placed in piles. They are cut open by hand with a knife and the seed and pulp are scraped into a container. Later the seeds are separated from the pulp either by fermentation, followed by washing or by washing alone. The seeds, after washing in water, are dried quickly in the sun.

Seed Yield

200 to 300 kg/ha.

Varieties/Hybrids Characteristics

Varieties/Hybrids Recommended by IARI, New Delhi

Pusa Madhuras

Fruit is roundish flat. The skin is pale green, sparsely netted with dark green stripes and salmon-orange flesh. Yield potential is 150 q/ha.

Pusa Rasraj

This variety has been developed by crossing of monoecious lines M-3 and Durgapur Madhu. Fruits are long with ridges on the skin. An average yield is 250 q/ha.

Pusa Sharbati

It is developed from the cross between Kutana and PM Resistant No. 6 of the USA. It is an early cultivar. Fruits are round having netted skin. Flesh is salmon orange, firm, thick. Yield potential is 150 q/ha.

Varieties/Hybrids Recommended by Karnataka State

Arka Jeet

It is a selection from 'Bati' strain of UP and released from IIHR, Bangalore. Fruits are small, round with flat ends. Skin is smooth and orange coloured. Flesh is white with medium soft and excellent flavours. The yield is 150 q/ha.

Arka Rajhans

It is selection from a local collection (IIHR-107) from Rajasthan and released by IIHR, Bangalore. It is a mid-season variety bearing large oval fruits. Flesh is thick, firm, white and sweet. This variety possesses good keeping. It is resistant to powdery mildew. The yield potential is 285 q/ha.

Durgapura Madhu

It is a selection from local material of Rajasthan. The fruits are oblong, and pale green rind. Flesh is very sweet, light green with dry texture.

Varieties/Hybrids Recommended by Punjab State

Hara Madhu

This is an inbred selection from the local material of UP, released by PAU, Ludhiana. Fruits are round, tapering toward the stalk-end. Rind is thin, smooth and pale yellow. Flesh is green, thick, and very juicy. It yields 125 q/ha.

Punjab Hybrid

This is an F_1 hybrid and released from PAU, Ludhiana. Its parents are a male sterile line (MS-1) and Hara Madhu. Fruits are globular with distinct sutures. Rind is quite thick, netted and creamy yellow. Flesh is thick, orange and juicy. It gives yield of 160 q/ha. It is moderately resistant to powdery mildew and fruit fly.

Punjab Sunehri

It is a derivative from the cross Hara Madhu x Edisto and released by the PAU, Ludhiana. Fruits are round to elliptical and devoid of distinct sutures at maturity. Its rind is thick, netted and dull yellow. Its flesh is thick, orange, fairly juicy. It has a long post harvest life and has excellent transportability. Its average yield is 160 q/ha.

Watermelon (*Citrullus lanatus* Thunb) 2n = 2x = 22

Land Requirement

The land should be free of volunteer plants. The soil of selected fields should be well drained and aerated.

Figure 79: Watermelon

Field Inspection

A minimum of three inspections should be done, the first before flowering, the second during flowering and fruiting stage and the third at mature fruit stage and prior to harvesting.

Field Standards

General Requirements

Isolation

Seed fields should be isolated from the contaminants shown in column 1 of the Table below by the distances specified in columns 2 and 3 of the said Table:

Contaminants	Minimum Distance (Meters)	
	Foundation	Certified
1	2	3
Fields of other varieties	1000	500
Fields of the same variety not conforming to varietal purity requirements for certification and from wild watermelon: *Citrullus colocynthis* L.	1000	500

Specific Requirements

Factor	Maximum Permitted (%)*	
	Foundation	Certified
Off types	0.10	0.20
** Objectionable weed plants	None	None

* Maximum permitted at and after flowering.

** Objectionable weed shall be: Wild watermelon): *Citrullus colocynthis* L.

Seed Standards

Factors	Standards for Each Class	
	Foundation	Certified
Pure seed (minimum)	98.0 per cent	98.0 per cent
Inert matter (maximum)	2.0 per cent	2.0 per cent
Other crop seeds (maximum)	None	None
Total Weed seeds (maximum)	None	None
Objectionable weed seeds (maximum)	None	None
Other distinguishable varieties (maximum)	5/kg	10/kg
Germination (minimum)	60 per cent	60 per cent
Moisture (maximum)	7.0 per cent	7.0 per cent
For vapour-proof containers (maximum)	6.0 per cent	6.0 per cent

Brief Cultural Practices

Time of Sowing

February to March.

Seed Rate

2.5-5 kg/ha.

Method of Sowing

Seeds are sown on both sides of the channels. The channels are irrigated and the seeds are sown on the ridges at a place above water level, so that capillary movement receives water. Two-three seeds may be sown on each hill, and subsequently thinned out.

Spacing

Two meters between channels and one meter between plants.

Fertilizers

Apply 250 to 300 q of farmyard manure per hectare in the soil at the time of preparation. A mixture of 50 kg nitrogen, 50 kg phosphorus, and 50 kg potash per hectare should be prepared and mixed at 500 g/pit. Top dress 50 kg nitrogen at the time of flowering.

Irrigation

Irrigate the furrows lightly soon after sowing, for good germination. A good moisture supply is essential for proper development of fruits. Irrigate once in 10 days during spring and every week during the hot months. When the plants have fully grown and the fruits are developed, a few more irrigations would be required for the rest of the season.

Interculture

Weeding and hoeing around the bases of the plants should be thoroughly attended to.

Insects and Diseases

Insects/Diseases	*Control Measures*
	Insects
Red pumpkin beetle (*Aulacophora foveicollis*)	Spraying of carbaryl 0.2 per cent or rogor 0.1 per cent.

Contd...

Insects/Diseases	*Control Measures*
Aphids (*Aphis spp.*)	Spraying malathion or metasystox @ 0.2 per cent.
Fruit fly (*Dacus spp.)*	Using light traps and poison baits. Spray of endosulfan @ 6 ml/ 4.5 litres of water.
	Diseases
Downy mildew and Powdery mildew	Ridomil (0.2 per cent) and blitox (0.2 per cent). Seed treatment with dithane M-45 (0.3 per cent).
Anthracnose (*Colletrotrichum lagenarianum*)	Seed treatment with diathane M-45 (2 g/kg seed). Spraying of blitox (0.125 per cent).
Watermelon mosaic (Vector: Aphids)	Spray with dimecron (0.05 per cent), or Mineral oil, krishi oil @ 1.5 per cent.

Roguing

Whatever has been said for cucumber and muskmelon applies equally well to watermelons. Such characters as fruit shape and exterior colour may often be determined relatively early. Prompt roguing of off type plants will reduce to some extent the contamination of surrounding plants by off type pollen. Again, the plant should be considered as a unit in the roguing operation. All the fruits on a plant should be considered as a group rather than choosen individually, without relation to plant.

Harvesting and Extraction of Seed

Fruit are ready for harvest when they reach edible maturity. However, there is no harm in delaying the harvesting for a few days, or even a week, or longer, unless it is desired to examine flesh quality and flavour. Only men familiar with plant and fruit types should do the fruit picking, so that any previously unrecognised rogue may be removed. The fruits are placed in piles after picking. The fruits are cut open individually by hand, seed scraped out and washed over a screen with clean water. Seed must be dried quickly after washing, and stored in a cool dry place.

Seed Yield

Average seed yield varies from 200 to 250 kg/ha.

Varieties/Hybrids Characteristics

Varieties/Hybrids Recommended by IARI, New Delhi

Sugar Baby

It is an early season American introduction which is released by IARI, New Delhi. The fruits are slightly smaller weighing 3 to 5 kg, round in shape with bluish black rind. Flesh is pink. This is a small seeded variety. It yields 150 q/ha.

Varieties/Hybrids Recommended by Karnataka State

Arka Jyoti

It is a F1 hybrid developed by crossing a local selection (IIHR-20) from Rajasthan with Crimson Sweet from USA and released by IIHR, Banglore. Fruit is oval, dark green with deep blue angular stripes, flesh bright crimson with good granular texture. The yield potental is 840 q/ha.

Arka Manik

Evolved by crossing IIHR-21 with cv. 'Crimson Sweet'and released by IIHR, Bangalore. Fruits are round to oval with green rind and dark green stripes, flesh deep crimson with granular texture. The yield is 600 q/ha. This variety is resistant to powdery mildew.

Varieties/Hybrids Recommended by Rajasthan State

Durgapura Kesar

It is a late variety released by ARS, Durgapura, Rajasthan. Fruit weight is 4-5 kg, green skin with stripes. The flesh is yellow and moderately sweet. Seeds are large and yellow.

Durgapura Meetha

It is a late maturing (125 days) variety released from ARS, Durgapura, Rajasthan. Fruits are round with light green having average fruit weight of 6 to 8 kg. Rind is thick with good keeping quality.

Bitter Gourd (*Momordica charantia* Linn.) 2n = 2x = 22

Land Requirements

Land requirements are sandy-to-sandy loam, but the land should be free of volunteer plants. The soil of selected fields should be well drained.

Figure 80: Bitter Gourd

Field Inspection

A minimum of three inspections should be done, the first before flowering, the second during flowering and fruiting stage, and the third at the mature fruit stage and prior to harvesting.

Field Standards

General Requirements

Isolation

Seed fields should be isolated from the contaminants shown in column 1 of the Table below by the distances specified in columns 2 and 3 of the said Table:

Contaminants	*Minimum Distance (Meters)*	
	Foundation	*Certified*
1	*2*	*3*
Fields of other varieties	1000	500
Fields of the same variety not conforming to varietal purity requirements for certification and from balsam apple *(mokha): Momordica balsamina* L.; *Bhat karela: (kakrol): M. cochinchinensis*; *Jangli karela: M. dioica.*	1000	500

Specific Requirements

Factor	*Maximum Permitted (%)**	
	Foundation	*Certified*
Off types	0.10	0.20
** Objectionable weed plants	None	None

* Maximum permitted at and after flowering.

** Objectionable weeds shall be: Balsam apple *(mokha)*, Bhat karela *(kakrol)*.

Seed Standards

Factors	*Standards for Each Class*	
	Foundation	*Certified*
Pure seed (minimum)	98.0 per cent	98.0 per cent
Inert matter (maximum)	2.0 per cent	2.0 per cent
Other crop seeds (maximum)	None	None
Total weed seeds (maximum)	None	None
Objectionable weed seeds (maximum)	None	None
Other distinguishable varieties (maximum)	5/kg	10/kg
Germination (minimum)	60 per cent	60 per cent
Moisture (maximum)	7.0 per cent	7.0 per cent
For vapour-proof containers (maximum)	6.0 per cent	6.0 per cent

Brief Cultural Practices

Time of Sowing

February to March.

Preparation of Land

Prepare the field to a good tilth by one deep ploughing, two to three harrowings followed by levelling.

Source of Seeds

Obtain nucleus/breeders/foundation seed from source approved by a seed certification agency.

Seed Rate

5 kg/ha.

Methods of Sowing

Seeds are sown either on one side or both sides of the channels. The channels are irrigated and the seeds are sown on the ridges at a place above water level, so that capillary movement receives water. In case of early sowings, sprouted seeds are sown on the ridges. Two–three seeds may be sown on each hill, and subsequently thinned out.

Planting Distance

Two meters between channels and one meter between plants.

Fertilizers

Farmyard manure @ 200-250 q/ha should be applied 15-20 days before sowing seeds in pits/furrows. Fertilizer recommendation in the ratio of 100:50:50 kg nitrogen, phosphorous and potash/ha respectively. Half quantity of nitrogen and full quantity of each phosphorus and potash should be applied in the pits or furrows at the time of sowing. Remaining quantity of nitrogen may apply 30 days after sowing or just before flowering.

Irrigation

The summer crop requires irrigation every third or fourth day. The number of irrigations for the rainy season would depend upon the number of rainy days.

Interculture

The beds should be kept free of weeds, especially in the early stages. Later on, rapidly spreading vines smother the weeds. Frequent hoeings in the early stages are also desirable.

Insects and Diseases

Insects/Diseases	*Control Measures*
	Insects
Aphids (*Aphis spp.*)	Spraying malathion 0.1 per cent.
Red pumpkin beetle (*Aulacophora foveicollis*)	Spraying of carbaryl 0.1–0.2 per cent or rogor 0.1 per cent.

Contd...

Insects/Diseases	*Control Measures*
Aphids (*Aphis spp.*)	Spraying malathion 0.1 per cent.
Fruit fly (*Dacus cucurbitae, Dacus dorsalis)*	Using light traps and poison baits. Spray of endosulfan @ 6 ml per 4.5 litres of water.
	Diseases
Downy mildew and Powdery mildew	Spray Ridomil (0.2 per cent). Seed treatment with dithane M-45 (0.3 per cent).
Anthracnose (*Colletrotrichum lagenarianum*)	Seed treatment with diathane M-45 (2 g/kg seed). Spraying of difolatan (0.2 per cent).
Bitter gourd mosaic (Vector: Aphids)	Spray the crop with nuvacron (0.05 per cent). Mineral oil, krishi oil @ 1.5 per cent
Bitter gourd witches broom	Application of furadan @ 1.5 kg a.i./ha at the time of sowing followed by 5-6 foliar spray of either nuvacron (0.05 per cent).

Roguing

Off type plants are usually detectable in early stage of growth. Bush type plants in a variety, which normally has runners, or *vice versa*, is easilly detected. Although some damage may have already resulted from cross-pollination, such off type plants should be immediately rogued.

Harvesting and Extraction of Seed

Seeds are ready to harvest when the fruits fully rippened and seeds in side the shell break readily from the pulp. Immature seeds adhere to pulp. Fruits are picked from plants and piled. The fruits are cut in half and the seeds scooped out and the contents dried in the sun. The seeds are removed from the flesh by rubbing over a sieve, and finally winnowed to remove light matter. Seeds may be washed, if necessary, and dried in trays and stored in a cool and dry place.

Seed Yield

300 to 500 kg/ha.

Varieties/Hybrids Characteristics

Varieties/Hybrids Recommended by IARI, New Delhi

Pusa Do Mausami

It is a selection from local collection, suitable for spring-summer and rainy seasons and released from IARI, New Delhi. Fruits are dark green, long, medium thick, 7-8 continuous ridges, and having 8-10 fruits in one kg.

Pusa Hybrid-1

This hybrid has been developed and released from IARI, New Delhi. The fruits are attractively green, medium-long and medium thick and an average fruit weight of 100 g. The fruits are suitable for making vegetable, pickles and also for dehydration purposes. Its average yield is 218 q/ha.

Pusa Vishesh

It is a dwarf vine variety suitable for higher plant density. Fruits are glossy green, medium long and thick, suitable as vegetable for pickling and dehydration. It is released by IARI, New Delhi.

Varieties/Hybrids Recommended by Uttar Pradesh State

BTH-165

Promising hybrid by cross combination KB-1 x VRBT-36 developed at IIVR, Varanasi. Fruits are long green with attractive colour. It gives an average yield of 1.84 kg/plant and sown tolerance against powdery and downy mildews.

BTH-7

This hybrid is cross combination between VRBT-3A and VRBT-82-1 and developed at IIVR, Varanasi. It bears long attractive green fruits of 20-25 cm length. The plant having field tolerance to powdery and downy mildews. It gives an average yield of 300 q/ha.

Kalyanpur Baramasi

This variety has been develped from Vegetable Research Centre, Kalyanpur (Kanpur). Fruits are 20-25 cm long, green with pointed on tips. An average yield of this variety is 100-125 q/ha.

Varieties/Hybrids Recommended by Tamil Nadu State

CO-1

It is a selection from a local type collected from Thudiyalur and released from TNAU, Coimbatore. The fruit is dark green and medium long (25-30 cm). The yield potential of this variety is 144 q/ha.

COBGH-1

It is a cross of MC-84 and MDU-1 and developed at TNAU, Coimbatore. Fruits are light green to creamy white. It gives an average yield of 52.39 t/ha.

Varieties/Hybrids Recommended by Other States

Arka Harit

It is a selection from Rajasthan collection and released by IIHR, Banglore. Fruits are spindles shaped attractive, glossy green with smooth regular ribs and thick flesh. Yield is 130 q/ha in 120 days.

MDU-1

It is an induced mutant developed by gamma irradiation of local cultivar MC 103. The fruits are long with a mean length of 40.34 cm and a girth of 17.54 cm. It yields 321.9 q/ha. Fruits also contain less seed.

Preethi

It is a high yielding variety developed at the College of Horticulture, Vellanikkara, Kerala. Fruits are white, spiny, medium and gives an average yield of 14-15 t/ha.

Punjab-14

This variety has been developed through selection from local material and released by the PAU, Ludhiana. It has smaller vines. Fruit large, oblong and green in colour. It yield is 125 q/ha. It is suitable for spring summer and rainy seasons.

Family–Compositae

Lettuce (*Lactuca sativa*) 2n = 2x = 18

Land Requirements

Land to be used for seed production should be free of volunteer plants. The soil of selected fields should be rich in organic matter.

Figure 81: Lettuce

Field Inspection

A minimum of three inspections should be made, the first before heads formed in heading types, and before full grown stage in non heading types, the second when heads have formed in heading types, and at full grown stage in non heading types and the third at flowering stage.

Field Standards

General Requirements

Isolation

Lettuce seed fields shall be isolated from the contaminants shown in column 1 of the Table below by the distances specified in columns 2 and 3 of the said Table:

Contaminants	*Minimum Distance (Meters)*	
	Foundation	*Certified*
1	*2*	*3*
Fields of other varieties	50	25
Fields of the same variety not conforming to varietal purity requirements for certification and wild lettuce (*Lactuca scariola* L).	50	25

Specific Requirements

Factor	*Maximum Permitted (%)**	
	Foundation	*Certified*
Off types	0.10	0.20
**Objectionable weed plants	0.010	0.200
*** Plants affected by seed borne diseases	0.10	0.50

* Maximum permitted at the final inspection.

** Objectionable weed shall be: wild lettuce (*Lactuca scariola* L).

*** Seed borne diseases shall be: Lettuce mosaic virus.

Seed Standards

Factors	*Standards for Each Class*	
	Foundation	*Certified*
Pure seed (minimum)	98.0 per cent	97.0 per cent
Inert matter (maximum)	2.0 per cent	2.0 per cent None
Other crop seeds (maximum)	None	10/kg
Total weed seeds (maximum)	5/kg	5/kg
Objectionable weed seeds (maximum)	2/kg	20/kg
Other distinguishable varieties (maximum)	10/kg	70per cent
Germination (minimum)	70 per cent	8 per cent
Moisture (maximum)	8.0 per cent	6.0 per cent
For vapour-proof containers (maximum)	6.0 per cent	

Brief Cultural Practices

Time of Sowing

For seed crops, sowing should be done in October.

Preparation of Field

Prepare the field to fine tilth by ploughing, three to four harrowings followed by levelling.

Source of Seeds

Obtain nucleus/breeders/foundation seed from source approved by a seed certification agency.

Seed Rate

0.5-0.75 kg/ha.

Method of Sowing

The seeds may be sown in raised nursery beds in rows 3-4 cm apart, and seedlings later transplanted at a distance of 30-45 cm x 20-30 cm, when they are 5-6 weeks old.

Fertilizers

Apply well-rotted farmyard manure @ 200-250 q/ha and 60kg/ha phosphorus at the time of preparation of land. Top dressing of 60 kg /ha nitrogen in two splits (early spring, and in May when the stalks emerge) is essential for good seed yield.

Irrigation

Water should be supplied to keep uniform moisture conditions in the soil. Drainage is as important for removing excess water from the field because of excess soil moisture may cause rotting.

Interculture

Frequent hoeing and weeding is necessary for proper aeration of soil and to keep down the weeds.

Tearing or Making Cut in Head Types

In the head types, a slight cut, or tearing the upper leaves by hand, helps the seed stalk to emerge easily.

Insects and Diseases

Insects/Diseases	*Control Measures*
	Insects
Cabbage looper, Aphids (*Aphis gossypii*, *Myzus persicae* and *Lipaphis erysimi*)	Spray malathion 0.1 per cent.
	Diseases
Damping off and Root rot (*Rhizoctonia solani* and *Phythium ultimum*)	Seed treatment with 2 per cent ceresan. Seedling should be sprayed with 0.2 per cent difoltan.

Contd...

Insects/Diseases	*Control Measures*
Downy mildew (*Bremia lactucae*)	Spraying of 0.2 per cent dithane Z-78.
Lettuce drop (*Sclerotinia sclerotiorum* and *S. minor*)	Soil sterilization with steam air mixture at 70-80 °C. Single spray of benomyl (3-4 kg/ha).
Lettuce mosaic (Vector: aphid)	Remove all affected plants from the field.

Roguing

Roguing should be completed before bolting starts in loose varieties, or when heading is complete in heading varieties. Direct bolters, ill formed plants, off-type plants, diseased plants affected by mosaic, yellow mosaic and wild lettuce plants should in variably be removed time to time as required.

Harvesting and Threshing

Lettuce seeds ripen unevenly, and fully mature seed shatter readily. It is advisable to harvest when 30 to 50 per cent of the seeds

Figure 82: Staked Plants

Figure 83: Shattering Seeds

in the heads show white pappus or fluff and a good number of heads after opening have turned brown or dark brown but have not shown out the pappus. The plants are cut by hand and left to dry on the ground for a few days. Latter the seeds are collected by gently shaking them.

Seed Yield

The leafy types usually produce 500 to 600 kg seed per hectare. However, the variety, Great Lakes produces only 100 to 125 kg per hectare.

Varieties/Hybrids Characteristics

Varieties/Hybrids Recommended by IARI, New Delhi

Great Lakes

It is resistant to " tip burn" and marginal burning of outer leaves. The heads are hard and slow to bolt, leaves are light green. The outer leaves are blistered.

Chines Yellow

The leaves are light green in colour, tender and crisp. It is highly productive, early in maturity and produces white seed.

Slow Bolt

It is better variety for kitchen gardens since it is very slow in bolting. The leaves are bread, frilled, slight yellowish green in colour.

Other Varieties/Hybrids

Imperial 859

It can be grown at high temperature. The head are medium to large in size, compact and well covered by heavily blistered outer leaves.

White Boston

It has soft solid heads, inner leave have oily, soft, greasy texture.

Dark Green

It has narrow leaves with elongated upright heads.

Chapter 10

Quality of Seeds

Seed is called the carrier of technology as it carries the total genetic potential of the plant it is going to produce. The expression of this genetic potential is influenced by the environment and the genotype x environment interaction. If the environment were optimized, the full expression of the genetic potential of the seed (or crop) can be released. Improved or quality seed is the cheapest input in crop production. It is the first link in crop production around which revolves the success of all other factors used for crop production. Seed is the single well recognized carrier of production technology facilitating our quest for higher and better crop yields.

Trueness to Variety (Genetic Purity)

To achieve maximum potential of an improved variety with respect to high yield, resistance to disease, pest or any desired characteristics, it is necessary that the seed must be free from genetic contaminants.

Methods of Testing for Genetic Purity

The following methods of testing are descriebd for genetic purity of seeds.

Morphological Test

Examination of morphological features of seeds, such as, length, width, thickness, shape, weight, helium colour, seed coat colour, etc., and comparing them with those of the authentic sample. The seeds may be examined with naked eye or magnified with a hand lens. The seeds may be examined with rapidly comparing length, width, thickness etc. of seed samples at a very reasonable cost. The structure of seed coat surface may be examined and compared with that of the authentic sample with the help of a scanning electron microscope. But the equipment is expensive and not suited for routine analysis.

KOH-Bleach Test

Seeds are immersed in a 1:5 (wt/vol.) solution of potassium hydroxide (KOH) and fresh household bleach (5.25 per cent NaOCl) for 10 (in case of seeds with brown coats) or 5 minutes (in case of seeds with white seed coat); the mixture is stirred gently periodically. Seeds are rinsed with tap water, dried on paper towel and observed for the presence of dark pigmented tannic acid and light pigmented testa.

NaOH Test

This test is used to distinguish red and white coloured seeds that have been treated with a fungicide or weather damaged. The seeds are immersed in a 5 per cent sodium hydroxide (NaOH) solution and observed for 60 minutes. Red seeds become bright yellow (just after immersion), then turn to dark orange (after 15 minutes) and finally become bright yellow on immersion; the colour fades to straw colour (after 10 minutes) and remain as such.

Peroxidase Test

Seed coats of individual seeds are removed and placed in separate test tubes. To each test tube the following reagents are added: 10 drops of 0.5 per cent guaiacol; after 10 minutes, one drop 0.1 per cent H_2O_2; 1 minute later, reddish brown colour of solution indicates the seed coat to be peroxidase positive, while a colourless solution reveals it to be peroxidase negative.

Phenol Test

Seeds are soaked in water for 16 hr. at 20 + 1°C, placed in a Petri dish lined with filter paper soaked with 1 per cent phenol solution and kept at 30 + 1°C, and the reaction noted after 4 hrs Phenol vapour reacts chemically (not enzymatically) with seeds to produce black, dark brown, brown, light brown colours or there may be no reaction. For a better resolution, the colour development may be observed every 30 minutes.

Grow Out Test

The seed sample is sown in a controlled condition with authentic sample. Genetic purity is determined on the basis of observations made on plant morphological characters with reference to authentic sample. Its genetic purity has to be worked out both at seed and plant level then the size of submitted sample is to be doubled and genetic purity is reported in percentage. The procedures for grow out test are as given below.

1. While raising the desired population, standard and recommended agronomic/cultural practices (*e.g.*, field preparation, size of the plot, row length, distance between plants, irrigation, fertilization etc.) in respect of individual crops are to be followed both for the unknown sample and its control.
2. The possibility to prove the genuineness of a cultivars by grow out test is based on heredity characteristics of the plants. Usually, the cultivars differences are more distinct if growth conditions are favourable. Crop should be so grown that the genetically differences express themselves as clearly as possible. In self-fertilizing species, the individual of cultivars may be theoretically identical, whereas, the individual of cultivars in cross-pollinated fertilizing species may not be genetically similar, but comprises a number of types. Therefore, it is easier to determine the cultivars purity in self-fertilizing species than in cross-fertilizing species where the examination for greater part is based on the mutual comparison between the samples to be tested and the standard sample. Hence, it is essential to sow the various samples of the same cultivars in succession and standard samples are sown at suitable intervals (for example one standard sample for every ten samples to be tested). The size of plots, row length etc. will differ crop to crop.

3. The seed rate may be adjusted depending on the germination percentage of individual samples and the sowing may be done by dibbling. Subsequent thinning is not recommended.
4. The test crop may be raised along with the control either in the areas recommended for the variety or in off-season nurseries. The authentic control sample from the originating plant breeder/breeding Institute is to be maintained by the testing station /agency following standard procedures. A minimum of two hundred plants from control sample will be raised along with the test crop.
5. All plants are to be studied keeping in view the distinguishing characters described for the cultivars both in the test crop as well as the control. Necessary corrections may be incorporated if the control is found to be heterogeneous.
6. The specification of the field plot and row length are determined. The number of plants required for taking observation is dependended on maximum permissible off types which are given in Table 1.

Table 1

Maximum Permissible Off Types (%)	*Minimum Genetic Purity (%)*
0.10	99.9
0.20	99.8
0.30	99.7
0.50	99.5
1.00 and above	99.0 and below

7. The analysis employed for conduction 'grow out test' should possess the basic qualification as identified under Seeds Rules, 1968. These are given in Table 2.

Greenhouse/Growth Chamber Test

The variety/crop is grown under controlled environmental condition in a specified growing medium, which helps in better development of distinct morphological features at a particular stage of development. Though this method is less time consuming, it is based on limited number of plants and also expensive (Payne, 1993).

Table 2

Standard	*Reject Numbers for Sample Sizes of*	
	800	*400*
99.5 (1 in 200)	8	*
99.0 (1 in 100)	16	8
95.0 (5 in 100)	48	24
90.0 (10 in 100)	88	44
85.0 (15 in 100)	128	64

* Indicates that the sample size is too small for a valid test.

Application of Molecular Markers-Proteins/Isozymes/DNA Based Markers

There are several markers/descriptors being used for identification of varieties and parental lines of hybrids on the basis of morphology, proteins and isozymes and recent techniques based on DNA polymorphism. As to overcome the innate constraints associated with certain technologies and to meet the day–to–day challenge of identifying and cataloguing newer and essentially derived varieties (EDVS), a research for better marker systems to establish the uniqueness of a particular variety is on. A suitable alternative for assessing genetic purity (Smith and Register, 1998) could be, PAGE/SDS PAGE of protein/isozymes, IEF/Ultra thin layer IEF of proteins and Electrophoresis of DNA based marker-RFLP/RAPD/AFLP/Microsatellites/minisatellite etc.

Sample Size for Genetic Purity Testing

A key step in the seed production process is the design of sampling and testing procedure used to evaluate seed lot purity. However, due to uncertainties in such methods, there is always a risk of incorrectly rejecting or accepting a seed lot. To manage this misclassification risk, especially with regard to the presence or absence of transgenic traits, certain sampling plan and formulae for determining the sample sizes necessary to reduce the size to the lowest possible and to control these miss classification errors when accepting or rejecting seed lots are to be worked out. In this regard, sequential sampling procedures would offer satisfactory solutions. Conventionally, for testing genetic purity of hybrid seed lot, for assessing one per cent off type plants, maximum permissible limit, at least 400 plants need to be tested under field plot grow out test (GOT). That case, an equal number of seeds (*i.e.* 400) therefore, need to be tested individually in electrophoresis analysis. However, for total replacement or supplementing fields grow–out test by electrophoretic analysis, such huge sample size will be a major constraint. The cost involved in electrophoretic analysis of 400 individual seed would be a limiting factor. However, sequential analysis technique may be followed for reducing the sample size to a batch of 20 or more, which would mean up to 95 per cent saving in work load, if specification is predetermined. Various research workers have worked out on the smallest individual seed samples for assessing genetic purity in different crops *viz;* 10 to 20 seeds in pepper, 25 to 30 in Chinese cabbage, 40 in tomato and 400 seeds in chicory have been reported to provide reliable assessment of genetic purity (Rao *et al;* 2004).

Table 3: Biochemical Markers Used in Identification of Parental Lines in Vegetable Crops

Crop	*Marker Systems and Purpose*
Tomato	ESTand ADH (Alcohol dehydrogenese).
Pepper	Total seed protein polymorphism (12 bands with 22-98 Kda and POX.
Radish and cabbage	EST, GOT (Glutamate oxalo-acetate trensaminase), PGM.
Pea	Protein poly morphism gave 70-96 per cent similarity among cultivars.
Beans	EST, GOT, 6-PGD.
Chinese cabbage	PGM, POX, EST, 6-PGD
Muskmelon	Perosidase, Shikimate dehydrogenase, Acid phosphatase.
Lettuce	Protein polymorphism.

Source: Rao *et al.* (2004).

Physical Purity

Physical purity of a seed lot refers to the physical composition of seed lots. A seed lot is composed of pure seeds-the seeds of same kind, inert matter, broken grains less than half in size, soil and dust particles, chaff etc., weed seeds and other crop seeds. Higher the content of pure seed the better would be the seed quality. The test is made upon a seed sample of prescribed minimum size, which is representative of the lot. The quantity of seed prescribed for purity seeds is based on a weight estimated to contain atleast 2500 seeds subject to a minimum of 0.5 g and a maximum of 1000g. In purity test the sample is analysed into three components *i.e.* pure seed, other seeds and inert matter.

Table 4: Physical Purity of Some Vegetable Seeds

Sl.No.	*Crop*	*Physical Purity (%)*
1.	Gourds	99
2.	Cabbage, Knol khol	98
3.	Onion	98
4.	Brinjal, tomato	98
5.	Chilli	98
6.	Bhendi (Lady's finger)	99
7.	Cauliflower	98
8	Methi	96
9.	Palak	98
10.	Carrot	95
11.	Radish	98

Viability (Germination)

Seed germination refers to the ability of a seed when planted under normal sowing conditions to give rise to a normal seedling. This is the most important attribute because unless the seed is alive and germinable, other factors become meaningless. It indicates the stand and plant producing potential of crop seed. Usually 400 seeds are tested in replicates of 100. The seeds are planted under prescribed conditions favourable for germination and early growth, *i.e.* provided with adequate moisture, a favourable temperature and a suitable sub stratum or other special requirement. Not only are the conditions of the test prescribed, but also the duration of the test is prescribed in days. Usually the germination test is evaluated at least two times (or more) during the test period. One of these is a final evaluation at the termination of the test. Germination test are evaluated as normal seedlings, abnormal seedlings, dead seeds and hard seeds.

Table 5: Seed Weight, Number of Seed, Germination Percentage, Temperature Required and Time Taken for Germination of Different Vegetables

Vegetable	*Seed*		*Germination*		
	Weight (g/1000 Nos.)	*Number/g*	*Per cent*	*Temperature Required (°C)*	*Time Taken (Days)*
Okra	50	18-20	65	20-30	4-7
Bitter gourd	250	3-5	60	25-30	4-7
Bottle gourd	240	3-6	60	25-30	4-7
Brinjal	4	176-229	70	25-30	3-7
Cabbage	3.69	212-300	70	8-25	3-7
Carrot	0.80	564-1164	60	8-18	3-10
Cauliflower	3.40	283-564	65	8-25	6-21
Capsicum	6.6	106-159	60	18-30	6-14
Chilli	6.6	106-159	60	18-30	6-14
Cucumber	36.1	35-63	60	25-30	3-7
Pea	150	2-6	75	4-18	5-8
Muskmelon	25	14-42	60	25-30	4-7
Methi	7.6	120-150	70	10-20	3-7
Onion	3.1	247-459	70	11-30	6-14
Radish	10	74-159	70	11-30	4-7
Tomato	3.2	282-564	70	18-30	5-14
Watermelon	113	4-7	60	20-30	4-14

Vigour

It indicates the " degree of aliveness of seed". A seed may be viable but too low in vigour that it may not be able to emerge under field conditions. Seed vigour is related to the capacity of seed to germinate and establish seedlings under unfavourable field conditions and produce healthy and vigorous plants.

Moisture Content

The seed moisture is the most critical factor in the maintenance of seed germination and viability during storage. The seed must be dried to safe moisture content.

Mechanical Damage

Most of the seed damage is caused by mechanical processes in harvesting, threshing, processing and handling of the seed, which results in broken, chipped, cut or internally damaged seeds, as well as seeds with cracked coats. Mechanical damage reduces the planting values of the seed.

Size and Appearance

The seed should be fully developed, matured and plump and bold in size, colour and good appearance of seed has more acceptability of consumers. Generally seed with higher volume weight, has the high seed quality. Low volume weights are an indication of immature, insect damage, drough damage, frost damage and sterilty in the seed.

Seed Health

The health of seed refers to the presence or absence of disease organisms/insect pests on seeds. The quality of a seed lot very much depends on its health.

Planting Value

Planting value is the real worth of a seed lot for raising the crop. It is determined by calculating pure live seed percentages as follows:

$$\text{Pure Live Seed (P.L.S.)} = \frac{\text{Pure seed (\%)}}{100} \times \frac{\text{Germination (\%)}}{100} \times 100$$

Causes of Unsatisfactory Quality of Seeds

The main problem of a vegetable grower is the availability of good seeds. It is needless to say that vegetable seeds available in the country did not come up to the expectation of the growers. This was due to:

1. General ignorance on the part of the seed growers as to the proper scientific method of vegetable seed production, processing and storage.
2. Lack of proper organization and coordination of available resources in improving quality of available varieties.

Not only this, there has been a market deterioration in the available seed quality, because in the absence of really good seed the ignorant farmers were saving the seed from their commercial vegetable crop by simple selection. The new improved varieties developed by the research institutions could not find adequate place in the market because of lack of proper organization in multiplying them and at the same time maintaining their quality.

Sale of Seeds

1. Any institution, agency, company or individual wishing to do business in seeds must obtain necessary permission for buying and selling seeds. Without permission such business cannot be carried out.
2. License holder must keep all seeds notified by the Central Govt. for sale.
3. The seeds to be sold to farmers must satisfy the minimum quality, standards of physical purity and germination percentage fixed under the Seed Act.
4. The bags or packets containing seeds for sale must have specifications fixed for that particular seed. These labels are of different colours for different seeds for the convenience of buying farmers:
   (*a*) Breeder's seed–Yellow colour
   (*b*) Foundation seed–White colour
   (*c*) Certified seed–Blue colour
5. The labels must contain following information on them.
   (*a*) Seed–Name and variety
   (*b*) Lot number

(*c*) Physical purity (in per cent) and germination (in per cent)

(*d*) Date of testing

(*e*) Weight of the seed in the bag/packet

(*f*) Seed class/category.

(*g*) Expiry date/viability period

(*h*) Name and address of seed producer.

(*i*) Signature of the Officer of seed certification agency.

Care by Farmers

Farmers should take following care while purchasing the seeds.

1. They should buy required seed from the authorized License holders only.
2. They should obtain printed receipt from the dealer/shopkeeper for the seed purchased and payment made. The receipt should contain important specifications of the seed written on the label *viz.* Name of the buyer, name of the crop and variety, plot numbers, producer's name, signature of shopkeeper, dealer, etc.
3. While buying the seed they should see carefully all the entries on the label particularly germination, percentage, date of seed test, etc.
4. They should ensure that the label is duly signed by the officer of Seed Certification agency.
5. The bag should be opened from the bottom side keeping the stitches of the upper side and the label intact. Empty bags with tag and the printed receipt should be preserved for seasonal period.
6. If there are two or more bags of seed, the seed from them should be sown separately without mixing.
7. The seed should be sown when there is optimum moisture and the date of sowing should be recorded.
8. If there is any complaint about the quality of seed, the matter should be reported to Seed Inspector or to Agricultural Development Officer of Zilla Parishad of the

concerned district. There is a committee at each district, which actually visits the plots of complaining farmers, sees condition of the plot and verifies the receipts and the label or tag. If the committee feels that the seed was not of genuine quality and the complaint was real, then the suitable action is taken against both seed producer and the dealer or shopkeeper. If there is financial loss to the farmer, he can approach the District Consumer Forum for compensation.

Chapter 11

Seed Germination and Seed Dormancy

Germination Process

The initiation of germination requires that three conditions must be fulfilled:

1. The seed must be viable; *i.e.* the embryo must be alive and capable of germination.
2. The seed must be nondormant and the embryo quiescent. There must be no dormancy inducing physiological, physical, or chemical barrier to germination.
3. The seed must be subjected to the appropriate environmental conditions: available water, proper temperature, a supply of oxygen, and sometimes light. Because of the complex interactions between the environment and specific dormancy conditions in the seed, these environmental requirements may change with time and methods of seed handling. Thus, the second requirement, avoiding dormancy, may sometimes be met by providing the appropriate environmental condition.

Stages of Germination

Stage 1: Awakening or Activation

Imbibition of Water

Water is absorbed by the dry seed and the moisture content increases rapidly at first, then levels off. Initial absorption involves the imbibition of water by colloids of the dry seed, which softens the seed coverings and causes hydration of the protoplasm. The seed swells and the seed coats may break. Imbibition is a physical process and can take place even in dead seeds.

Synthesis of Enzymes

Enzymes activity begins very quickly after the start of germination as the seed becomes hydrated. Activation results in part from reactivation of stored enzymes previously formed during development of the embryo and in part from synthesis of new enzymes as germination starts.

Seed development and germination represent a biological system in which the metabolic machinery of the cell is turned "on" and "off" by control of the flow of genetic information from DNA in the cell. Simply stated the process involves two basic steps. One is the transcription of genetic instructions from the DNA to form specific messenger RNA molecules (mRNA). The second step is the translation of this information by means of another group of transfer RNA molecules (tRNA) to synthesize specific proteins. These proteins are enzymes that control the complex biochemical reactions involved in metabolism and growth. Energy for these processes is obtained from the high-energy phosphate bonds in adenosine triphosphate (ATP) present in mitochondria.

Synthesis of enzymes during germination then requires RNA molecules, which are present in the embryo axis. The concept has been advanced that the initiation of germination results from the activation of unique germination controlling mRNA molecules in which genetic information for germination was transcripted during seed development but the translation is delayed until imbibitions of the seed. Some of the ATP to provide energy was formed during seed development, preserved in the dormant seed, and then reactivated with cell hydration.

Cell Elongation and Emergence of the Radicle

The first visible evidence of germination is emergence of the radicle, which results from the elongation of cells rather than from cell division. In a nondormant seed emergence of the radicle can occur within a few hours or a few days after planting and is usually considered to mark the end of stage 1.

Stage 2: Digestion and Translocation

Fats, proteins and carbohydrates are stored in the endosperm, cotyledon, perisperm, or female gametophyte (conifers). The compounds are digested to simpler chemical substances, which are translocated to the growing points of the embryo axis. The metabolic patterns in seeds of different species differ with the type of chemical reserves in the seed. Fats and oils the major food constitutes in the seeds of higher plants are converted enzymatically to fatty acid and eventually the sugar. Storage proteins present in most seeds are a source of amino acid and nitrogen essential to the growing seedling. Starch present in many seed as an energy source, is converted to sugar. The metabolic patterns that occur during germination involved the activation of specific enzymes in the proper sequence and the regulation of their activity. Control may be exercised by various biochemical processes within cells and may depend on the presence of specific chemicals or hormones. Water uptake and respiration now continue at a steady rate. The existing cell systems have been activated and the protein synthesizing system is functioning to produce new enzymes, structural materials, regulator compounds and nucleic acid to carry on the cell functions and synthesize new materials.

Stage 3: Seedling Growth

In the third stage, development of the seedling plant results from continued cell division in the separate growing points of the embryo axis, followed by the expansion of the seedling structures. The initiation of the cell division in the growing points appears to be independent of the initiation of cell elongation. Once growth begins from the embryo axis, fresh weight and dry weight of the new seedling plant increase but total weight of storage tissue decreases. The respiration rate as measured by oxygen uptake, increases steadily with advanced in growth. Storage tissues of the seed eventually cease to be involved in metabolic activities except in plants of which the cotyledons emerge from the ground and become active in photosynthesis. Water absorption increases steadily as new roots explore the germination medium and the fresh weight of the seedling plant increases.

As germination proceeds, the structure of the seedling soon becomes evident. The embryo consists of an axis bearing one or more seed leaves, or cotyledons. The growing point of the root, the radicle, emerges from the base of the embryo axis. The growing point of the shoot, the plumule, is at the upper end of the embryo axis, above the cotyledons. The seedling stem is divided into the section below the cotyledons–the hypocotyls and the section above the cotyledons–the epicotyl. The initial growth of the seedling follows one of two patterns. In one type–epigeous germination–the hypocotyls elongates and raises the cotyledons above the ground. In the other type–hypogeous germination–the lengthening of the hypocotyls does not raise the cotyledons above the ground, and only the epicotyl emerges.

Dormancy

Various types of seed dormancy results from different germination controlling mechanisms with the seed. Classification has been developed to explain the biological mechanisms involved and suggest ways to overcome dormancy. The first system was formulated by Crocker, who described seven types of seed dormancy. More recently Nicolaeva has defined a system based predominantly on the physiological causes of dormancy. The following categories are based on both physiological and morphological differences, primarily following Nicolava's classification.

Physical Dormancy

Physical dormancy is characteristic of a large number of plant families in which the seed coats and sometimes hardened sections of other seed coverings are impermeable to water. These include Leguminoseae, Malvaceae, Cannaceae, Geraniaceae, Chenopodiaceae, Convolvulaceae, and Solanaceae. The embryo is quiescent but is sealed inside a water impermeable covering that can preserve the seed at low moisture contents for many years even at warm temperatures. Among cultivated crops hardseedness is found chiefly in the herbaceous legumes, including clover, alfalfa, etc., as well as many woody legumes.Hardseedness depends on the genetic nature of the species and cultivar, environmental conditions during seed maturation, and environmental conditions during seed storage. Drying at high temperatures during ripening will increase hardseedeness.

Impermeability of the seed coat is due to a layer of palisade–like macrosclereid cells, especially thick-walled on their outer surfaces and coated with a layer of waxy, cuticular substances. Disintegration of the "caps" of such cells, or mechanical stress separating the cells, may allow water to enter and produce germination. In some legume species, the point of attachment of the seed acts as a one-way valve during ripening by opening to allow water to escape in a dry atmosphere but closing in a moist atmosphere to prevent water uptake. In such species as *Albizia lophantha* a small opening (strophiole) near the hilum is sealed with a cork-like plug which can be dislodged with vigorous shaking or impaction or by exposure to dry heat as in a fire. In nature, seed coverings are softened by various agents of the environment, including mechanical abrasion, alternate freezing and thawing, attack by soil microorganisms, passage through the digestive tracts of birds or mammals, or fire. Softening of seed coverings by fungi and bacteria is most effective at temperatures above 10°C or when seeds are planted in a nonsterile medium.

Mechanical Dormancy

Refers to seed coverings that are too hard to allow the embryo to expand during germination. This factor probably is not the sole cause of dormancy in any species, but combines with other factors to delay germination. In general, once water is absorbed by the seed the expanding force of germination ruptures the seed coats and breaks apart any outer covering. The shells of walnuts and the pits of stone fruits or other hard fruit coverings can delay germination, however. The hard endocarp of peach seeds slows water absorption and delays the leaching of germination inhibitors out of the seed. It has been difficult to demonstrate in many species that such mechanical effects are a primary cause of dormancy. Any effects of mechanically resistant coats can be overcome by the same treatments used for impermeable seed coats in native or cultivated plants.

Chemical Dormancy

Chemicals that act as seed germination inhibitors have been extracted from various plant parts and identified. Such chemicals are produced and accumulate in the fruit as well as the seed coverings. Fleshly fruits or juices from them can strongly inhibit seed germination. This occurs in citrus, cucurbits, stone fruits apples, pears, grapes, and tomatoes. Likewise, dry fruits and fruit coverings such as the hulls of guayule, wheat as well as the capsules of mustard can inhibit germination. When the fruit coverings remain with the seed, such inhibition can persist into the germination period. Some of the substances associated with inhibition are various phenols, coumarin, and abscisic acid.

Morphological Dormancy

Morphological dormancy is found in plant families that characteristically have seeds in which the embryo is not fully developed at the time of ripening. Additional embryo growth is necessary after separation of the seed from the plant and prior to germination. As a general rule embryo enlargement is favored by warm temperatures, but the response may be complicated by the presence of other dormancy mechanisms. There are two groups within this category:

Rudimentary Embryo

Plants with rudimentary embryos produce seeds with little more than a proembryo embedded in a massive endosperm at the time of fruit maturation. This type occurs in various families, including some that are important as herbaceous flower crops, such as Ranunculaceae, Papaveraceae, and Araliaceae. Germination inhibition chemicals also occur in the endosperm, and become particularly active at high temperatures. Effective aids for inducing germination include: (*a*) exposure to temperatures of 15° or below; (*b*) exposure to alternating temperatures; (*c*) treatment with chemical additives such as potassium nitrate or gibberellic acid.

Underdeveloped Embryo

At fruit maturity some seeds have partially developed, torpedo-shaped embryos that may attain a size up to one half that of the seed cavity. Further embryo enlargement occurs prior to germination. Some important families and species in this category include Umbelliferae, Ericaceae, Primulaceae and Gentianaceae. Other conditions such as semipermeability of the inner seed coats and internal germination inhibitors, may be involved in freshly harvested seed. A temperature of about 20° is favorable for germination as are gibberellic acid treatments.

Internal Dormancy

Several categories include seeds in which dormancy is internally controlled within the living tissues of the seed. Two separate phenomena are involved in the internal control of germination. The first is the control exerted by the semipermeability of seed coverings. The second is a dormancy present within the embryo that is overcome by exposure to moist-chilling.

Treatments to Overcome Seed Dormancy

Scarification

It is any process of breaking, scratching, mechanically altering, or softening the seed coverings to make them permeable to water and gases.

Mechanical Scarification

Mechanical scarification is simple and effective with seeds of many species if suitable equipment is available. The seeds are dry after such treatment and may be stored or planted immediately by mechanical seeders. Scarified seeds are more susceptible to injury from pathogenic organisms, however, and will not store as well as comparable nonscarified seeds. Chipping hard seed coats by rubbing with sandpaper, cutting with a file, or cracking with a hammer or a vise are simple methods useful for small amounts of relatively large seed. For large–scale mechanical operations, special scarifiers are used. Small seeds of legumes, such as alfalfa and clover, are often treated in this manner to increase germination. Seeds may be tumble in drums lined with sandpaper or in concrete mixtures, combined with coarse sand or gravel. The sand or gravel should be of a different size than the seed to facilitate subsequent separation. Scarification should not proceed to the point at which the seeds are injured. To determine the optimum time, a test lot can be germinated, the seeds may be soaked to observe swelling, or the seed coats may be examined with a hand lens. The seed coats generally should be dull but not so deeply pitted or cracked as to expose the inner parts of the seed.

Hot Water Scarification

Drop the seeds into four times their volume of hot water 77° to 100 °C.The heat source is immediately removed, and the seed soaked in the gradually cooling water for 12 to 24 hours. Following this the unswollen seeds can be separated from the swollen ones by suitable screens and either retreated or subjected to some other treatment. The seeds should usually be planted immediately after the hot water treatment; some kinds of seed have been dried and stored for later planting without impairing the germination percentage, although the germination rate is reduced.

Acid Scarification

Dry seeds are placed in containers and covered with concentrated sulfuric acid in a ratio of about one part seed to two parts acid. The amount of seed treated at any one time should be restricted to no more than about 110 kg to avoid uncontrollable heating. Containers should be glass, earthenware, or wood–not metal or plastic. The mixture should be stirred cautiously at intervals during the treatment to produce uniform results and to prevent accumulation of the dark, resinous material from the seed coats *i.e.*, sometimes present. Since stirring tends to raise the temperature, vigorous agitation of the mixture should be avoided to prevent injury to the seeds. The time of treatment may very from as little as ten minutes for some species to six hours or more for other species. Since treatment time may vary with different seed lots, making a preliminary test on a small lot is recommended prior to treating to large lots. With thick-coated seeds that require long periods, drawing out samples at intervals and checking the thickness of the seed coat may follow the progress of the acid treatment. When it becomes paper thin, the treatment should be terminated immediately. At the end of the treatment period the acid is poured off, and the seeds are washed to remove the acid. Placing seeds in a large amount of water with a small amount of baking soda will neutralized any adhering acid; or the seeds can be washed for 10 minutes in running water. The acid-treated seeds can either be planted immediately when wet, or dried and stored for later planting. Large seeds of most legumes respond to the simple acid treatment, but variations are required for some species.

Warm Moist Scarification

Keeping seeds in a nonsterile, moist, warm medium for several months can soften seed coats through microorganism activity. This treatment can be provided by planting hard seeds in summer or early fall while soil temperatures are warm. Usually it is most effective for seeds having double dormancy where the warm moist scarification precedes cold stratification during the winter.

High Temperature Scarification

Seeds of certain species of native plants with hard seed coats germinate extensively after a forest of range fire. Seed coats are modified by the high temperatures. Closed cone species of pine also respond to high temperatures in fires by melting the enclosing resins that seal the cone. The action allows release of the seeds which are then able to germinate.

Harvesting Immature Fruits

Extracting seeds from immature fruits improves germination in certain tree species by avoiding the development of hard seed coats. Such seeds must be planted immediately without drying.

Stratification

Stratification is a method of handling dormant seeds in which the imbibed seeds are subjected to a period of chilling to after-ripen the embryo. The term originated because nursery men placed seeds in strartified layers interspersed with a moist medium, such as soil or sand, in out–of-doors pits during winter. The term moist-chilling has been used as a synonym for stratification.

Refrigerated Stratification

Dry seeds should be fully imbibed with water prior to refrigerated stratification. Twelve to 24 hours of soaking at warm temperatures may be sufficient for seeds without hard seed coats or coverings. Longer periods are required, with aeration, for seeds enclosed in hard endocarp or pericarp; soaking for three days to a week or more may be necessary. Leaching seeds in running water also can be used, or alternately soaking seeds for 12 hours, then draining for 12 hours. After soaking, seeds are usually mixed with a moisture-retaining medium for the stratification period. Almost any medium that holds moisture, provides aeration, and contains no toxic substances is suitable. These include well-washed sand, peat moss, chopped or screened with smaller parts discarded sphagnum moss, vermiculite, and composed sawdust. Fresh sawdust may contain toxic substances. A good material is a mixture of one part sand and one part peat moss, moistened and allowed to stand 24 hours before use. Any medium used should be moist but not so wet that can be squeezed out.

Where refrigerated storage is not available, storing outdoors may do stratification, eighter in pit several feet deep or in raised beds enclosed in wooden frames. Essentially the same preparation procedures are used but outdoor winter chilling and natural rainfall provide the required chilling temperature and moisture. However seed may be protected against freezing, drying, and rodents.

Technique of Storing Seed

Seed is viable and has a potential to develop into a new plant. When the seed is produced it goes into dormancy for some time. This dormancy has to be preserved till the seed is used in the next season, next year or after some years.

The problem of storing seeds in an appropriate manner to ensure proper germination in the next season is as old as agriculture itself. The discovery that drying of seeds led to fairly satisfactory preservation of quality from the consumption point of view must have led to the adoption of drying as a seed storage practice, on the assumption that good table quality automatically ensures good viability and vitality. This is not surprising since even the earlier scientific theories on seed viability postulated that the loss in viability was due to embryo starvation brought about by respiratory depletion of food reserves. The fact that non-viable seeds appear as rich in food reserves as the viable ones and that seeds with abundant food reserves deteriorate more rapidly than those with lesser amounts goes against the Embryo Starvation Theory. The exact mechanism responsible for deterioration in seed quality is even today not fully understood. A decrease in the enzymatic activity of glutamic acid decarboxylase and the mutagenic effects of the end products of respiration under adverse storage conditions (leading to aberrations in embryonic tissue and normal cell division) are considered to be the possible causes.

Though respiratory depletion of food reserves is not held to be the cause of seed quality deterioration, the theories advanced for seeds becoming non-viable are related to the respiratory process as the primary cause. Again, the secondary effects of respiration appear to become operational under adverse conditions of storage and after a considerable lapse of time. The requirements for satisfactory preservation of seed quality, therefore, appear to be those that would result in a marked reduction in seed respiration and its maintenance at very low rates over long periods, say about 2 years. This is in view of the need to use the seeds in the next season and to keep a sufficient stock, in case of total failure of seed crop in the ensuing season. The respiration rate is very greatly influenced by the moisture content of seeds and the temperature conditions.

Seeds dried to a moisture content at 4 per cent level show negligible respiratory activity while in those with over 10 per cent moisture level seed respiration begins to increase rapidly. Unlike other plant parts where the decline in respiration may set in at about 30°C, seed respiration continues to increase up to 50 °C. Under high relative humidity (90 per cent) and high temperature (more than 32°C) most of the seeds become non-viable in a short period of time. Low temperature and low humidity enhance the life of the seed.

Table 1: Approximate Moisture Content of Vegetable Seeds in Equilibrium with Air at Different Relative Humidity at Room Temperature (approximately 25°C)

Seed	*Relative Humidity (%)*					
	10	*20*	*30*	*45*	*60*	*75*
Cabbage	3.2	4.6	5.4	6.4	7.6	9.6
Carrot	4.5	5.9	6.8	7.9	9.2	11.6
Cucumber	2.6	4.3	5.6	7.1	8.4	10.1
Eggplant	3.1	4.9	6.3	8.0	9.8	11.9
Lettuce	2.8	4.2	5.1	5.9	7.1	9.6
Okra	3.8	7.2	8.3	10.0	11.2	13.1
Onion	4.6	6.8	8.0	9.5	11.2	13.4
Pea	5.4	7.3	8.6	10.1	11.9	15.0
Pepper	2.8	4.5	6.0	7.8	9.2	11.0
Radish	2.6	3.8	5.1	6.8	8.3	10.2
Spinach	4.6	6.5	7.8	9.5	11.1	13.2
Tomato	3.2	5.0	6.3	7.8	9.2	11.1
Turnip	2.6	4.0	5.1	6.3	7.4	9.0
Watermelon	3.0	4.8	6.1	7.6	8.8	10.4

Many of the high-yielding varieties have non-dormant seeds and hence run the risk of loss of viability by quiscent germination and absorption of atmospheric moisture in storage. For example, the vitality of the second crop varieties of paddy suffer a significant loss in viability even during the period of 9 months from harvest to next sowing season, if stored in ordinary gunny bags after air drying for a fortnight after harvest. The viability came down to 30 per cent and the loss in viability was stepped up during rainy and winter months, since moisture absorption was facilitated by high relative humidity (R.H) and low temperature. Drying of the seeds to a moisture level of 11 per cent is vital for maintaining viability. Thus, initial stages of seed multiplication, the need for keeping every kilo of seed in top condition of viability is apparent.

Table 2: Moisture of Seed Packaged

Family	*Crops*	*Maximum % Seed Moisture*
Liliaceae	Onion	6.5
Chenopodiaceae	Spinach	8.0
Cruciferae	Cabbage, cauliflower, turnip, raddish	5.0
Leguminosae	Pea	7.0
Umbelliferae	Carrot	7.0
Solanaceae	Tomato	5.5
	Pepper	4.5
	Egg-plant	6.0
Cucurbitaceae	Cucumber, muskmelon	6.0
	Watermelon	6.5
Compositae	Lettuce	5.5

Air drying of seeds even in open does not reduce the moisture content of seeds to a low level for ever. Seeds being hygroscopic tend to attain a moisture content which is in equilibrium with the R.H. of the air to which they are exposed. Moisture equilibria values, on per cent wet weight basis, for seeds of various crops and vegetables, at different relative humidities and a constant temperature of 25°C given by James (1967) are set out in Tables 1 and 2.

Chapter 12

Economics of Hybrids Seed Production

Cost of Hybrid Seed Production

Production of hybrid seed in vegetables is a complex and expensive venture. It requires about three to four times more capital than ordinary commercial vegetable cultivation. The cost arises from major three components *viz.*

1. Raising of healthy vegetable crop for hybrid seed production.
2. Cost of emasculation and pollination.
3. Post harvest operations of seed crop like harvesting, seed extraction, packing, storage and marketing of hybrid seeds.

Cost on Raising of Healthy Crop for Hybrid Seed Production

Cost of raising crop varies depending upon type of vegetable, season, region, availability of inputs and facilities like irrigation, good soil, labour, etc. Various cost included in this component are.

1. Cost of seed of parental lines.
2. Cost of land rent, land preparation and labour
3. Cost of seed sowing, nursery raising and planting.
4. Cost on fertilizing, weeding and pesticides.
5. Cost of cultural practices and other special operations of seed production such as rouging, staking, stackling preparations over wintering, etc.
6. Cost of hiring of farm implements and machinery and other material cost involved in crop raising.

The cost of cultivation although vary depending upon place, crop and availability of labour and inputs. The cost of just raising of crop for seed production under Kashmir conditions ranged from approximately Rs. 0.5lakh in okra to Rs. 1.0 lakh in tomato (Ahmed and Narayan 2005). In chilli, capsicum and brinjal it worked out to be around 0.8 lakh while in cucumber and bottle gourd it was about 0.9 lakh including stacking. This cost of raising crop was much less than the cost worked out for emasculation and pollination performed in hybrid seed production.

Cost of Emasculation and Pollination

The production of hybrid seed involves two basic procedures namely emasculation and pollination. These procedures are carried out either manually or by employing certain mechanisms available in crop plants. The hand emasculation and pollination of individual flowers is most expensive but it is economical in crops like tomato, brinjal, chilli, capsicum and cucurbits where single fruit gives large number of seeds and where small quantity of seed is required for planting of one hectare land. Such a procedure is not worth while in other vegetables like cabbage, cauliflower, radish and turnip where single pollination results in limited number of seeds and in crops like onion and carrot where floral parts are so arranged that selection of florets in umbels and their emasculation is difficult.

To ease the procedure and to produce hybrid seed at an economically acceptable level, a number of morphological and physiological mechanisms operating in plants such as sex expression, male sterility, self incompatibility, heterostyly etc and certain pollinator were exploited commercially to over come expensive emasculation and hand pollination. Removal of male plants in dioecious crops like spinach and asparagus or pinching of male flower in monoecious cucurbits before flowering facilitates large scale emasculation and subsequent natural pollination results in production of hybrid seed in female plants or flowers while in crops like cabbage, cauliflower, kale, chinese cabbage, turnip, radish etc. where self incompatibility mechanism is operating both the male and female parental lines are producing hybrid seed by pollinating each other. In some crops like tomato, brinjal capsicum, onion, carrot and melon, the male sterility arises either due to genes or the cytoplasm or induced through gameticides is being used as a potent self emasculating mechanism. The hybrid seed here is produced by planting alternatively male sterile female parent and with pollinating with fertile counter part male parent either by hand as in case of tomato, brinjal, chilli, capsicum or by natural pollination through pollinators as in onion, carrot or melons. Even heterostyly condition has been used in production of hybrids by pollinating long styled flower of some varieties of tomato and brinjal.

Cost of Postharvest Operations of Seed Crop Like Harvesting, Seed Extraction, Drying, Packing Labelling, Storage and Marketing of Hybrids Seeds

Hybrid seed production is a highly capital intensive activity, requires utmost care from harvesting hybrid seeds to till it is marketed. Fruits after attaining physiological maturity are harvested carefully from female parent and seeds extracted using standard procedures applicable under different crops. After extraction, seeds are dried either in sun or through mechanical driers up to a moisture percentage of about 7-8 percent depending upon crops. After drying, seeds are properly treated with fungicides are vaccume packed preferably in small units. Packaging in small units makes identification, transportation, handling, marketing and distribution earier. A wide range of material and types of container are in use for seed packing. The choice of material and size of seed packing depends on seed, value of seed /kg/unit, destination, type of market, mode of distribution, protection required from biotic and abiotic hazards. Moisture proof material is generally used for the purpose. Labelling is must on a seed package containing all particulars as per seed certification standards like name of the crop hybrid, percent germination, date of packing, date of expiry, weight of seeds etc. packed and labelled seeds are then stored in different structures like ordinary well ventilated stores, air cooled storages, dehumidifiers or in vapour proof containers on short or long term basis depending upon the requirement and than, marketed through authorized whole sale and retail outlets on a prefixed prizes. All the post harvest operations of hybrid seed production are labour intensive and are carried out based on modern techniques and therefore the cost of post harvest operations are generally more than the ordinary seeds resulting in additional costs. The total cost of production of hybrid seed is thus very high and therefore are sold at exauburant rates.

Methods of Hybrid Seed Production and their Production Cost Analysis

On the basis of sex expression, flower structure, flower position, pollination mechanisms, male sterility and self incompatibility and the ease with which seed can be produced the following methods of hybrids seed production are described along with economics of production of hybrid seed.

Hybrid Seed Production Using Hand Emasculation and Hand Pollination

In bisexual crops like tomato, brinjal, chilli, capsicum and okra and in romonocious varieties of musk melon the male part in bisexual flower is emasculated and pollinated manually by hand. This method was first practiced in Japan in 1935 for production of hybrid seed in brinjal and subsequently in tomato by various workers (Nishi, 1955; Hojby 1958). It involved 5 different steps like emasculation, bagging or caging, removal of bags / cages, pollination and subsequent bagging which takes about 50 to 60 seconds by a trained worker (Nishi, 1967). In a day it is possible to handle only about 300 to 350 flowers, therefore, the hybrid seed produced by this method is very laborious and expensive.

Obra *et al.* (1945) produced 28 g of hybrid tomato seed from 3-4 hours of pollination which was sufficient for planting in half an acre of land. Hojby (1958) obtained 24000 seeds from 8.1 hour while Hafen and Stevenson (1956) produced 4000 hybrid seeds from 51 minutes through emasculation and pollination. It reduced to 12 minutes when male sterile plants were used as female parents. Hojby (1960) again worked out economics of hybrid seed production in tomato variety Pearson through natural pollination, hand pollination and hand emasculation and hand pollination with pollen of long fruited San Marzano variety. The maximum hybrid seed quantity (108.0 lbs) was obtained with hand emasculation and hand pollination than either hand pollination or natural pollination when male sterile seed parent was used. Both percent hybrid seed as well as quantity of hybrid seed however reduced in fertile seed parent than male sterile. Ahmed and Narayan (2005) worked out labour requirement and economics of hybrid seed production in chilli, capsicum, tomato, brinjal, okra, cucumber and bottle gourd following hand emasculation and hand pollination. The number of flowers emasculated and pollinated per hour per day was 360 for chilli, 320 for capsicum, 400 for tomato, 340 for brinjal, 280 for okra and 240 each for cucumber and bottle gourd. The approximate numbers of plants which one labour can handle was about 14 for chilli 53 for capsicum, 20 for tomato, 34 for brinjal, 28 for okra, 48 each for cucumber and bottle gourd (Table 1). On the basis of number of plants a labour can handle, the total number of labours required per hectare for doing emasculation and pollination were worked out as 1580 for chilli, 474 for capsicum, 741 for tomato, 436 for brinjal, 1058 for okra, 124 for cucumber and 93 for bottle gourd (Table-2). The total cost of hybrid seed production per hectare which includes labour cost of emasculation and pollination and the cost of raising of crop in one hectare of land ranged from 1.0389 lac for bottle gourd to 3.17 for chilli (Table 2). For capsicum it was Rs. 1.711 lacs, and for tomato, brinjal, okra and cucumber it was 2.115, 1.454, 2.087 and 1.086 lacs respectively. The hybrid production cost was maximum for chilli followed by tomato. Involvement of more labour hours for emasculation and pollination per hectare resulted in high production cost as compared to cucurbits or brinjal whose labour requirement is almost 5 to 10 times lesser. Nearly 74 to 76 per cent of the total cost in chilli and okra and 42 to 52 per cent in capsicum, brinjal and tomato goes towards the cost of human labour who do the work of emasculation and pollination. For cucumber and bottle gourd the cost of emasculation and pollination was only 13 to 17 per cent of total cost of production. It is because, the labour requirement for pinching of male flowers and pollination of large female flowers is comparatively less, easy to handle and economical. The hybrid seed production using hand emasculation and hand pollination in other than cucurbits is therefore very expensive requiring two to three fold more capital than ordinary commercial seed production.

Table 1: Emasculation and Pollination per Day/Labour (4 hrs. Emasculation + 3 hrs. Pollination + 1 hr. Preparation)

Sl.No.	*Crop*	*Av. No. of Flowers Emasculated/Bagged and Tagged in 4 hrs.*	*No. of Flowers Pollinated in 3 hrs*	*No. of Flowers Emasculated and Pollinated per Labour per day*	*Approximate No. of Plants Handled/ Day/Labour*
1.	Chilli	360	360	360	15.0
2.	Capsicum	320	320	320	50.00
3.	a) Tomato (Fertile)	400	400	400	20
	b) Tomato (Male sterile)	600	600	600	30
4.	Brinjal	340	340	340	34
5.	Okra	280	280	280	28
6.	Cucumber	240	240	240	48
7.	Bottle gourd	240	240	240	48

(Ahmed and Narayan, 2005).

Table 2: Labour Requirement, Labour Cost and Cost of Cultivation per Hectare

Sl.No.	*Crop*	*Approx. No. of Plants/ha at 4:1 (F:M) Ratio*	*No. of Labours/ha Required for E and P*	*Labour Cost Rs. 150/Labour for E and P (lakhs)*	*Cost of Cultivation/ha (lakhs)*	*Total Cost of Prod/ha (lakhs)*	*Per cent of Total Cost for Emasculation and Pollination*
1.	Chilli	23700 (F) 5925 (M)	1580	2.37	0.8	3.17	74.76
2.	Capsicum	23700 (F) 5925 (M)	474	0.711	1	1.711	41.55
3.	Tomato	14815 (F) 3704 (M)	741 494 (MS)	1.1115 0.741	1 1	2.115 1.741	52.55 42.56
4.	Brinjal	14815 (F) 3704 (M)	436	0.654	0.8	1.454	44.97
5.	Okra	29628 (F) 7407 (M)	1058	1.5872	0.5	2.087	76.05
6.	Cucumber	5924 (F) 1481 (M)	124	0.186	0.9	1.086	17.12
7.	Bottle Gourd	4444 (F) 1111 (M)	93	0.1389	0.9	1.0389	13.36

(Ahmed and Narayan, 2005).

Hybrid Seed Production Using Hand Pollination Without Emasculation

This method is followed in crops where two sexes namely male and female flowers are separate as in cucurbits or crops where male part of the bisexual flower is non functional i.e male sterility which is exploited commercially in number of crops like capsicum, tomato, brinjal, squash, muskmelon, cucumber, pumpkin etc. For production of hybrid seed, in monoecious cucurbits, the female flowers of the seed parent and pollen flowers of the male parents are bagged separately before opening and hand pollination is done subsequently after opening of female flowers of seed parent. No pinching of male flower from seed parent in this method saves lot of labour resulting in low cost seed production. Nishi (1955) worked out number of workers for pollination in cucumber and reported handling of about 350-400 flowers per day as against 240 flowers in cucumber and bottle gourd when both hand pollination and pinching of male flowers was done (Ahmed and Narayan (2005). It means 30-40 per cent additional labour hours are required when hand emasculation i.e pinching of male flowers of seed parent is carried out. But this labourious emasculation can be overcome if some self emasculating genes like male sterile genes are transferred to commercial seed parents using conventional methods or biotechnological techniques and later exploited for producing hybrids commercially at much lower cost using different types of male sterility, hybrid seed production is being commercialized in crops like tomato, muskmelon, chilli, onion and carrot use of male sterility not only improves the labour efficiency but also quality of hybrid seed when efficient and stable male sterile systems are used. Ahmed and Narayan (2005) compared the efficiency of hybrid seed production in tomato using hand emasculation and hand pollination and no emasculation using functional male sterility and hand pollination. With the use of male sterility it was possible to handle 600 flowers by one labour as against 400 when both hand emasculation and hand pollination is used. Male sterility therefore not only increase labour efficiency by about 50 per cent but also reduce cost of production per kg hybrid seed from Rs. 1133 to Rs. 904 which was equivalent to 25.33 per cent reduction (Tables 2 and 3).

Table 3: Cost Benefit Analysis of Hybrid Seed Production of Different Vegetables Under Kashmir Condition

Sl.No.	*Crop*	*Av. No. of Seeds/ Fruit*	*Seed wt./Fuit (g)*	*Seed Number (g)*	*No. of Fruits/ Plant*	*Seed Wt./ Plant (g)*	*Seed Yield/ha (kg)*	*Total Cost of Prod. of Hybrid Seed/ ha*	*Av. Cost of Prod./kg Hybrid Seed (Rs.) Excluding Cost on Processing and Marketing*	*Average Sale Price of Hybrid Seed/kg in India (Rs)*	*Cost Benefit Ratio*
1.	Chilli	87	0.59	147	25	14.75	349.57	3.17	906.82	10000	01:11.0
2.	Sweet Pepper	281	2.72	103	6	16.32	386.78	1.711	442.37	15000	01:33.9
3.	Tomato (F)	195	0.63	309	20	12.6	186.67	2.115	1133.01	12000	01:10.6
	Tomato (MS)	203	0.65	298	20	13	192.59	1.741	903.99		01:13.3
4.	Brinjal	782	3.89	201	10	38.9	576.3	1.454	252.29	3500	01:13.9
5.	Okra	50	3.5	14	10	35	1036.98	2.087	201.25	500	01:02.5
6.	Cucumber	355	7.8	46	5	39	231.04	1.086	470.04	2000	01:04.3
7.	Bottle Gourd	426	52	8	5	260	1155.44	1.039	89.92	1800	01:20.0

(Ahmed and Narayan, 2005).

Hybrid Seed Production Through Emasculation and Wind Pollination

This method is generally used in dioecious crops like spinach and asparagus and in monoecious sweet corn. The abundant quantity of light weight pollen produced in male parent is carried to female line by wind and the seed collected from female parent is the hybrid seed. All these procedures in this method are carried in isolation. As compared to hand emasculation and hand pollination, the hybrid seed production through this method is easy and economical as no labour is required for pollination. However, identification and removal of 50 percent male plants of seed parent in dioecious crops and detassaling of male part of seed parent of monoecious sweet corn before flowering are the only major labour involving procedures. No laborious bagging and tagging are also involved as in hand emasculation and pollination. The efficiency can be further improved if some marker trait is associated either with male or female plants of a line of dioecious crop and it is identifiable before flowering as it facilitates easy identification and removal of males from female parent in a hybrid seed production plot. The efficiency also depends upon the pollen producing capacity of the male parent, the prevailing weather conditions especially dry sunny day, direction and speed of wind. Planting of male first followed by seed parent in the direction of wind is very important for obtaining higher and better seed set. Due to very less labour requirement the cost of production of hybrid seed through this method is very cheap as compared to other methods. The cost of production of one lb hybrid seed was just $11-18 in sweet corn and $12-13 in spinach as against $1917-2109 in tomato and $1246-1534 in pepper through hand emasculation and hand pollination (Table 4). This method although is very easy and economical but limited only to vegetables crops which are wind pollinated.

Hybrid Seed Production Through Hand Emasculation and Use of Insect Pollination

This method is used mostly in monoecious cucurbits like cucumber, melons, squashes and gourds where male and female parents are planted alternatively in commercial hybrid seed production plots in the ratio of 1:2 and male flower of female parent are removed before opening regularly and allowed insect pollination in isolation in which the pollen of male parent is transferred to

Table 4: Cost Comparison Analysis of Hybrid Seeds (Produced Through Different Methods/Techniques) Vs Non-hybrid Seeds (Open–Pollinated or Pweeline Cultivars)

Sl.No.	*Crop*	*Method*	*Non-Hybrid Cost /lb ($)*	*F_1 Hybrid Cost/lb ($)*
1.	Onion	CMS, Insect pollination	100–208	167–638
2.	Spinach	Hand rouging of male plants	11	12–13
3.	Muskmelon	Hand emasculation, Insect pollination	–	255–639
4.	Cabbage	Self incompatible, Insect pollination	79–102	351–682
5.	Tomato	Hand emasculation and hand pollination	534–829	1,917–2,109
6.	Pepper	Hand emasculation and hand pollination	134–421	1,246–1,534
7.	Eggplant	Hand emasculation and hand pollination	–	509–1085
8.	Carrot	CMS, Insect pollination	63–104	175–534
9.	Cucumber		–	–
(*a*)	Slicer	Hand emasculation or ethrel treatment	43	97–127
(*b*)	Pickler	Ethrel treatment or hand emasculation	60	–
(*c*)	Slicer	Gynoecious flowers	–	20
(*d*)	Pickler	Gynoecious flowers	–	206–271
10.	Pumpkin	Ethrel elimination of males	47–143	95–319
11.	Watermelon	Diploid	57–58	278–444
	Triploid	–	700–780	

female parent through insect pollinators and the seed harvested from female is collected as hybrid seed. It is one of the cheapest method which does not involve laborious hand emasculation and hand pollination. The identification and pinching of male flowers of female parent can be done very easily in the bud stage and before opening by very few labourers. In cucumber as reported by Nishi (1955) only thirty workers are required in one hectare for carrying out emasculation work as against 124 and 93 labourers required in cucumber and bottle gourd respectively when both emasculation and pollination is carried out (Ahmed and Narayan 2005). This method save about 50 to 75 per cent of labour cost involved in emasculation and pollination and the cost of hybrid seed thus produced is very cheap. The cost of production of hybrid musk melon ranged from 255-639 per lb while for cucumber it was 97-127 per lb. The cost would further reduce if hand pinching is replaced by elimination of males through application of ethrel or by using gynoecious lines. By use of gynoecious lines and ethrel elimination one can produce only one pound hybrid cucumber seed in just $ 20 as against $ 97-127 when hand emasculation is used (Table 4).

Recently gynoecious lines and ethrel in cucumber, mostly ethrel in pumpkin and squash are being commercially exploited for economical production of hybrid seed world over.

Hybrid Seed Production Through Free Insect or Wind Pollination Without Emasculation

It is one of the most important method particularly in crops where emasculation is tedious as in onion and carrot due to umbellate inflorescence and in crops where each pollination results in only a small number of seeds as in cole crops and radish. The presence of self-incompatibility in cruciferous vegetables and male sterility in onion, carrot and beet have been exploited to overcome laborious emasculation. While insect pollinator and wind have been used for carrying out pollination. Since both emasculation and pollination have been done utilizing plants inherent mechanisms like self incompatibility and male sterility and natural pollinating agencies like wind and insect, the hybrid seed produced is very economical in comparison to methods where manual emasculation and pollinations are carried out. The cost of production of hybrid seed of onion and carrot using male sterility and free insect pollination per lb ranged from $167-638 and $ 175-534 respectively as compared to cost by hand emasculated and hand pollinated seeds of tomato or pepper whose production cost for one lb seed ranged between $ 1246, to 2109 (Table 4).

Cost Benefit Analysis of Vegetable Hybrid Seed Production

The hybrid seed yield per hectare varies from one crop to another crop and from one variety to another variety and also from one area to another depending upon the quality of soil, topography weather conditions and the labour availability and quality. Hojby (1960) recorded hybrid seed yield ranging from 95.1 in fertile seeded parent to 107.5 lb/acre in male sterile tomato under South African conditions. Similarly Ahmed and Narayan (2005) under temperate Kashmir condition recorded 186.67 kg/ha hybrid seed yield in fertile seeded parent to 192.59 kg/ha in functional male sterile tomato indicating that Kashmir climate is more suitable for producing hybrid seed of tomato. The availability of cheap labour, fertile soil and congenial weather the cost of production of per kg hybrid seed under Kashmir condition was just Rs. 906.82 for chilli, 442.37 for sweet pepper, 904 to 1133 for tomato, 252 for brinjal, 201 for okra, 470 for cucumber and only 90 for bottle gourd. On the basis of average sale price the cost benefit ratio of hybrid seed production for important crops as reported by Ahmed and Narayan (2004) under Kashmir conditions, ranged from 1:2.48 for okra to 1:33.90 for sweet pepper. For chilli it was 1:11.02, tomato 1:10.59 to 1:13.27, brinjal 1:13.87, cucumber 1:4.25 and bottle gourd 1:20.01. This indicated that except okra, the hybrid seed production of rest of the vegetables is highly remunerative with sweet pepper hybrid seed production being highly profitable followed by hybrid seed production of bottle gourd, brinjal tomato, chilli and cucumber. For every one rupee spent on cost of production we can get net returns of Rs. 33.90, 20.01, 13.87, 13.27, 11.02 and 4.25. It is the promise of these profits that attracts number of private Indian and multinational companies to venture in this highly profitable seed business.

There is a buoyant market for hybrid seeds. Unlike seeds of open pollinated varieties, hybrid seeds are used for only one crop and therefore have to be purchased every year. Moreover, in India vegetables are major commercial crops occupying about 6 million, hectares the demand for hybrid seed is increasing both in India and also in international markets. To meet the growing demand, new hybrid seed production zones have to be identified. Temperate hills in this regard can compliment in meeting the requirement of hybrid seed. Ideal cool summers, fertile soil, adequate irrigation, cheap labour and seed production expertise available in temperate hills could be exploited for production of high quality hybrid seed at much lower cost than other traditional seed growing areas of Southern India. Hybrid vegetable seed production in temperate hills, beside contributing to national seed pool can help in reviving sick seed industry in this area and also helps in generating employment to rural men and women.

Chapter 13

Biotechnology in Vegetable Seed Production

Biotechnology, in the simplest and broadest sense, is the utilization of living organisms or their components to provide useful products or processes. This definition encompasses essentially all of agriculture, since agriculture is based on the production of plants and animals to provide food, fiber and other products for human use. It also includes familiar uses of microorganisms, such as the yeast used in brewing or baking or the symbiosis of legume plants with nitrogen-fixing bacteria (such as commercial *Rhizobium* inoculants for peas). Thus, humans have been employing biotechnology for at least 10,000 years since the origin of agriculture. Recently, however, the term 'biotechnology' has become synonymous with 'genetic engineering.' The cloning and movement of genes among life forms as diverse as bacteria, tomatoes or sheep is now possible. In this context, biotechnology refers to a wide range of enabling technologies that allow the alteration of heritable traits outside of the living organism and the subsequent reintroduction of the new trait into an organism for specific purposes. These new varieties are called transgenic because they require the recombination of genetic materials from different organisms to cause the desired change.

Some of these enabling techniques are applied without resulting in a transgenic plant. These advanced technologies are sophisticated enhancements of methods that have long been common practice. For example, potatoes are propagated primarily by planting the buds, or "eyes," present on the tubers. This vegetative reproduction ensures that the new plants are identical to their parents. As one tuber can give only a few new plants, one for each "eye," several rounds of seed increase are needed to produce commercial quantities of potato "seed." Unfortunately, this also allows the transmission of disease-causing bacteria or viruses that may have infected the parent plant. Micro propagation, the mass production of identical plants from tiny buds of the parent plant, is a biotechnique that can eliminate these pathogens from the progeny plants while retaining the advantages of vegetative reproduction. Similarly, the individual cells of a plant can be separated, multiplied, and regenerated into whole plants through a process known as tissue culture. In this way, thousands of copies of a single plant can be made and certified pest-free. These techniques are valuable for propagating plants and are an essential component of more advanced methods of genetic modification.

Major improvement in vegetable crops has been achieved through conventional breeding. However, it is an established fact that conventional breeding methods have limited for creating super cultivars/ hybrids of vegetable crops. Biotechnology in the recent years have created unprecedented opportunities, not only for the manipulation of biological systems for the benefit of man kind but also to undertake for better understanding of the fundamental life process. Consequently, it has become the fastest and most rapidly growing technology in the world. In the coming years, biotechnological tools have played very crucial role in various ways including development of specific parental lines for hybrids in vegetables. Some of the potential thrust areas where bio-technology can have additional aid to the conventional hybrid breeding are as follows (Kalloo and Singh, 2001):

1. Molecular markers can be used to identify genic male sterile lines at seedling stage, identified male sterile plants can serve the purpose of female parent in F_1 seed production field. For this purpose, polymerase chain reaction (PCR) based markers seem feasible particularly in vegetables having less plant population per hectare (*e.g.* cucurbits). However before utilization, cost effectiveness of such techniques needs to be worked out.
2. For developing multiple biotic stress resistant lines, already available molecular markers having established linkage with stresses may be used to enhance breeding programme for development of parental lines.
3. In vegetable crops where male sterility system is not available, *e.g.*, okra, male sterility system can be developed through genetic transformation. There are a number of molecular strategies, which can be used to develop male sterility system depending upon the availability of gene constructs and plant materials (Kumar *et al.*, 2000).
4. Anther culture technique can be utilized for development and maintenance of self-incompatible lines in Cole vegetables and also to develop inbred lines in other cross-pollinated vegetables.
5. Protoplast culture and fusion technique may be used to transfer traits of interest, *e.g.* male sterility cytoplasm, resistant gene (s) for biotic stresses inside the parental lines from cross incompatible genus/texa. Embryo culture technique can also be used for the same purpose.
6. Most of the genic male sterility is controlled by single recessive gene, however, dominant gene controlling male sterility are also available in some vegetables *e.g.* cabbage. Advancement of biotechnology has made possible to

utilize such dominant male sterility for F_1 seed production through micro propagation in those vegetable crops where vegetative part is consumed. In the first step, male sterile line with homozygous dominant gene can be multiplied by clones and then crossed as female sterile line. In second step, seeds of male sterile lines can be used to raise female parent seedlings for hybrid seed production programme (Fang *et al.*, 1997).

7. Genetic transformation technique can be used to develop trait specific transgenic parental lines for hybrids.

Micro Propagation

Some breeding materials *i.e.* male sterile lines, self incompatible lines, inbred lines and haploids etc; are not easily propagated by seed, therefore, the tissue culture method is employed for their propagation. The tissue culture method for regeneration, supplements the conventional breeding methods. Plants can be propagated asexually on a large scale under in vitro conditions either directly from apical, auxillary and adventitious buds or indirectly through somatic embryogenesis. Micro-propagation is very useful especially to propagate superior manually produced hybrids for several generations without any genetic change. It might be suitable for clonal propagation of F_1 hybrids in crops like cabbage and their parental *SI* or *CMS* lines on a large scale. Until recently, auxillary shoot production has been found the most advanced method of micro-propagation but since the mid 80's interest in somatic embryogenesis has increased tremendously with the success achieved in many crops. Now production of somatic embryos and their conversion into artificial or synthetic seed appears as an efficient system of propagation of true to type hybrids. By propagating male sterile and male fertile or self-sterile parental lines through synthetic seeds, a pollination control system can be introduced in the plant population. A small number of hybrids can be produced by hand pollination and subsequently mass multiplied by artificial seed technology. Thus, making it possible to exploit hybrid vigour in crops like beans where the production of hybrid seed is very difficult and reduces the cost of hybrid seed.

Table 1: Achievements of Tissue Culture for Improvement of Vegetable Crops

Crop	*Technique Applied for*
Anther/microspore/ovule culture	
Brassica oleracea	Haploid production
Capsicum annum	Haploid production
Lycopersicon esculentum	Haploid production
Solanum melongena	Haploid production
Cucurbita pepo	Haploid production
Brassica rapa	Haploid production
Embryo rescue	
L. esculentum x *L. peruvianum*	To rescue the hybrid embryo
P. vulgaris x *P. angustissimus*	To rescue the hybrid embryo
Vigna pubescens x *V. unguiculata*	To rescue the hybrid embryo
S. melongena x *S. torvum*	To rescue the hybrid embryo
Meristem culture	
Allium sativum	Elimination of onion yellow dwarf virus
A. ascalonicum	Elimination of leek yellow dwarf virus
Pisum sativum	Elimination of pea seed borne mosaic virus
Solanum melongena	EMCV
L. esculentum	Cryopreservation
P. sativum	Cryopreservation
Protoplast fusion	
S. melongena + *S. integerefolium*	Resistance to *Pseudomonas*
S. melongena + *S. saintwongesi*	Resistance to *Pseudomonas*
B. oliracea + *B. compestris*	Disease resistance
B. oleracea + *B. napus*	Atrazine resistance
B. oleracea + *Armoraia*	Asymmetric hybridization
L. esculentum x *L. peruvianum* + *L. lycopersicoides*	Fertile somatic hybrids
L. esculentum + *L. peruvianum*	Somatic hybrids with intermediate morphology
Somaclonal variation	
L. esculentum	Tolerance to *Fusarium oxysporum* culture filtrate

Source: Singh *et al.* (2001).

Anther Culture

Guha and Maheshwari proposed the first haploid plant through anther culture in 1964. In anther or pollen culture, haploid plants may be obtained from pollen grains by placing anthers or isolated pollen grains on a suitable culture medium. Anthers may be taken from plants grown in the field or in pots, but ideally, these plants should be grown under controlled temperature, light and humidity. The optimum condition may differ from species to species. Often, the capacity for haploid production declines with the age of donor plants. In some species, however, a brief exposure of the anthers to a high temperature is reported to have a promontory effect, *e.g.* at 35^{0}C for 24hours in *Brassica campestris*. The medium requirements may vary with the vegetable crops. Anther cultures are generally maintained in altering period of light (12–18 h; 5000–10,000 lux m^2) at 28 ^{0}C and darkness (12-6h) at 22 ^{0}C, but the optimum conditions vary with the species. The wall of responsive anthers turn brown and after 3-8 weeks they burst open due to the developing callus or embryos. After seedling (from embryos) or shoots (from callus) become 3-5cm long, they are transferred to soil in the same way as other in *vitro*–regenerated plantlets. Pollen grains of the cultured anthers show remarkable cytological changes during the first 6-12 days, called the *inductive period*.

Genetic Engineering

Genetic (recombinant DNA) engineering is the manipulation of DNA molecules to produce modified plants. A plant in which a gene has been transferred through genetic engineering is called a transgenic plant and the gene to so transfer is called transgenic. The necessary techniques for genetic engineering in plants are being constantly refined, but the available techniques have already demonstrated the power and the usefulness of the approach. The general approach for genetic engineering in plants may be outlined as follows: (*i*) introduction of the gene of interest into the cells of concerned plant species. (*ii*) Integration of this gene into the nuclear/ organellar genome of the plant cells, (*iii*) Expression of the transferred gene in the new genetic background. (*iv*) Regeneration of whole plants from the genetically modified cells, and finally (*v*) Transmission of the transferred gene to the sexual progeny of these plants. It may be kept in mind that any gene in *E. coli* or by amplifying the gene using PCR (polymerase chain reaction). It may further be noted that gene transfers in plants are primarily based on tissue culture (therefore, a reliable protocol for the regeneration of whole plants from tissue cultures of the concerned plant species is a prerequisite for such work) and that the integration and expression of the introduced genes must be stable to be transmitted through the sexual process. The various gene transfers for development of hybrids may be grouped into the following categories.

Virus Resistance

There are several approaches, which have been used to engineer plants resistant to viruses. The first approach aims at introducing genes encoding viral coat protein to produce transgenic plants resistant to plant virus infection. The viral coat protein produced by the transgenic plants appear to block a receptor in the plant cell that is required for uncoating the viral nucleic acid, or to interfere with viral replication or expression of viral genes. Resistance is manifested by delay in symptom development or escape from infection altogether. A second approach to engineering virus resistance in plants has centered on satellite RNA's. It has been observed that expression of satellite RNA's in transgenic plants can reduce infection with cucumber mosaic virus. The third approaches aims at using virus RNA to govern transgenic resistance. Following the successful field-testing of coat protein mediated resistance against TMV in tomato plants (Nelson *et al.*, 1988), additional testing of a number of plant lines has occurred. A better understanding of the cellular and molecular mechanisms that confer transgene mediated resistance may lead to second generation viral genes that improve resistance beyond current levels and extend its application to still more viruses and crop plants. The transgene (s) in vegetable crops against viruses is given Table 2.

Fungal Resistance

One on the major approach for engineering plants resistant to fungal diseases includes use of *chitinase* genes of plant and/or bacterial origin. *Chitinase* enzyme degrades *chitin* rich walls of fungi and delay symptom development. Other antifungal protein genes include osmotin, *glucanase thionin, peroxidase* etc. Recently, an effective antifungal protein from barley seeds called *ribosome inactivating protein* (RIP) has been isolated. RIP acts by inhibiting protein synthesis via N–glycosidase modification of fungal 28 S r DNA. Transgenic crop plants expressing RIP proteins are now being tested against many fungi (Sharma and Kumar, 1995).

Table 2: Examples of Engineered Vegetables Against Viruses

Purpose of Genetic Manipulation	*Transgenic Product*	*Transformed Plant*	*Origin of Transgene*
Potato virus X (PVX)	Viral coat protein	Potato	PVX
Potato virus Y (PVY)	Viral coat protein	Potato	PVY
Tomato yellow leaf curl virus	Coat protein	Tomato	TYLCV
Tomato spotted wilt virus	N-gene	Tomato	TSWV
Tomato yellow leaf curl virus	Truncated version of CI gene of TYLCV	Tomato	TYLCV
Tomato mosaic virus	Antigens RNA	Tomato	TOMV
Cucumber mosaic virus	Satellite RNA	Tomato	CMV
Cucumber mosaic virus	Viral coat protein	Cucumber	CMV

Source: Singh *et al.* (2001).

Bacterial Resistance

Most of the crop plants are susceptible to bacterial diseases, which have been difficult to control. Bactericides are not a complete solution because bacteria quickly evolve resistance to them. Using molecular biological tool, a novel approach for engineering resistance against bacteria has been undertaken to isolate the *tabtoxin* resistance genes (*ttr*) from bacteria *Pseudomonas syringae* pv. *tabaci* and subsequently to introduce into susceptible plants for transgenic resistance (Yoneyyama and Anazi, 1991). In a similar way *Phaseolotoxin–insensitive OCTase* (Ornithine carbamoyl transferase) gene was isolated from *Pseudomonas syringae* pv. *phaseolicola* and introduced to beans to observe total resistance(Herrera Estrella and Simpson, 1995). This simple and useful strategy for molecular breeding of disease resistant plants by detoxification of pathogenic toxins has a great potential to be exploited against bacterial and fungal diseases in vegetable crops. The successful transgene(s) in vegetables against bacteria are given in Table 3.

Table 3: Examples of Successful Engineered Vegetables Against Bacteria

Resistance Against	*Transgene Product*	*Origin of Transgene*	*Transformed Plant*
Erwinia carotovora	Lysozyme	T4 bacteriophage	Potato
Erwinia carotovora	Tachyplesin	Horseshoe crab	Potato
Pseudomonas syringae pv. *Phaseolicola*	Phaseolotoxin insensitive OCTase	*Pseudomonas syringae* pv. *Phaseolicola*	Bean
Erwinia carotovora	Pestate lyase	*Erwinia carotovora*	Potato
Erwinia carotovora	Glucose oxidase	*Aspergillus niger*	Potato

Source: Singh *et al.* (2001).

Insect-pest Resistance

Crop protection from insects plays a very important role in the agricultural productivity. Progress in developing insect control in transgenic plants has been initially achieved through the expression of insect control protein genes of *Bacillus thuringensis*. The insecticidal protein gene from *Bt* conferred for reaching resistance to classes of certain lepidopteron species when transgenic tomato, potato, maize and cotton plants were produced. There are several *Bt* genes encoding proteins that are largely insecticidal. These proteins are not harmful to mammals but kill insect by interfering with the ion transport in the midgut. Different types of proteins are available that selectivity kill specific groups of insects raising the possibility of producing plants resistant to particular pests.

Plants also have endogenous defense mechanisms against insect attack. Several plants have been shown to synthesize systematically protease inhibitors in response to wounding. These proteins are believed to provide some defense against feeding insects. Recently, works have shown that cowpea seeds containing a high concentration of chymotrypsin inhibitor are resistant to insect attack (Hilder *et al.*, 1987). Transgenic tobacco plants have been generated that express constitutively different amounts of cowpea seed chymotrypsin inhibitor. Feeding experiments have shown that there is a direct relationship between the amount of chymotrypsin inhibitor produced and the degree of protection afforded against a range of insects. The transgene(s) expressing resistant to insects are given in Table 4.

Table 4: Engineered Vegetables Expressing Insect Resistant Gene

Transgenic Products	*Organic of Transgene*	*Target Insects*	*Transformed Plants*
Cry 1 AB	*Bacillus thuringiensis*	Lepidoptera	Tomato
Cry 1 AB	*Bacillus thuringiensis*	Lepidoptera	Tomato
Cry 1 H	*Bacillus thuringiensis*	Lepidoptera	Corn
Cry 3 A	*Bacillus thuringiensis*	Coleptera	Eggplant
CpT1 (cowpea trypsin inhibitor)	Cowpea	Coleoptera Lepidoptera	Tomato
Tomato proteinase inhibitor I	Tomato	Lepidoptera	Tomato
Tomato proteinase inhibitor II	Tomato	Lepidoptera	Tomato
Alpha-amylase inhibitor of the common bean	Common bean	Coleoptera	Common bean
Snowdrop lectin (GNA)	-	Lepidoptera	Tomato
Tobacco anionic peroxidase)		Lepidoptera	Tomato
Bovine pancreatic trypsin inhibitor	-	Lepidoptera	Lettuce

Source: Singh *et al.* (2001).

Parthenocarpic Plants

In plants which are able to develop fruits in the absence of fertilization are said to be parthenocarpic. In such fruits seeds are absent which can increase fruit acceptance by consumers. Parthenocarpy has been achieved by genetic engineering in eggplant and tomato. For the parthenocarpic development of fruit the heterologous gene def H9-iaaM has been used in transformation experiments to increase the concentration of IAA in the ovule. Line transformed with this gene gave normal sized parthenocarpic fruits even from the emasculated flowers. It has been observed that in a protected cultivation trial, three transgenic plant hybrids yielded more than the corresponding non transgenic control even when the latter used treated with plant growth regulators (Rotino *et al.*, 1998).

The transgenic vegetables for parthenocarpic fruits are given in Table 5.

Table 5: Successful Engineered Vegetables for Parthenocarpic Fruits

Transgenic Against	*Transgene*	*Origin of Transgene*	*Transformed Plants*
Tomato	Rol B gene	*A. rhizogens*	Tomato
Brinjal	*Iaah gene*	*Ps. Syringae* pv. *savas*	Brinjal

Source: Singh *et al.* (2001).

Postharvest Traits

In tomato plants have been engineered for reduced synthesis of polygalacturonase, which is responsible for softening the cell wall during ripening of some fruits. Polygalacturonase degraded the pectin fraction of the fruit cell wall thus plays a role in fruit softening. It has been shown that a single antisense gene could inhibit expression of tomato Pgu gene during ripening. Using the antisense technology, antisense RNA from polygalacturonase has been inserted. When normal PGU mRNA is synthesized the antisense RNA binds to it and prevents synthesis of *Pgu.* Antisense RNA against ACC synthase (or ACC oxidase). It has been observed that when it expressed in tomato plants and this inhibited by ethylene biosynthesis by 99.5 per cent production of which appears to be involved in the initiation, modulation and co-coordination of expression of many of the genes required for the ripening process. Genetically engineered tomatoes have become a reality of commercial company collagen is set to market genetically engineered *'Flvr-savr'* tomatoes using antisense RNA technology. These tomatoes have extended shelf life and increased resistance to brushing. The successful transgene(s) in vegetable crops for post harvest traits are given in Table 6.

Male Sterility

The advent of recombinant DNA methods and the identification and characterization of genes involved in pollen development have led to the development of molecular male sterility system. The best example of genetically engineered male sterility restorer system is based on the suppression of pollen development by expressing a

Ribonuclease (R Nase) gene under tapetum specific TA 29 promotor, which removes the tapetum, the tissue that supports pollen development. This system known as "*Barnase-Bar Star*" systems has been proved to be successful in tomato and corn.

Table 6: Examples of Successful Engineered Vegetables for Postharvest Traits

Purpose of Genetic Manipulation	*Transgene Product*	*Origin of Transgene*	*Transformed Plants*
Improved storage/ shelf life	Antisense polygalacturonase	Tomato	Tomato
Ripening	Antisense ACC oxidase	Tomato	Tomato
	Antisense ACC synthase	Tomato	Tomato
Fruit pigmentation	Phytoene synthase gene	Tomato	Tomato

Source: Singh *et al* (2001).

Herbicide Tolerance

Pre and post emergence herbicides are used to control weeds in vegetable crops. However, certain herbicides are not selective to the crop. Therefore tissue culture technique has been used to isolate herbicide tolerant mutants. A paraquat tolerant mutant was identified in *Lycopersicon*. In this case an interspecific hybrid strain L2 was used. The callus cells of this genotype were kept on an agar medium containing a lethal concentration of paraquat. The tolerant clone grew under these conditions. Plants produced have about 30-fold increase in paraquat tolerance as compared to strain L2 (Thomas and Pratt, 1981). Similarly cultures have been used to select and isolate mutants tolerant to various chemicals *e.g.* metribuzin in *Lycopersicon* and *D. carota* and 2,4-D resistant cells in *D. carota*.

Salt Tolerance

Salt is one of the most important limiting factor for the production of vegetable crops. A wide range of variation in the plants for tolerance to salt has been reported. Traditional breeding is expensive time consuming and some tmes may not be successful. Therefore the plant tissue culture technique may be used to obtain salt tolerance cell lines. Salt tolerant lines have been isolated in

Lycopersicon by the tissue culture technique. *L. peruvianum* and *L. pennelli* have been found tolerant to sodium chloride. Similarly applying selection pressure to suspension cultures has isolated salt tolerant lines of *Capsicum annuum*. Such lines could grow in liquid media containing 1-2 w/v sodium chloride.

Use of Molecular Markers

Gene Tagging

The most interesting application of molecular markers at present is the ability to facilitate the method of "Conventional" gene transfer. In plant breeding gene tagging is a term commonly applied where a genetic marker is used as the selection criteria for a closely linked agronomic trait. Example of some markers linked to disease resistance genes are summarized in Table 7. A classic example resulted from the discovery of tight linkage between the isozyme markers *Aps 1* and *Mi*, a gene encoding root knot nematode resistance in tomato (Rick and Fobes, 1977). As a result, many seed companies now use isozyme analysis as the preferred method for the incorporation of this trait into elite tomato lines. More than 1000 markers have been mapped onto the tomato molecular linkage maps and these markers cover a total of 1276 map units (Tanksley *et al.*, 1992). Many of the markers of the tomato map have also been mapped onto the potato and pepper molecular maps. All these solanaceous species share the same chromosome number and by having molecular maps based on the same RFLP probes; it has been possible to determine chromosomal homologies with a fairly high degree of precision (Bonierbale *et al.*, 1988; Tanksley *et al.*, 1988; Tanksley *et al.*, 1992).

Another area where molecular markers facilitate gene transfer is the analysis of linkage drag. Linkage drag is frequently encountered when wild species are used as a trait donor source; while backcrossing the trait, considerable excess linked chromosomal DNA may inadvertently be incorporated. A study by Young and Tanskley (1989) demonstrates the power of RFLP in these areas. The tomato cultivar "Craigella" was found to carry approximately 50 cM piece of DNA derived from the disease resistant donor parents following 11 backcross generations. In a single generation of selection, Young and Tanksley (1989) were able to reduce the size of this fragment to less than 7 cM.

QTL Mapping

Many quantitative traits of economic value are under polygenic control and are selected for directly. Such a selection is often ineffective, since the effect of each gene is small and is also influenced by the environment. Therefore, one would welcome a procedure for indirect selection, which is not influenced by the environment. For this purpose, linkage between quantitative traits and major marker genes should be known. Such marker genes involving morphological traits have been identified and mapped in several crop plants. Isozyme marker loci linked to quantitative traits of economic value had also been earlier identified in tomato and maize. However, in recent years it became possible to identify and map quantitative traits loci (QTLs) using RFLPs markers particularly in some crop plants like tomato and maize. With QTL analysis, hypothesis about the effects of a locus on a quantitative trait are tested at discrete positions along a map of genetic markers. The presence of a QTL is inferred when a significant amount of phenotypic variance can be attributed to estimate QTL parameters. This type of procedure was first described as "Simplified Interval mapping (SIM)" several variations of "SIM" have been described (Knapp *et al.*, 1990; Haley and Knot 1992). All the SIM procedures search for a single "Target QTL" at positions through out a mapped genome.

When multiple QTL segregate, the sampling error associated with inferences about a target the effects of other QTL may inflate QTL. Furthermore, linked QTL can cause biased estimates of QTL position. These problems have been recognized for sometime (Knapp *et al.*, 1990), and have stimulated efforts to develop methods of QTL analysis that can simultaneously account for multiple QTL. Recently, Zeng (1993) described a method called "Composite interval mapping" (CIM) which permits searching for one QTL at a time while simultaneously accounting for the effect of other segregating QTL. This is accomplished by filling parameters for a target QTL in one interval while simultaneously fitting partial regression coefficients for "background" markers, outside of that interval. In principle, the genetic variance caused by QTL other than the target is absorbed by regression coefficients of the background markers. Other authors (Jansen and Stain, 1994) have described similar methods based on this principle.

Another facet of QTL analysis concerns the fact that QTL effects may depend on environment. Methods of analysis of QTL x R interactions are not well developed. Tinker and Mather (1995) proposed a procedure of simplified composite interval mapping (SCM) for the analysis of QTL replicated in multiple environments.

Table 7: Identification of Markers Linked to Disease Resistance Genes in Vegetable Crops

Host	*Pathogen*	*Gene*	*Type of Markers*
Tomato	*Meloidogyne incognita*	Mi	RAPD
	Cladosporium fulvum	*Cf9*	AFLP
	Meloidogyne incognita	*Mi3*	RAPD, RFLP
	Leveillula taurica	*Lv*	RFLP, RAPD
	Phytopthora infestans	*Ph2*	RFLP
	Verticillium dahlie	*Ve*	RAPD
	Yellow leaf curl virus	*Ty1*	RFLP
	Tospovirus	*SW-5*	RFLP
Potato	*Phytopthora infestans*	*R1* and *R3*	RFLP
Lettuce	*Bremia lactuceae*	*Dm 17* and *18*	RAPD, SCAR
		Dm 8 and *10*	RAPD, SCAR
	Plasmopara lactucae-radicis	*Plr*	RAPD, RFLP
	Turnip mosaic virus	*Tu*	RAPD, RFLP
Bean	Common bean mosaic virus	*I*	RAPD
Pea	Pea common mosaic virus	*Mo*	RFLP
	Erysiphe polygone	*Er*	RAPD
	Fusarium oxysporum	*Fw*	USAT
Chinese cabbage	*Plasmodiophora brassicae*		RAPD

Source: Singh *et al.* (2001).

Gene Cloning

The advent of genome mapping at the DNA level has provided a method for localizing genes of economic importance to specific chromosomal positions. The ability to map any gene of economic importance to a defined chromosomal site opens the possibility of isolating genes via chromosome walking. This method is sometimes referred to as " map based cloning" or the " top down approach since it starts with an interesting phenotype observed at the whole plant level (*e.g.*; diseases resistance or high yield) and proceeds phenotype observed at the genetic mapping and chromosome walking (Table 8) Chromosome walking presents serious difficulties in crop species they tend to have relatively large genome. For example, the "average crop species has a nuclear genome 10-100 times larger than yeast, *Drosophila* or *Arabidopsis*. Nevertheless, the problems with chromosome walking the genome is rapidly being overcome by new methods for manipulating large pieces of DNA (e.g; pulsed field gel electrophoresis and yeast artificial chromosome library potentially provide the foundation on which to initiate map based gene cloning for genes underlying any trait that can be genetically mapped (Martin *et al;* 1992). A critical prerequisite for map–based gene cloning is that target gene must be localized definitively with respect to markers in that region of the map. With the small genetic distances being considered, analysis of very large progenies is often required for such fine mapping. New protocols for rapid DNA isolation as well as new strategies for high resolution genetic mapping should enhance our ability to meet this prerequisite for map based cloning (Churchill and Doerge, 1994). It is anticipated that, in the future, the molecular linkage map will play a key role in cloning genes for a variety of economically valuable and/ or biologically interesting characters including disease resistance, growth habit, fruit ripening behaviour and even yield.

Table 8: Gene Mapping in Vegetable Crops for Various Polygenic Traits

Crop Plants	*Traits*	*Molecular Markers*
Tomato	I-2 *Fusarium oxysporum*	RFLP
	Male sterile 14 gene and a hotspot for recombination	RFLP
	Regulation of iron–metabolism	RFLP
	Sequencing of cDNA clone	RFLP
	Integration of the classical and molecular linkage maps of chromosome-6	RAPD
	Soluble solids, fruit weight, fruit pH, fruit size and shape, fruit ripening and yield	RFLP

Contd...

Table 8–Contd...

Crop Plants	*Traits*	*Molecular Markers*
	Ripening inhibitor and non ripening loci	RFLP
	Root knot nematode resistance locu	AFLP and RFLP
	Salt tolerance	RAPD
	Golden nematode resistance gene (*Hero)*	RFLP
	Lateral suppressor (*ls)* locus	RFLP
	Tandemly repeated telomeric DNA sequences	RFLP
	Physical location of *cf–2* gene	RFLP
	Integration of the classical and molecular linkage maps of tomato chromosome-6	RAPD
	Close linkage between the Cf-2/ Cf-5 and *Mi* resistance loci	RAPD and RFLP
	Carotenoid loci	RAPD
Potato	High resolution map of the vicinity of the R1 locus of chromosome V of potato	RFLP and AFLP
	Gene pool similarities of potato cyst nematode population	AFLP
	Quantitative resistance to blight	AFLP
	High resolution map of chromosome VII harbouring the nematode resistance gene	RAPD and AFLP
	Mapping of simple sequence repeat loci	RFLP
Cabbage	Linkage group alignment	RFLP
	Location of self–incompatibility locus	RFLP
	Comparative genome mapping	RFLP
	Linkage group alignment from four independent–RFLP maps	RFLP
Cucumber	Plant height, sex expression and lateral branches	RFLP
Melon	Construction of molecular map	AFLP
Hot pepper	*Xanthomonas compestris*	RFLP and RAPD
Egg plant	Bacterial wilt resistance	RAPD
Onion	Downey mildew and onion leaf blight	RFLP and RAPD

Source: Singh *et al.* (2001).

Transpoon Tagging

Transpoon tagging might be more straight forward approach for the molecular isolation of useful genes. Application of endogenous plant transposable elements has shown to be obtaining plant mutants. The insertion of a transposable element in to genetic locus can be given rise to mutation. Using a transposable element from maize, the *Ac*–element, Fedoroff *et al.* (1984), were able to clone a locus, which until then was only characterized genetically. Their result indicates that in principle any locus in maize, identifiable with transposable element, can be cloned and analyzed at the molecular level. This approach, called transpoon tagging, has successfully tag applied for the isolation of various maize genes. In vegetable crops, researches on transpoon tagging are going on.

Transgenic

China was the first country to commercialize transgenic in the early nineties with the introduction of virus–resistant tobacco, which was later followed by a virus–resistant tomato (James, 1998). Subsequently, the global area of transgenic increased 4.5 fold from 2.8 million ha in 1996 to 12.8 million ha, with seven crops in six countries. Commercially released transgenic vegetable varieties are given in Table 9. For years, seed companies have been investing huge amounts of money in research on genetically modified plants. An overview of developmental and notification status of genetically modified vegetable is given in Table 10. In India, genetically altered seed has yet not been commercialized as the transgenic varieties developed are under going field trials and evaluation. The private sector is emerging as a major player in the area of transgenic research and commercial field releases in view of the fact that most Indian companies have now tie ups with multinational companies involved transgenic research. Pro Agro–PGS have obtained permits to carry out controlled field trials with male sterile cauliflower and transgenic tomato, brinjal, cauliflower and cabbage carrying *Bt* gene. Rallis India Ltd. have given permission by DBT to carry out transformation of chillies and bell pepper with GNA gene, pepper vein banding virus (PVBV) coat protein gene and its modified sequences and also to conduct green house trials of these transformants. Transgenic brinjal carrying the *Bt* gene developed by the National Research Center for Plant Biotechnology is also being evaluated. Besides, several programme on developing transgenic vegetable are also going in government funded research laboratories in India. These include (a) development of insect–resistant brinjal, tomato, cauliflower and cabbage through transfer of Bt gene, (b) Development of transgenic tomato resistant to TLCV, CMV, TSWV and early blight. (c) Development of transgenic watermelon resistant to watermelon bud necrosis virus (d) Genetic manipulation of ripening in tomato. (e) Development of transgenic amaranths with low oxalate and nitrate level.

Table 9: Commercial Release of Transgenic Vegetable Varieties

Traits	*Crops*	*Name*	*Company*	*Product status*
Quality (vine–ripened favour, shelf–life)	Tomato	Flavr–Savr	Calgene	Released 1994
Quality (vine–ripened favour, shelf–life)	Tomato	Endless Summer	DNA Plant Technology	Blocked by patent claims
Quality (paste consistency)	Tomato	-	Zenca	Released 1995
Virus resistance	Tomato, capsicum	-	China	Released 1993-94
Virus resistance	Squash	Freedom II	Asgrow	Released 1995

Table 10: Details of the Notification of Field Released of Genetically Modified Vegetable Crops

Company	*Crop*	*Trait*
Agritope	Melon	Fruit ripening altered; BCTV resistance.
Hunt–Wesson	Tomato	Pectinesterase level reduced
Lipton	Tomato	Antioxidant enzyme increased

Transgenic with Antifungal Protein Genes

Among the antifungal proteins, PR proteins have been found to express the resistance to several fungal diseases. Over expression of single PR–protein genes coding for *chitinase and glucanase* have been used to engineer fungal resistance in vegetable crops (Table 11). It has also been shown constitutive expression of PR2 and PR3 genes resulted in an increase in resistance against several fungi, which could be explained only by synergy between the gene products.

Table 11: Single PR–Protein Genes Used in Vegetable Crops Against Resistance to Fungal Diseases

PR–protein	*Source*	*Recipient*	*Fungus tested*
PR-2 (Class I glucanase)	Tobacco	Tomato	*Fusarium oxysporum* f. sp. *lycopersicon*
PR-3 (Class I chitinase)	Tobacco	Tomato	*Fusarium oxysporum* f. sp. *lycopersicon*
PR-5	Tobacco	Carrot	Erysiphe heraclei

Source: Does and Cornelissen, 1996.

Transgenic Through Pathogen Induced Cell Death

This approach is being with the aim to produce disease resistant transgenics with a more durable resistance to fungal pathogen in comparison with above described system. Strittmatter *et al.* (1995) developed a molecular strategy to engineer controlled cell death in pathogen infected plants utilizing *barnase* gene producing cytotoxic protein with RNAse activity and its in activator gene (Barstar). The *barmase* gene was placed under the control of pathogen specific (inducible) promoter sequence (*prp1-1* from potato) and to avoid cell death due to unwanted expression of banase gene, the Barstar gene was constitutively expressed in all the tissues. Therefore, in transgenic plants expressing *barnase* encoded cytotoxic protein, upon pathogen attack, cell death resulted (a kind of HR response). A number of transgenic potato plants expressing both genes showed severe necrosis of leaf tissue after inoculation with *Phytophthora infestans* spores due to the induced HR reaction in infected cells mediated through the expression of *barnase* gene under pathogen inducible promoter.

Transgenic with R and Avr Genes

As described earlier, many plant pathogen interactions fit in the gene for gene model. In such cases, transgenic plants are also produced with an *R* gene and for corresponding *Avr* gene under the control of pathogen inducible promoter. In this case, resistance reaction can be activated upon pathogen attack provided all components necessary to complete the signal transduction pathway leading to resistance are present. Honnee *et al.* (1995) showed that constitutive expression of the elicitor gene *Avr9* (from *Cladosposium fulvum*), in tomato plants carrying the corresponding resistance gene (*Cf9*) resulted in delayed growth, necrosis and complete cell death. In recent years, a number of *R* genes have been isolated which may be introduced in heterologous system (unrelated species) to engineer disease resistance.

Table 12: Examples of Development of Virus Resistant Transgenic in Vegetables

Crop/Class of Trans Gene	*Resistant to*
Tomato	
Cp gene	Tomato mosaic virus
Cp gene	Tomato yellow leaf curl virus
Antisense RNA	Tomato mosaic virus
Satellite RNA	Cucumber mosaic virus
N gene	Tomato spotted wilt virus
Truncated C1 gene*	Cucumber mosaic virus
Two *Cp* gene**	Cucumber mosaic virus
Cucumber	
Cp gene	Cucumber mosaic virus

Source: Singh *et al.* (1998).

Transgenic Resistant to Viral Disease

Ever since the first report on the development of resistant transgenic tobacco plants to virus using coat protein gene by Powell Abel in 1986, several strategies have been described and executed to produce virus resistant transgenic plants. Presently, virus resistant transgenics are being produced utilizing following gene constructs: (i) virus resistance gene from one plant species to other, (ii) sequence corresponding to ameliorative satellite RNAs, (iii) gene (s) that encode antiviral proteins or compounds, (iv) antisense RNA gene corresponding to segments of viral genome and (v) viral genes like coat protein genes etc. some of these approaches have been utilized to develop virus resistant transgenic plants in vegetable crops. Recently, four transgenics tomato lines possessing single copy of four different coat protein (*Cp*) genes were evaluated under field condition in various parts of Italy. In general, CMV resistance was confirmed, although it was less effective than that in growth chamber experiment (Tomassoli *et al.*, 1999). Against bacteria, major strategy to develop resistant transgenics is detoxification of pathogenic bacterial toxins. Isolation and characterization of genes for toxins resistance in recent past has opened the path of developing bacteria resistant transgenic crops (Table 12).

Transgenic with Delayed Ripening

Use of transgenic in vegetable crops is an other area for longer shelf life. The strategies adopted so for the delay ripening have been to engineer the crops to express low levels of polygalacturonase (PG), ACC synthase and ACC oxidase using the respective antigens RNAs or to over express ACC deaminase and SAM decarboxlase which lead to decrease ethylene synthesis. These approaches have been successfully adopted in tomato and melons.

Chapter 14

National Seed Policy

Seed is the most important determinant of agricultural production potential on which the efficiency of other agriculture inputs is dependent. Seeds of appropriate characteristics are required to meet the demand of diverse agro-climatic conditions and intensive cropping systems. Sustained increase in agriculture production and productivity is dependent to a large extent on the development of new and improved varieties of crops and an efficient system for timely supply of quality seeds to farmers. The seed sector has made impressive progress over the last three decades. The area under certified seeds has increased from less than 500 hectares in 1962-63 to over 5 lakh hectares in 1999-2000. The quantum of quality seeds has crossed 100 lakh quintals. The Seeds Act, 1966 and Seeds Control Order promulgated thereunder, and the New Policy on Seeds Development, 1988, form the basis of promotion and regulation of the Seed Industry. Far-reaching changes, however, have taken place in the national economic and agricultural scenario and in the international environment since the enactment of the existing seed legislation and the announcement of the 1988 Policy.

Aims and Objectives

It has become evident that in order to achieve the food production targets of the future a major effort will be required to enhance the seed replacement rates of various crops. A major increase in the production of quality seeds is required in which the private sector is expected to play a significant role. At the same time, private and Public Sector Seed Organizations at both Central and State levels are expected to adopt economic pricing policies that would seek to realize the true cost of production. The creation of a facilitative climate for growth of a competitive and localized seed industry, encouragement of import of useful germplasm and boosting of exports are the core elements of the agricultural strategy of the new millennium.

Biotechnology will be a key factor in agricultural development in the coming decades. Genetic engineering/modification techniques hold enormous promise in developing crop varieties with a higher level of tolerance to biotic and abiotic stresses. A conducive atmosphere for application of frontier sciences in varietal development and for enhanced investments in research and development is a pressing requirement. At the same time, concerns relating to possible harm to human and animal health and bio-safety, as well as interests of farmers, must be addressed.

Globalization and economic liberalization have opened up new opportunities as well as challenges. The main objectives of the National Seeds Policy, therefore, are the provision of an appropriate climate for the seed industry to utilize available and prospective opportunities, safeguarding of the interests of Indian farmers and the conservation of agro-biodiversity. While unnecessary regulation needs to be dismantled, it must be ensured that gullible farmers are not exploited by unscrupulous elements. A regulatory system of a new genre is, therefore, needed, which will encompass quality assurance mechanisms coupled with facilitation of a vibrant and responsible seed industry.

Varietal Development and Plant Variety Protection

1. The development of new and improved varieties of plants and availability of such varieties to Indian farmers is of crucial importance for a sustained increase in agricultural productivity.

2. Appropriate policy framework and programmatic interventions will be adopted to stimulate varietal development in tune with market trends, scientific-technological advances, suitability for biotic and abiotic stresses, locational adaptability and farmers' needs.
3. An effective sui generis system for intellectual property protection will be implemented to stimulate investment in research and development of new plant varieties and to facilitate the growth of the Seed Industry in the country.
4. A Plant Varieties and Farmers' Rights Protection (PVP) Authority will be established which will undertake registration of extant and new plant varieties through the Plant Varieties Registry on the basis of varietal characteristics.
5. The registration of new plant varieties by the PVP Authority will be based on the criteria of novelty, distinctiveness, uniformity and stability.
6. The criteria of distinctiveness, uniformity and stability could be relaxed for registration of extant varieties, which will be done within a specified period to be decided by the PVP Authority.
7. Registration of all plant genera or species as notified by the Authority will be done in a phased manner.
8. The PVP Authority will develop characterisation and documentation of plant varieties registered under the PVP Act and cataloguing facilities for all varieties of plants.
9. The rights of farmers to save, use, exchange, share or sell farm produce of all varieties will be protected, with the proviso that farmers shall not be entitled to sell branded seed of a protected variety under the brand name.
10. The rights of researchers to use the seed/planting material of protected varieties for bonafide research and breeding of new plant varieties will be ensured.
11. Equitable sharing of benefit arising out of the use of plant genetic resources that may accrue to a breeder from commercialisation of seeds/planting materials of a new variety, will be provided.

12. Farmers/groups of farmers/village communities will be rewarded suitably for their significant contribution in evolution of a plant variety subject to registration. The contribution of traditional knowledge in agriculture needs to be highlighted through suitable mechanisms and incentives.
13. A National Gene Fund will be established for implementation of the benefit sharing arrangement, and payment of compensation to village communities for their contribution to the development and conservation of plant genetic resources and also to promote conservation and sustainable use of genetic resources. Suitable systems will be worked out to identify the contributions from traditional knowledge and heritage.
14. Plant Genetic Resources for Food and Agriculture Crops will be permitted to be accessed by Research Organisations and Seed Companies from public collections as per the provisions of the 'Material Transfer Agreement' of the International Treaty on Plant Genetic Resources and the Biological Diversity Bill.
15. Regular interaction amongst the Private and Public Researchers, Seed Companies/Organisations and Development Agencies will be fostered to develop and promote growth of a healthy seed industry in the country.
16. To keep abreast of global developments in the field of Plant Variety Protection and for technical collaboration, India may consider joining Regional and International Organisations.
17. The PVP Authority may, if required, resort to compulsory licensing of a protected variety in public interest on the ground that requirements of the farming community for seeds and propagating material of a variety are not being met or that the production of the seeds or planting material of the protected variety is not being facilitated to the fullest possible extent.

Seed Production

1. To meet the Nation's food security needs, it is important to make available to Indian farmers a wide range of seeds of

superior quality, in adequate quantity on a timely basis. Public Sector Seed Institutions will be encouraged to enhance production of seed towards meeting the objective of food and nutritional security.

2. The Indian seed programme adheres to the limited three generation system of seed multiplication, namely, breeder, foundation and certified seed. Breeder seed is the progeny of nucleus seed.
3. Nucleus seed is the seed produced by the breeder to develop the particular variety and is directly used for multiplication as breeder seed.
4. Breeder seed is the seed material directly controlled by the originating or the sponsoring breeder or Institution for the initial and recurring production of foundation seed.
5. Foundation seed is the progeny of breeder seed. Foundation seed may also be produced from foundation seed. Production of foundation seed stage-I and stage-II may thus be permitted, if supervised and approved by the Certification Agency and if the production process is so handled as to maintain specific genetic purity and identity.
6. Certified seed is the progeny of foundation seed or the progeny of certified seed. If the certified seed is the progeny of certified seed, then this reproduction will not exceed three generations beyond foundation stage-I and it will be ascertained by the Certification Agency that genetic identity and genetic purity has not been significantly altered.
7. Public Sector Seed Production Agencies will continue to have free access to breeder seed under the National Agriculture Research System. The State Farms Corporation of India and National Seeds Corporation will be restructured to make productive use of these organisations in the planned growth of the Seed Sector.
8. Private Seed Production Agencies will also have access to breeder seed subject to terms and conditions to be decided by Government of India.
9. State Agriculture Universities/ICAR Institutes will have the primary responsibility for production of breeder seed as per the requirements of the respective States.

10. Special attention will be given to the need to upgrade the quality of farmers' saved seeds through interventions such as the Seed Village Scheme.
11. Seed replacement rates will be raised progressively with the objective of expanding the use of quality seeds.
12. DAC, in consultation with ICAR and States, will prepare a National Seed Map to identify potential, alternative and non-traditional areas for seed production of specific crops.
13. To put in place an effective seed production programme, each State will undertake advance planning and prepare a perspective plan for seed production and distribution over a rolling (five to six year) period. Seed Banks will be set up in non-traditional areas to meet demands for seeds during natural calamities.
14. The 'Seed Village Scheme' will be promoted to facilitate production and timely availability of seed of desired crops/ varieties at the local level. Special emphasis will be given to seed multiplication for building adequate stocks of certified/quality seeds by providing foundation seed to farmers.
15. For popularising newly developed varieties and promoting seed production of these varieties, seed minikits of pioneering seed varieties will be supplied to farmers. Seed exchange among farmers and seed producers will be encouraged to popularise new/non-traditional varieties.
16. Seeds of newly developed varieties must be made available to farmers with minimum time gap. Seed producing agencies will be encouraged to tie up with Research Institutions for popularization and commercialization of these varieties.
17. As hybrids have the potential to improve plant vigour and increase yield, support for production of hybrid seed will be provided.
18. Seed production will be extended to agro-climatic zones which are outside the traditional seed growing areas, in order to avoid unremunerative seed farming in unsuitable areas.

19. Seed Banks will be established for stocking specified quantities of seed of required crops/varieties for ensuring timely and adequate supply of seeds to farmers during adverse situations such as natural calamities, shortfalls in production, etc. Seed Banks will be suitably strengthened with cold storage and pest control facilities.
20. The storage of seed at the village level will be encouraged to facilitate immediate availability of seeds in the event of natural calamities and unforeseen situations. For the storage of seeds at farm level, scientific storage structures will be popularised and techniques of scientific storage of seeds will be promoted among farmers as an extension practice.
21. Seed growers will be encouraged to avail of Seed Crop Insurance to cover risk factors involved in production of seeds. The Seed Crop Insurance Scheme will be reviewed so as to provide effective risk cover to seed producers and will be extended to all traditional and non-traditional areas covered under the seed production programme.

Quality Assurance

1. The Seeds Act will be revised to regulate the sale, import and export of seeds and planting materials of agriculture crops including fodder, green manure and horticulture and supply of quality seeds and planting materials to farmers throughout the country.
2. The National Seeds Board (NSB) will be established in place of existing Central Seed Committee and Central Seed Certification Board. The NSB will have permanent existence with the responsibility of executing and implementing the provisions of the Seeds Act and advising the Government on all matters relating to seed planning and development. The NSB will function as the apex body in the seed sector.
3. All varieties, both domestic and imported varieties, that are placed on the market for sale and distribution of seeds and planting materials will be registered under the Seeds Act. However, for vegetable and ornamental crops a simple

system of varietal registration based on "breeders declaration" will be adopted.

4. The Board will undertake registration of kinds/varieties of seeds that are to be offered for sale in the market, on the basis of identified parameters for establishing value for cultivation and usage (VCU) through testing/trialling.

50 Registration of varieties will be granted for a fixed period on the basis of multilocational trials to determine VCU over a minimum period of three seasons, or as otherwise prescribed as in the case of long duration crops and horticultural crops. Samples of the material for registration will be sent to the NBPGR for retention in the National Gene Bank.

6. Varieties that are in the market at the time of coming into force of the revised Seeds Act, will have to be registered within a fixed time period, and subjected to such testing as will be notified.

7. The NSB will accredit ICAR, SAUs, public/private organisations to conduct VCU trials of all varieties for the purpose of registration as per prescribed standards.

8. The NSB will maintain the National Seeds Register containing details of varieties that are registered. This will help the Board to coordinate and assist activities of the States in their efforts to provide quality seeds to farmers.

9. The NSB will prescribe minimum standards (of germination, genetic characteristics, physical purity, seed health, etc.) as well as suitable guidelines for registration of seed and planting materials.

10. Provisional registration would be granted on the basis of information filed by the applicant relating to trials over one season to tide over the stipulation of testing over three seasons before the grant of registration.

11. Government will have the right to exclude certain kinds or varieties from registration to protect public order or human, animal and plant life and health, or to avoid serious prejudice to the environment.

12. The NSB will have the power to cancel the registration granted to a variety if the registration has been obtained

by misrepresentation or concealment of essential data, the variety is obsolete and has outlived its utility and if the prevention of commercial exploitation of such variety is necessary in the public interest.

13. Registration of Seed Processing Units will be required if such Units meet the prescribed minimum standards for processing the seed.
14. Seed Certification will continue to be voluntary. The Certification tag/label will provide an assurance of quality to the farmer.
15. The Board will accredit individuals or organisations to carry out seed certification including self-certification on fulfillment of criteria as prescribed.
16. To meet quality assurance requirements for export of seeds, Seed Testing facilities will be established in conformity with ISTA and OECD seed certification programmes.
17. The State Government, in conformity with guidelines and standards specified by the Board, will establish one or more State Seed Testing Laboratories or declare any Seed Testing Laboratory in the Government or non-Government Sector as a State Seed Testing Laboratory where analysis of seeds will be carried out in the prescribed manner.
18. Farmers will be encouraged to use certified seeds to ensure improved performance and output.
19. Farmers will retain their right to save, use, exchange, share or sell their farm seeds and planting materials without any restriction. They will be free to sell their seed on their own premises or in the local market without any hindrance provided that the seed is not branded. Farmers' right to continue using the varieties of their choice will not be infringed by the system of compulsory registration.
20. Stringent measures would be taken to ensure the availability of high quality of seeds and check the sale of spurious or misbranded seeds.

Seed Distribution and Marketing

1. The availability of high quality seeds to farmers through an improved distribution system and efficient marketing

set-up will be ensured to facilitate greater security of seed supply.

2. For promoting efficient and timely distribution and marketing of seed throughout the country, a supportive environment will be provided to encourage expansion of the role of the private seed sector. Efforts will be made to achieve better coordination between State Governments to facilitate free Inter-State movement of seed and planting material through exemption of duties and taxes.
3. Private Seed Sector will be encouraged and motivated to restructure and reorient their activities to cater to non-traditional areas.
4. A mechanism will be established for collection and dissemination of market intelligence regarding preference of consumers and farmers.
5. A National Seed Grid will be established as a data-base for monitoring of information on requirement of seed, its production, distribution and preference of farmers on a district-wise basis.
6. Access to term finance from Commercial Banks will be facilitated for developing efficient seed distribution and marketing facilities for growth of the seed sector.
7. Distribution and marketing of seed of any variety, for the purpose of sowing and planting will be allowed only if the said variety has been registered by the National Seeds Board.
8. National Seeds Board can direct a dealer to sell or distribute seeds in a specified manner in a specified area if it is considered necessary to the public interest.

Infrastructure Facilities

1. To meet the enhanced requirement of quality/certified seeds, creation of new infrastructure facilities along with strengthening of existing facilities, will be promoted.
2. National Seed Research and Training Center will be set up to impart training and build a knowledge base in various disciplines of the seed sector.

3. The Central Seed Testing Laboratory will be established at the National Seed Research and Training Center to perform referral and other functions as required under the Seeds Act.
4. Seed processing capacity will be augmented to meet the enhanced requirement of quality seed.
5. Modernisation of seed processing facilities will be encouraged in terms of modern equipment and latest techniques, such as seed treatment for enhancement of performance of seed, etc.
6. Conditioned storage for breeder and foundation seed and aerated storage for certified seed would be created in different regions.
7. A computerized National Seeds Grid will be established to provide information on availability of different varieties of seeds with production agencies, their location, quality etc. This network will facilitate optimum utilisation of available seeds in every region.
8. Initially, seed production agencies in the public sector would be connected with the National Seed Grid, but progressively the private sector will be encouraged to join the Grid for providing a clear assessment of demand and supply of seeds.
9. State Governments, or the National Seeds Board in consultation with the concerned State Government, may establish Seed Certification Agencies.
10. State Governments will establish appropriate systems for effective execution and implementation of the objectives and provisions of the Seeds Act.

Transgenic Plant Varieties

1. Biotechnology will play a vital role in the development of the agriculture sector. This technology can be used not only to develop new crops/varieties, which are tolerant to disease, pests and abiotic stresses, but also to improve productivity and nutritional quality of food.
2. All genetically engineered crops/varieties will be tested for environment and bio-safety before their commercial

release, as per the regulations and guidelines of the Environment Protection Act (EPA), 1986.

3. The EPA, 1986, read with the Rules, 1989 would adequately address the safety aspects of transgenic seeds/planting materials. A list will be generated from Indian experience of transgenic cultivars that could be rated as environmentally safe.
4. Seeds of transgenic plant varieties for research purposes will be imported only through the National Bureau of Plant Genetic Resources (NBPGR) as per the EPA, 1986.
5. Transgenic crops/varieties will be tested to determine their agronomic value for at least two seasons under the All India Coordinated Project Trials of ICAR, in coordination with the tests for environment and bio-safety clearance as per the EPA before any variety is commercially released in the market.
6. After the transgenic plant variety is commercially released, its seed will be registered and marketed in the country as per the provisions of the Seeds Act.
7. After commercial release of a transgenic plant variety, its performance in the field, will be monitored for at least 3 to 5 years by the Ministry of Agriculture and State Departments of Agriculture.
8. Transgenic varieties can be protected under the PVP legislation in the same manner as non-transgenic varieties after their release for commercial cultivation.
9. All seeds imported into the country will be required to be accompanied by a certificate from the Competent Authority of the exporting country regarding their transgenic character or otherwise.
10. If the seed or planting material is a product of transgenic manipulation, it will be allowed to be imported only with the approval of the Genetic Engineering Approval Committee (GEAC), set up under the EPA, 1986.
11. Packages containing transgenic seeds/planting materials, if and when placed on sale, will carry a label indicating their transgenic nature. The specific characteristics including the agronomic/yield benefits, names of the

transgenes and any relevant information shall also be indicated on the label.

12. Emphasis will be placed on the development of infrastructure for the testing, identification and evaluation of transgenic planting materials in the country.

Import of Seeds and Planting Material

1. The objective of the import policy is to provide the best planting material available anywhere in the world to Indian farmers, to increase productivity, farm income and export earnings, while ensuring that there is no deleterious effect on environment, health and bio-safety.
2. While importing seeds and planting material, care will be taken to ensure that there is absolutely no compromise on the requirements under prevailing plant quarantine procedures, so as to prevent entry into the country of exotic pests, diseases and weeds detrimental to Indian agriculture.
3. All imports of seeds will require a permit granted by the Plant Protection Advisor to the Government of India, which will be issued within the minimum possible time frame.
4. All import of seeds and planting materials, etc. will be allowed freely subject to EXIM Policy guidelines and the requirements of the Plants, Fruits and Seeds (Regulation of import into India) Order, 1989 as amended from time to time. Import of parental lines of newly developed varieties will also be encouraged.
5. Seeds and planting materials imported for sale into the country will have to meet minimum seed standards of seed health, germination, genetic and physical purity as prescribed.
6. All importers will make available a small sample of the imported seed to the Gene Bank maintained by NBPGR.
7. The existing policy, which permits free import of seeds of vegetables, flowers and ornamental plants, cuttings, saplings of flowers, tubers and bulbs of flowers by certain specified categories of importers will continue. Tubers and bulbs of flowers will be subjected to post-entry quarantine.

8. After the arrival of consignments at the port of entry, quarantine checks would be undertaken; which may include visual inspection, laboratory inspection, fumigation and grow-out tests. For the purpose of these checks, samples will be drawn and the tests will be conducted concurrently.

Export of Seeds

1. Given the diversity of agro-climatic conditions, strong seed production infrastructure and market opportunities, India holds significant promise for export of seeds.
2. Government will evolve a long term policy for export of seeds with a view to raise India's share of global seed export from the present level of less than 1 per cent to 10 per cent by the year 2020.
3. The export policy will specifically encourage custom production of seeds for export and will be based on long term perspective, dispensing with case to case consideration of proposals.
4. Establishment and strengthening of Seeds Export Promotion Zones with special incentives from the Government will be facilitated.
5. A data bank will be created to provide information on the International Market and on export potential of Indian varieties in different parts of the world.
6. A data base on availability of seeds of different crops to assess impact of exports on domestic availability of seeds will be created.
7. Promotional programmes to improve the quality of Indian seeds to enhance its acceptability in the International Market will be taken up.
8. Testing and certification facilities will be established in conformity with international requirements.

Promotion of Domestic Seed Industry

1. Incentives will be provided to the domestic seed industry to enable it to produce seeds of high yielding varieties and hybrid seeds at a faster pace to meet the challenges of domestic requirements.

2. Seed Industry will be provided with a congenial and liberalized climate for increasing seed production and marketing, both domestic and international.
3. Membership to International Organisations and Seed Associations like ISTA, OECD, UPOV, ASSINSEL, WIPO, at the National level or at the level of individual seed producing agencies, will be encouraged.
4. Emphasis will be given to improving the quality of seed produced and special efforts will be directed towards improving the quality of farmers' saved seeds.
5. Financial support for capital investment, working capital and infrastructure strengthening will be facilitated through NABARD/ Commercial Banks/Cooperative Banks.
6. Tax rebate/concessions will be considered on the expenditure incurred on in-house research and development of new varieties and other seed related research aspects. In order to develop a competitive seed market, the States will be encouraged to remove unnecessary local taxation on sales of seeds.
7. To encourage seed production in non-traditional areas including backward areas, special incentives such as transport subsidy will be provided to seed producing agencies operating in these marginalized areas.
8. Reduction of import duty will be considered on machines and equipment used for seed production and processing which are otherwise not manufactured in the country.

Strengthening of Monitoring System

1. The Department of Agriculture and Cooperation (DAC) will supervise the overall implementation and monitoring of the National Seeds Policy.
2. The physical infrastructure in terms of office automation, communication facilities, etc., in DAC will be augmented in a time bound manner.
3. The technical capacity of DAC need to be augmented and strengthened to undertake the additional work relating to implementation of National Seeds Policy, implementation

of PVP&FR Bill, Seeds Act, Import and Export of Seeds, etc.

4. Capacity building, including National and International training and participation in Seminars/Workshops will be organized for concerned officials.

The Government of India trusts that the National Seeds Policy will receive the fullest support of State Governments/Union Territory Administrations, State Agricultural Universities, plant breeders, seed producers, the seed industry and all other stakeholders, so that it may serve as a catalyst to meet the objectives of sustainable development of agriculture, food and nutritional security for the population, and improved standards of living for farming communities. The National Seeds Policy will be a vital instrument in attaining the objectives of doubling food production and making India hunger free. It is expected to provide the impetus for a new revolution in Indian agriculture, based on an efficient system for supply of seeds of the best quality to the cultivator. The National Seeds Policy will lay the foundation for comprehensive reforms in the seed sector. Significant changes in the existing legislative framework will be effected accompanied by programmatic interventions. The Policy will also provide the parameters for the development of the seed sector in the Tenth and subsequent Plans. The progress of implementation of the Policy will be monitored by a High Level Review Committee.

Chapter 15

Certification of Seeds

The introduction of hybrid and high yielding varieties of different crops have further given incentive to the farmers for use of quality seed in crop husbandry. The demand for certified seed is therefore on increase particularly for hybrid crops. The National Seeds Corporation started functioning as Seed Certification Agency for the whole country initially. The Government of India subsequently made provision under section 8 of the Seed Act 1966 to have an independent Seed Certification Agency.

Seed certification is a legally sanctioned system for quality control of seed multiplication and production. A well-organized seed certification system helps in accomplishing the following three primary objective of seed programme:

1. The systematic increase of superior varieties.
2. The identification of new varieties and their rapid increase under appropriate and generally accepted names.
3. Provision of a continuing supply of comparable materials by careful maintenance.

Purpose of Seed Certification

The main function of seed certification is to encourage the production of adequate quantities of genetically pure and good quality seeds, particularly high potential, improved and hybrid varieties of different crops. In order to achieve this requirement the Seed Certification Agency carries out the certification work various seed notified under Indian Seeds Act 1966 and Seed Rules 1968.

Certification Standards

To ensure the quality seed material to the farmers, minimum seed certification standards are most important. The standards, so prescribed should be achievable by the grower and at the same time high enough to meet the needs of the farmer–consumer. The certification standards in force can broadly be divided in to 2 groups:

1. Indian Minimum Seed Certification Standards (IMSCS)
2. Seed Certification Procedure (SCP)

The Indian Minimum Seed Certification Standards

1. General Seed Certification Standards requirements, which are basic and are applicable to all crops under certification.
2. Specific seed certification Standards–requirements that are applicable only to a crop or a group of crops.

Phases of Seed Certification

Certification shall be completed in five broad phases listed as under.

1. Receipt and scrutiny of application including verification and establishment of seed source and other requirements.
2. Field inspections to verify conformity to the prescribed field standards.
3. Supervision at post-harvest stages including processing and packing.
4. Seed sampling and analysis, including genetic purity test and/or seed health test, if any, in order to verify conformity to the prescribed standards, and
5. Grant of certificate and certification tags, tagging and sealing.

General Seed Certification Standards

The General Seed Certification standards are applicable to all crops, which are eligible for certification, and with field and seed standards for the individual crops, shall constitute the Minimum Seed Certification Standards.

Eligibility Requirements for Certification of Crop Varieties

All the varieties notified under section 5 of the Seed Act would be accepted for certification provided that Minimum Seed Certification Standards are prescribed for these crops and / or varieties by Central Seed Committee New Delhi and it meets minimum requirements laid down hereinafter or those notified time to time.

Classes and Sources of Seed

1. Breeder seed
2. Foundation seed stage-1
3. Foundation seed stage-2
4. Certified seed stage-1
5. Certified seed stage-2

Breeder Seed

Breeder seed is a seed or vegetative propagating material directly controlled by the originating or sponsoring plant breeder of the breeding programme or institution and or seed whose production is personally supervised by a qualified plant breeder and which provides the source for the initial and recurring increase of Foundation Seed.

Breeder seed is not covered under certification. However the breeder seed crop is being inspected by a monitoring-team consisting of the Breeder, representative of National Seeds Corporation (not less than the rank of Deputy Manager) and representative of Seed Certification Agency (not less than the rank of Deputy Director of Seed Certification) and nominees of crop coordinator (not less than the officer of S-II rank). The crops shall be inspected at appropriate stages.

Breeder seed shall be genetically so pure as to guarantee that in the subsequent generation, i.e., certified foundation class seeds shall conform to the prescribed standards of genetic purity. The other quality factors of breeder seed such as physical purity, inert matter, germination etc. shall be indicated on the label on actual basis. Breeder seed shall not be used directly for producing certified class seed skipping foundation class.

Certified Seed

Certified seed shall be the seed certified by Certification Agency notified under section 8 of the Seeds Act, 1966. Certified seed shall consist of two classes, namely,

1. Foundation class seed
2. Certified class seed

Foundation Class Seed

Foundation class seed shall be the progeny of Breeder seed, or be produced from foundation seed, which can be clearly traced to Breeder seed. The production shall be supervised and approved by the Certification Agency to conform the required certification standards viz. genetic purity, Physical purity etc. Thus, foundation seed can even be produced from foundation seed. During the production of Foundation seed, the following guidelines shall be observed:

1. Foundation class seed produced directly from Breeder seed shall be designated as Foundation Seed Stage-1.
2. Foundation seed produced from Foundation Seed Stage-I shall be designated as Foundation Seed Stage-II.
3. Foundation Seed Stage-II will not be used for further increase of Foundation Seed and shall be used only for production of Certified seed class.
4. Minimum Seed Certification standards shall be the same for both Foundation Seed Stage-I and Stage-II.

Certified Class Seed

Certified class seed shall be the progeny of Foundation Seed and its production shall be so handled as to maintain specific genetic identity and purity according to standards prescribed for the crop being certified.

Certification tags issued for certified class shall carry the word "CERTIFIED" for certification stage-I and "CERTIFIED STAGE–II" for certified stage II Class. At the discretion of the certification agency (when considered necessary to maintain adequate seed supplies) certified seed may be the progeny of certified seed provided this reproduction may not exceed three generations and provided further that it is determined by the seed certification agency, that the genetic purity will not be significantly altered.

Seed Standards for Genetic Purity

All Certified Seed Lots Should Conform to the Following Minimum Standards for Genetic Purity Unless Otherwise Specified

Class	*Standards for Minimum Genetic Purity (%)*
Foundation	99.00
Certified	
A. Varieties, Composites, Synthetics and Multilines	98.00
B. Hybrids	95.00

Specific Genetic Purity Standards

Class	*Standards for Minimum Genetic Purity (%)*
Certified hybrid seeds of vegetables	90.00

Unit Area for Certification

State Seed Certification has fixed certain norms for the area to be certified. Area offered for seed production should not be less than one hect. except in case of some vegetables. Unit area must have a well-defined boundary to locate the seed plot properly. Minimum 25 acres area in a village for seed production is essential. In case of Potato Crop the Unit area in a village should be 5 acres. Area offer for seed certification should be easily approachable by public conveyance in all season.

Rejection of Seed Farms and Seed Lots

In general, the seed certification agency aims in large-scale production of certified seeds. In that process, elimination of seed farm below the minimum seed certification standards prescribed becomes inevitable. Any seed field or lot found to be not conforming to the IMSCS is subjected to rejection. In addition to non-compliance of IMSCS July 1988 prescribed, L.F.R shall be issued for the following reasons also.

1. When no crop is found in the seed farm registered.
2. When unable to carryout the minimum number of inspections as prescribed in Minimum Seed Certification Standards.
3. Seed farm without uniform male female ratio through out the field .
4. In case of lodging more than one third part of seed farm:
   (*a*) Prior to or during flowering phase and rendering inspection difficult in cross pollinated crops.
   (*b*) Having no scope to stand up again making inspection impossible in self pollinated crops.
5. Seed farm affected by flood drought and very poor crop husbandry.
6. Non compliance of Seed Certification officer's harvest and post harvest instructions leading to doubt of admixture.
7. Seed lots failing to comply with required standard in other variety seed (O.V.S.) and other crop seed (O.C.S.) test, before processing.
8. If the produce brought for processing exceeds more than 20 per cent of the estimated yield for all crops except hybrid cotton and its parents for which the percentage allowed is 10 per cent .
9. Seed lots losing its identity before the tagging.
10. Seed lots, which has lost the lusture or appearance.
11. Seed farm shown for inspection varies with the one mentioned in sowing report.
12. If the variety of the seed crop inspected is other than the variety registered then seed certification officer can straight away issue L.F.R. without taking counts quoting improper seed source as reason.
13. Preventing the SCO from entering into the Seed farm for counting.

14. Seed farms organised with area more than the area registered by 10 percent for hybrids and by 20 percent for varieties. Under such circumstances the SCO shall recommend for rejection quoting improper seed source due to excess area as reason after taking required counts.

L.F.R. reports should be written in the field inspection reports and sent to the various destinations within 24 hours. On receipt of L.F.R. reports, the Deputy Director of Seed Certification should take a decision and communicate within 5 days from the date of receipt. In case of rejection the same may be communicated to the producer with a copy to SCO and ADSCN in the proforma prescribed.

Certification Stages

Registration of Seed Programme

All those interested in certified seed production are required to submit an application on prescribed profarma to the concerned state seed certification agency alongwith an application of Rs 25. This fee is per season and irrespective of area and crops offered for certification in the application. The registration fee is non-refundable .In case of late registration the area may be accepted after due date alongwith late fee.

The seed certification agency upon receipt of the application is required to verify the following:

1. That the variety/varieties are eligible for seed certification, only notified kinds/varieties are eligible.
2. That the source seed of crop to be sown authentic and is in accordance with the conditions laid in the minimum seed certification standards.
3. That there would be no difficulties in reaching the field for carrying out timely field inspections.
4. That the seed produce is able to provide requisite isolation and the field meets the land requirement mentioned under the minimum seed certification standards.
5. That are seed processing facilities are available to the applicant.

Seed Certification Fee

If the application basis of verification made above is accepted by the certification the applicant (certified seed producer) is required to pay certification fee.

Field Inspection

Indian Minimum Seed Certification Standard have prescribed the field inspection according to growth habit of the crop. It may varies 2 to 4 inspections. Generally the first inspection is to be carried out before flowering (after second at the subsequent third inspection is to be carried out within 48 hours in case of highly cross pollinated crops such as hybrid jowar). The last inspection is to be carried out at the time of crop maturity. Concern officer is giving expected estimation of the yield, later on after harvesting the threshed material is duly sealed and proposed to carry on processing plant for further processing.

Processing/Sampling

The harvested and threshed produced is then brought to the approved seed processing plant where the material is graded by using specified sieves. The grading is important to meet the Minimum Seed Certification Standards for physical purity, inert matter, weed seed etc. The graded seed material is then bagged and sealed, representative samples of this baged material are then drawn as per prescribed procedures. Three samples of each lot are bagged and sealed by the Seed Certification Officer in the presence of seed producer. One sample is send to the laboratory (by giving code numbers). The second is kept by the processor and the third is given to the producer as a verification sample the results of the seed samples of these lots when received are decoded and the lots which made minimum prescribed standards of moisture, germination, physical purity etc. are eligible for tagging. These bags are labelled with certification tags (according to stage of seed) and sealed before handing over the material to the producer as certified seed.

Field Testing

In order to maintain high standard genetical purity of the hybrids are further subjected to field test similarly foundation seed for all varieties of all crops are also subjected to field test before they are released to certified seed. The lot samples of processed good seed are drawn and taken for field test to enables to confirm on field the genetic purity within prescribed limits. Those samples of seed lots confirmed the actual genetical character as given by breeder which are confirmed in field test are eligible for certification.

Revalidation

Fresh seed is having 9 months validity period from date of testing so unsold material from market are brought by different agencies for validation of seed. After accepting the material grading is carried out and sample send to STL for getting physical purity and germination results. After getting satisfactory results the seed is tagged and released. After validation subsequent period of validation is prescribed up to 6 months only.

Certification Tags

Each certified seed bag is attached with certification tag and truthful label. Tags and seals shall provide by Seed Certification Agency. For supply of truthful labels seed producers are responsible. The certification tags are rectangular in shape with proportion of 2:1 in length and breath. Colour of tags are as given below:

Sl.No.	*Seed Stage*	*Tag / Label Colour*
1.	Breeder Seed	Yellow Golden
2.	Foundation seed tag	White
3.	Truthful Label	Opale Green
4.	Certified Seed tag	Azure Blue

Appellate Board

If the decision of the seed certification agency is not acceptable to the seed grower as per section 9 and 10 of seed Act 1966, he may appeal to the Appellate Board within 30 days from the date of decision given by agency. The application obtained within 30 days board should give the decision as early as possible. The decision of Appellate Board is Final to both seed certification agency and grower.

Chapter 16

Seed Act and Seed Rules

The Seeds Act, 1966
(Act No.54 of 1966) [29th December, 1966]

An Act to provide for regulating the quality of certain seeds for sale, and for matters connected therewith.

BE it enacted by Parliament in the Seventeenth Year of the Republic of India as follows:

Short Title, Extent and Commencement

1. (1) This Act may be called the Seeds Act, 1966.

   (2) It extends to the whole of India.

   (3) It shall come into force on such date as the Central Government may, by notification in the Official Gazette, appoint, and different dates may be appointed for different provisions of this Act, and for different States or for different areas thereof.

Definitions

2. In this Act, unless the context otherwise requires,

1. "Agriculture" includes horticulture;
2. "Central Seed Laboratory" means the Central Seed Laboratory established or declared as such under sub-section (1) of section 4;
3. "Certification agency" means the certification agency established under Section 8 or recognized under Section 18;
4. "Committee" means the Central Seed Committee constituted under sub-section (1) of Section 3;
5. "Container" means a box, bottle, casket, tin, barrel, case, receptacle, sack, bag, wrapper or other thing in which any article or thing is placed or packed;
6. "Export" means taking out of India to a place outside India;
7. "Import" means bringing into India from a place outside India;
8. "Kind" means one or more related species or sub-species of crop plants each individually or collectively known by one common name such as cabbage, maize, paddy and wheat;
9. "Notified kind or variety" , in relation to any seed, means any kind or variety thereof notified under Section 5;
10. "Prescribed" means prescribed by rules made under this act;
11. "Seed" means any of the following classes of seeds used for sowing or planting–
    I. Seeds of food crops including edible oil seeds and seeds of fruits and vegetables;
    II. Cotton seeds;
    III. Seeds of cattle fodder; and includes seedlings, and tubers, bulbs, rhizomes, roots, cuttings, all types of grafts and other vegetatively propagated material, of food crops or cattle fodder;
12. "Seed Analyst" means a Seed Analyst appointed under section 12;

13. "Seed Inspector" means a Seed Inspector appointed under section 13;
14. "State Government", in relation to a Union territory, means the administrator thereof;
15. "State Seed Laboratory", in relation to any State, means the State Seed Laboratory established or declared as such under sub-section (2) of section 4 for that State; and
16. "variety" means a sub-division of a kind identifiable by growth, yield, plant, fruit, seed, or other characteristic.

Central Seed Committee

3. (1) The Central Government shall, as soon as may be after the commencement of this Act, constitute a Committee called the Central Seed Committee to advise the Central Government and the State Governments on matters arising out of the administration of this Act and to carry out the other functions assigned to it by or under this Act.

(2) The Committee shall consist of the following members, namely:

i. A Chairman to be nominated by the Central Government;
ii. Eight persons to be nominated by the Central Government to represent such interests that Government thinks fit, of whom not less than two persons shall be representatives of growers of seed;
iii. One person to be nominated by the Government of each of the States.

(3) The members of the Committee shall, unless their seats become vacant earlier by resignation, death or otherwise, be entitled to hold office for two years and shall be eligible for renomination.

(4) The Committee may, subject to the previous approval of the Central Government, make bye-laws fixing the quorum and regulating its own procedure and the conduct of all business to be transacted by it.

(5) The Committee may appoint one or more sub-committees, consisting wholly of members of the Committee or wholly of other persons or partly of members of the Committee

and partly of other persons, as it thinks fit, for the purpose of discharging such of its functions as may be delegated to such sub-committee or sub-committees by the Committee.

(6) The functions of the Committee or any sub-committee thereof may be exercised notwithstanding any vacancy therein.

(7) The Central Government shall appoint a person to be the secretary of the Committee and shall provide the Committee with such clerical and other staff as the Central Government considers necessary.

Central Seed Laboratory and State Seed Laboratory

4. (1) The Central Government may, by notification in the Official Gazette, establish a Central Seed Laboratory or declare any seed laboratory as the Central Seed Laboratory to carry out the functions entrusted to the Central Seed Laboratory by or under this Act.

(2) The State Government may, by notification in the Official Gazette, establish one or more State Seed Laboratories or declare any seed laboratory as a State Seed Laboratory where analysis of seeds of any notified kind or variety shall be carried out by Seed Analysts under this Act in the prescribed manner.

Power to Notify Kinds or Varieties of Seeds

5. If the Central Government, after consultation with the Committee, is of opinion that it is necessary or expedient to regulate the quality of seed of any kind or variety to be sold for purposes of agriculture, it may, by notification in the Official Gazette, declare such kind or variety to be a notified kind or variety for the purposes of this Act and different kinds or varieties may be notified for different States or for different areas thereof.

Power to Specify Minimum Limits of Germination and Purity, etc.

6. The Central Government may, after consultation with the Committee and by notification in the Official Gazette, specify-

(*a*) The minimum limits of germination and purity with respect to any seed of any notified kind or variety;

(*b*) The mark or label to indicate that such seed conforms to the minimum limits of germination and purity specified under clause (*a*) and the particulars which such mark or label may contain.

Regulation of Sale of Seeds of Notified Kinds or Varieties

7. No person shall, himself or by any other person on his behalf, carry on the business of selling, keeping for sale, offering to sell, bartering or otherwise supplying any seed of any notified kind or variety, unless:

(*a*) Such seed is identifiable as to its kind or variety;

(*b*) Such seed conforms to the minimum limits of germination and purity specified under clause (a) of section 6;

(*c*) The container of such seed bears in the prescribed manner, the mark or label containing the correct particulars thereof, specified under clause (b) of section 6; and

(*d*) He complies with such other requirements as may be prescribed.

Certification Agency

8. The State Government or the Central Government in consultation with the State Government may, by notification in the Official Gazette, establish a certification agency for the State to carry out the functions entrusted to the certification agency by or under this Act.

Grant of Certificate by Certification Agency

9. (1) Any person selling, keeping for sale, offering to sell, bartering or otherwise supplying any seed of any notified kind or variety may, if he desires to have such seed certified by the certification agency, apply to the certification agency for the grant of a certificate for the purpose.

(2) Every application under sub-section (1) shall be made in such form, shall contain such particulars and shall be accompanied by such fees as may be prescribed.

(3) On receipt of any such application for the grant of a certificate, the certification agency may, after such enquiry as it thinks fit and after satisfying itself that the seed to which the application relates conforms to the minimum limits of germination and purity specified for that seed under clause (a) of section 6, grant a certificate in such form and on such conditions as may be prescribed.

Revocation of Certificate

10. If the certification agency is satisfied, either on a reference made to it in this behalf or otherwise, that-

(*a*) The certificate granted by it under section 9 has been obtained by misrepresentation as to an essential fact; or

(*b*) The holder of the certificate has, without reasonable cause, failed to comply with the conditions subject to which the certificate has been granted or has contravened any of the provisions of this Act or the rules made there under;

then, without prejudice to any other penalty to which the holder of the certificate may be liable under this Act, the certification agency may, after giving the holder of the certificate an opportunity of showing cause, revoke the certificate.

Appeal

11. (1) Any person aggrieved by a decision of a certification agency under section 9 or section 10, may, within thirty days from the date on which the decision is communicated to him and on payment of such fees as may be prescribed, prefer an appeal to such authority as may be specified by the State Government in this behalf:

Provided that the appellate authority may entertain an appeal after the expiry of the said period of thirty days if it is satisfied that the appellate was prevented by sufficient cause from filing the appeal in time.

(2) On receipt of an appeal under sub-section (1), the appellate authority shall, after giving the appellant an opportunity of being heard, dispose of the appeal as expeditiously as possible.

(3) Every order of the appellate authority under this section shall be final.

Seed Analysts

12. The State Government may, by notification in the Official Gazette, appoint such persons as it thinks fit, having the prescribed qualifications, to be Seed Analysts and define the areas within which they shall exercise jurisdiction.

Seed Inspectors

13. (1) The State Government may, by notification in the Official Gazette, appoint such persons as it thinks fit, having the prescribed qualifications, to be Seed Inspectors and define the areas within which they shall exercise jurisdiction.

(2) Every Seed Inspector shall be deemed to be a public servant within the meaning of section 21 of the Indian Penal Code (45 of 1860) and shall be officially subordinate to such authority as the State Government may specify in this behalf.

Powers of Seed Inspector

14. (1) The Seed Inspector may:

(*a*) Take samples of any seed of any notified kind or variety from:

(*i*) Any person selling such seed; or

(*ii*) Any person who is in the course of conveying, delivering or preparing to deliver such seed to a purchaser or a consignee; or

(*iii*) A purchaser or a consignee after delivery of such seed to him;

(*b*) Send such sample for analysis to the Seed Analyst for the area within which such sample has been taken;

(*c*) Enter and search at all reasonable times, with such assistance, if any, as he considers necessary, any place in which he has reason to believe that an offence under this Act has been or is being committed and order in writing the person in possession of any seed in respect of which the offence has been or is being committed, not to dispose of any stock of such seed for a specific period not exceeding thirty days or, unless the alleged offence is such that the defect may be removed by the possessor of the seed, seize the stock of such seed;

(*d*) Examine any record, register, document or any other material object found in any place mentioned in clause (*c*) and seize the same if he has reason to believe that it may furnish evidence of the commission of an offence punishable under this Act; and

(*e*) Exercise such other powers as may be necessary for carrying out the purposes of this Act or any rule made thereunder.

(2) Where any sample of any seed of any notified kind or variety is taken under clause (a) of sub-section (1), its cost, calculated at the rate at which such seed is usually sold to the public, shall be paid on demand to the person from whom it is taken.

(3) The power conferred by this section includes power to break-open any container in which any seed of any notified kind or variety may be contained or to break-open the door of any premises where any such seed may be kept for sale:

Provided that the power to break-open the door shall be exercised only after the owner or any other person in occupation of the premises, if he is present therein, refuses to open the door on being called upon to do so.

(4) Where the Seed Inspector takes any action under clause (a) of sub-section (1), he shall, as far as possible, call not less than two persons to be present at the time when such action is taken and take their signatures on a memorandum to be prepared in the prescribed form and manner.

(5) The provisions of the Code of Criminal Procedure, 1898 (5 of 1898), shall, so far as may be, apply to any search or

seizure under this section as they apply to any search or seizure made under the authority of a warrant issued under section 98 of the said Code.

Procedure to be followed by Seed Inspectors

15. (1) Whenever a Seed Inspector intends to take sample of any seed of any notified kind or variety for analysis, he shall-

(*a*) Give notice in writing, then and there, of such intention to the person from whom he intends to take sample;

(*b*) Except in special cases provided by rules made under this Act, take three representative samples in the prescribed manner and mark and seal or fasten up each sample in such manner as its nature permits.

(2) When samples of any seed of any notified kind or variety are taken under sub-section (1), the Seed Inspector shall-

(*a*) Deliver one sample to the person from whom it has been taken;

(*b*) Send in the prescribed manner another sample for analysis to the Seed Analyst for the area within which such sample has been taken; and

(*c*) Retain the remaining sample in the prescribed manner for production in case any legal proceedings are taken or for analysis by the Central Seed Laboratory under sub-section (2) of section 16, as the case may be.

(3) If the person from whom the samples have been taken refuses to accept one of the samples, the Seed Inspector shall send intimation to the Seed Analyst of such refusal and thereupon the Seed Analyst receiving the sample for analysis shall divide it into two parts and shall seal or fasten up one of those parts and shall cause it, either upon receipt of the sample or when he delivers his report, to be delivered to the Seed Inspector who shall retain it for production in case legal proceedings are taken.

(4) Where a Seed Inspector takes any action under clause (c) of sub-section (1) of section 14:

(*a*) He shall use all despatch in ascertaining whether or not the seed contravenes any of the provisions of section 7 and if it is ascertained that the seed does not so contravene, forthwith revoke the order passed under the said clause or, as the case may be, take such action as may be necessary for the return of the stock of the seed seized;

(*b*) If he seizes the stock of the seed, he shall, as soon as may be, inform a magistrate and take his orders as to the custody thereof;

(*c*) Without prejudice to the institution of any prosecution, if the alleged offence is such that the defect may be removed by the possessor of the seed, he shall, on being satisfied that the defect has been so removed, forthwith revoke the order passed under the said clause.

(5) Where as Seed Inspector seizes any record, register, document or any other material object under clause (d) of sub-section (1) of section 14, he shall, as soon as may be, inform a magistrate and take his orders as to the custody thereof.

Report of Seed Analyst

16. (1) The Seed Analyst shall, as soon as may be after the receipt of the sample under sub-section (2) of section 15, analyse the sample at the State Seed Laboratory and deliver, in such form as may be prescribed, one copy of the report of the result of the analysis to the Seed Inspector and another copy thereof to the person from whom the sample has been taken.

(2) After the institution of a prosecution under this Act, the accused vendor or the complainant may, on payment of the prescribed fee, make an application to the court for sending any of the samples mentioned in clause (a) or clause (c) of sub-section (2) of section 15 to the Central Seed Laboratory for its report and on receipt of the application, the court shall first ascertain that the mark and the seal or fastening as provided in clause (b) of sub-section (1) of section 15 are intact and may then despatch the sample under its own seal to the Central Seed Laboratory which shall thereupon send its report to the court in the prescribed form within one month from the

date of receipt of the sample, specifying the result of the analysis.

(3) The report sent by the Central Seed Laboratory under sub-section (2) shall supersede the report given by the Seed Analyst under sub-section (1).

(4) Where the report sent by the Central Seed Laboratory under sub-section (2) is produced in any proceedings under Section 19, it shall not be necessary in such proceedings to produce any sample or part thereof taken for analysis.

Restriction on Export and Import of Seeds of Notified Kinds or Varieties

17. No person shall, for the purpose of sowing or planting by any person (including himself), export or import or cause to be exported or imported any seed of any notified kind or variety, unless:

    (*a*) It conforms to the minimum limits of germination and purity specified for that seed under clause (a) of section 6; and

    (*b*) Its container bears, in the prescribed manner, the mark or label with the correct particulars thereof specified for that seed under clause (b) of section 6.

Recognition of Seed Certification Agencies of Foreign Countries

18. The Central Govt. may, on the recommendation of the Committee and by notification in the Official Gazette, recognise any seed certification agency established in any foreign country, for the purposes of this Act.

Penalty

19. If any person:

    (*a*) contravenes any provision of this Act or any rule made there under; or

    (*b*) prevents a Seed Inspector from taking sample under this Act;or (*c*) prevents a Seed Inspector from exercising any other power conferred on him by or under this Act;

he shall, on conviction, be punishable-

(*i*) for the first offence with fine which may extend to five hundred rupees, and

(*ii*) in the event of such person having been previously convicted of an offence under this section, with imprisonment for a term which may extend to six months, or with fine which may extend to one thousand rupees, or with both.

Forfeiture of Property

20. When any person has been convicted under this Act for the contravention of any of the provisions of this Act or the rules made thereunder, the seed in respect of which the contravention has been committed may be forfeited to the Government.

Offences by Companies

21. (1) Where an offence under this Act has been committed by a company, every person who at the time the offence was committed was in charge of, and was responsible to the company for the conduct of the business of the company, as well as the company, shall be deemed to be guilty of the offence and shall be liable to be proceeded against and punished accordingly:

Provided that nothing contained in this sub-section shall render any such person liable to any punishment under this Act if he proves that the offence was committed without his knowledge and that he exercised all due diligence to prevent the commission of such offence.

(2) Notwithstanding anything contained in sub-section (1), where an offence under this Act has been committed by a company and it is proved that the offence has been committed with the consent or connivance of, or is attributable to any neglect on the part of, any director, manager, secretary or other officer of the company, such director, manager, secretary or other officer shall also be

deemed to be guilty of that offence and shall be liable to be proceeded against and punished accordingly.

Explanation. For the purpose of this section,

(*a*) "company" means any body corporate and includes a firm or other association of individuals; and

(*b*) "director", in relation to a firm, means a partner in the firm.

Protection of Action Taken in Good Faith

22. No suit, prosecution or other legal proceeding shall lie against the Government or any officer of the Government for anything which is in good faith done or intended to be done under this Act.

Power to Give Directions

23. The Central Government may give such directions to any State Government as may appear to the Central Government to be necessary for carrying into execution in the State any of the provisions of this Act or of any rule made thereunder.

Exemption

24. Nothing in this Act shall apply to any seed of any notified kind or variety grown by a person and sold or delivered by him on his own premises direct to another person for being used by that person for the purpose of sowing or planting.

Power to Make Rules

25. (1) The Central Government may, by notification in the Official Gazette, make rules to carry out the purpose of this Act.

(2) In particular and without prejudice to the generality of the fore-going power, such rules may provide, for:

(*a*) The functions of the Committee and the travelling and daily allowances payable to members of the Committee and members of any sub-committee appointed under sub-section (5) of section 3;

(*b*) the functions of the Central Seed Laboratory;

(*c*) the functions of a certification agency;

(*d*) the manner of marking or labelling the container of seed of any notified kind or variety under clause (c) of Section 7 and under clause (b) of section 17;

(*e*) the requirements which may be complied with by a person carrying on the business referred to in section 7;

(*f*) the form of application for the grant of a certificate under section 9, the particulars it may contain, the fees which should accompany it, the form of the certificate and the conditions subject to which the certificate may be granted;

(*g*) the form and manner in which and the fee on payment of which an appeal may be preferred under section 11 and the procedure to be followed by the appellate authority in disposing of the appeal;

(*h*) the qualifications and duties of Seed Analysts and Seed Inspectors;

(*i*) the manner in which samples may be taken by the Seed Inspector, the procedure for sending such samples to the Seed Analyst or the Central Seed Laboratory and the manner of analysing such samples;

(*j*) the form of report of the result of the analysis under sub-section (1) or sub-section (2) of section 16 and the fees payable in respect of such report under the said sub-section (2);

(*k*) the records to be maintained by a person carrying on the business referred to in section 7 and the particulars which such records shall contain; and

(*l*) any other matter which is to be or may be prescribed.

(3) Every rule made under this Act shall be laid as soon as may be after it is made, before each House of Parliament while it is in session for a total period of thirty days which may be comprised in one session or in two successive sessions, and if, before the expiry of the session in which it is so laid or the session immediately following, both Houses

agree in making any modification in the rule or both Houses agree that the rule should not be made, that rule shall, thereafter have effect only in such modified form or be of no effect, as the case may be; so however, that any such modification or annulment shall be without prejudice to the validity of anything previously done under that rule.

The Seeds (Amendment) Act, 1972 (No. 55 of 1972) (9th September, 1972)

An Act to amend the Seeds Act, 1966

Be it enacted by Parliament in the Twenty-third Year of the Republic of India as follows:

Short Title and Commencement

1. (1) This Act may be called the Seeds (Amendment) Act, 1972

   (2) It shall come into force on such date as the Central Government may, by notification in the Official Gazette, appoint.

Amendment of Section 2 of 1966

2. In section 2 of the Seeds Act, 1966 (hereinafter referred to as the principal Act), in clause (11), after sub-clause (iii), the following sub-clause shall be inserted, namely:-

   "(iv) jute seeds,"

Insertion of New Sections 8A to 8E

3. After section 8 of the principal Act, the following sections shall be inserted, namely:

The Central Seed Certification Board

"8A. (1) The Central Government shall, by notification in the Official Gazette, establish a Central Seed Certification Board (hereinafter referred to as the Board) to advise the Central Government and the State Governments on all matters relating to certification and to co-ordinate the functioning of the agencies established under section 8.

(2) The Board shall consist of the following members, namely:-

(*i*) a Chairman, to be nominated by the Central Government;

(*ii*) four members, to be nominated by the Central Government from out of the persons employed by the State Governments as Directors of Agriculture;

(*iii*) three members, to be nominated by the Central Government from out of the persons employed by the Agricultural Universities as Directors of Research;

(*iv*) thirteen persons, to be nominated by the Central Government to represent such interests as that Government thinks fit, of whom not less than four persons shall be representatives of seed producers or tradesmen.

(3) A member of the Board shall, unless his seat becomes vacant earlier by resignation or otherwise, be entitled to hold office for two years from the date of his nomination:

Provided that a person nominated under clause (ii) or clause (iii) of sub-section (2) shall hold office only for so long as he holds the appointment by virtue of which his nomination was made.

Other Committees

8B. The Board may appoint as many Committees as it deems fit consisting wholly of the members of the Board or wholly of other persons or partly of members of the Board and partly of other persons as it thinks fit to exercise such powers and perform such duties as may be delegated to them, subject to such conditions as it may think fit, by the Board.

Proceedings of Board or Committee not to be invalid by reason of any vacancy therein

8C. No proceeding of the Board or any Committee thereof shall become invalid merely by reason of the existence of any vacancy therein or any defect in the constitution thereof.

Procedure for Board

8D. The Board may, subject to the previous approval of the Central Government, make bye-laws for the purpose of regulating its own procedure and the procedure of any Committee thereof and the conduct of all business to be transacted by it or such Committee.

Secretary and Other Officers

8E. The Central Government shall:

(*i*) appoint a person to be the Secretary of the Board, and

(*ii*) provide the Board with such technical and other staff as the Central Government considers necessary."

Amendment of Section 9

4. In section 9 of the principal Act,–

(*i*) in sub-section (3), for the words, brackets, letter and figure "minimum limits of germination and purity specified for that seed under clause (a) of section 6", the words " prescribed standards" shall be substituted;

(*ii*) to sub-section (3), the following proviso shall be added, namely:

"Provided that such standards shall not be lower than the minimum limits of germination and purity specified for that seed under clause (a) of section 6."

Amendment of Section 25

1. In section 25 of the principal Act,

(*a*) in sub-section (2), after clause (f), the following clause shall be inserted, namely:

"(*ff*) the standards to which seeds should conform,";

(*b*) in sub-section (3), for the words "in two successive sessions, and if, before the expiry of the session in which it is so laid or the session immediately following", the words "in two or more successive sessions, and if, before the expiry of the session immediately following the session or the successive sessions aforesaid" shall be substituted.

The Seeds Rules, 1968 under Seed Act, 1966 (Act No. 54 of 1966)

Part I: Preliminary

1. Short title. – These rules may be called the Seeds Rules, 1968.
2. Definitions. – In these rules, unless the context otherwise requires,–
   (*a*) "Act" means the Seeds Act, 1966 (54 of 1966);
   (*b*) "advertisement" means all representations other than those on the label, disseminated in any manner or by any means relating to seed for the purposes of the Act;
   (*c*) "certification sample" means a sample of seed drawn by a certification agency or by a duly authorised representative of a certification agency established under section 8 or recognized under section 18 of the Act;
   (*d*) "certification tag" means a tag or label of certain design to be specified by the certification agency and shall constitute the certificate granted by the certification agency;
   (*e*) "certified seed" means seed that fulfils all requirements for certification provided by the Act and these rules and to the container of which the certification tag is attached;
   (*f*) "certified seed producer" means a person who grow or distributes certified seed in accordance with the procedure and standards of the certification agency;
   (*g*) "complete record" means the information which relates to the origin, variety, kind, germination and purity of seed of any notified kind or variety offered for sale, sold or otherwise supplied;
   (*h*)"form" means a form appended to these rules;

(*i*) "origin" means the State, Union Territory or foreign country where the seed is grown and in case seeds of different origin are blended the label shall show the percentage of seed of each origin;

(*j*) "processing" means cleaning, drying, treating, grading and other operations which would change the purity and germination of the seed and thus requiring re-testing to determine the quality of the seed, but does not include operations such as packaging and labelling;

(*k*) "section" means a section of the Act;

(*l*) "service sample" means a sample submitted to the Central Seed Laboratory or to a State Seed Laboratory for testing, the results to be used as information for seeding, selling or labelling purposes;

(*m*) "treated" means that the seed has been subjected to an application of a substance or process in such a manner as to reduce, control or repel certain disease organisms, insects, or any other pests attacking such seeds or seedlings growing therefrom and for other purposes.

Part II: Central Seed Committee

3. Functions of the Central Seed Committee. – In addition to the functions entrusted to the Committee by the Act, the Committee shall,–

(*a*) recommend the rate of fees to be levied for analysis of samples by the Central and State Seed Testing Laboratories and for certification by the certification agencies;

(*b*) advise the Central or State Governments on the suitability of seed testing laboratories;

(*c*) send its recommendations and other concerning records to the Central Government;

(*d*) recommend the procedure and standards for certification, tests and analysis of seeds; and (*e*) carry out such other functions as are supplemental, incidental or consequential to any of the functions conferred by the Act or these rules.

4. Travelling and daily allowances payable to members of the committee and its sub-committees. – The members of the committee and its sub-committees shall be entitled to draw travelling and daily allowances as specified below when they are called upon to attend a meeting of the committee or a sub-committee thereof:

(*a*) An official member of the committee or its sub-committees shall be entitled to draw travelling and daily allowances in accordance with the rules of the Government under which he is for the time being employed and from the same source from which his pay and allowances are drawn.

(*b*) A non-official member shall be allowed travelling and daily allowances in accordance with the general orders issued in this behalf by the Central Government from time to time.

Part III: Central Seed Laboratory

5. *Functions.* – In addition to the functions entrusted to the Central Seed Laboratory by the Act, the Laboratory shall carry out the following functions, namely:

(1) initiate testing programmes in collaboration with the State Seed Laboratories designed to promote uniformity in test results between all seed laboratories in India.

(2) collect data continually on the quality of seeds found in the market and make this data available to the Committee; and

(3) carry out such other functions as may be assigned to it by the Central Government from time to time.

Part IV: Seed Certification Agency

6. *Functions of the Certification Agency.* – In addition to the functions entrusted to the certification agency by the Act, the Agency shall–

(*a*) certify seeds of any notified kinds or varieties;

(*b*) outline the procedure for submission of applications and for growing, harvesting, processing, storage and labellilng of seeds intended for certification till the end to ensure that seed lots finally approved for certification are true to variety and meet prescribed standards for certification under the Act or these rules;

(*c*) maintain a list of recognised breeders of seeds;

(*d*) verify, upon receipt of an application for certification that the variety is eligible for certification, that the seed source used for planting was authenticated and the record of purchase is in accordance with these rules and the fees have been paid;

(*e*) take sample and inspect seed lots produced under the procedure laid down by the certification agency and have such samples tested to ensure that the seed conforms to the prescribed standards of certification;

(*f*) inspect seed processing plants to see that the admixtures of other kinds and varieties are not introduced;

(*g*) ensure that action at all stages, e.g. field inspection, seed processing plant inspection, analysis of samples taken and issue of certificates (including tags, marks, labels and seals) is taken expeditiously;

(*h*) carry out educational programmes designed to promote the use of certified seed including a publication listing certified seed growers and sources of certified seed;

(*i*) grant certificates (including tags labels, seals etc.) in accordance with the provisions of the Act and these rules;

(*j*) maintain such records as may be necessary to verify that seed plants for the production of certified seed were eligible for such planting under these rules;

(*k*) inspect fields to ensure that the minimum standards for isolation, roguing (where applicable) use of male sterility (where applicable) and similar factors are maintained at all times, as well as ensure that seedborne diseases are not present in the field to a greater extent than those provided in the standards for certification.

Part V: Marking or Labelling

7. Responsibility for marking or labelling. – When seed of a notified kind or variety is offered for sale under section 7, each container shall be marked or labelled in the manner hereinafter specified. The person whose name appears on the mark or label shall be responsible for the accuracy of the information required to appear on the mark or label so long as seed is contained in the unopened original container:

   Provided, however, that such person shall not be responsible for the accuracy of the statement appearing on the mark or label if the seed is removed from the original unopened container, or he shall not be responsible for the accuracy of the germination statement beyond the date of validity indicated on the mark or label.

8. Contents of the mark or label. – There shall be specified on every mark or label–

   (*i*) particulars, as specified by the Central Government under clause (b) of section 6 of the Act;

   (*ii*) a correct statement of the net content in terms of weight and expressed in metric system;

   (*iii*) date of testing;

   (*iv*) if the seed in container has been treated–

      (*a*) a statement indicating that the seed has been treated;

      (*b*) the commonly accepted chemical or abbreviated chemical (generic) name of the applied substance; and

      (*c*) if the substance of the chemical used for treatment, and present with the seed is harmful to human beings or other vertebrate animals, a caution statement such as "Do not use for food, feed or oil purposes". The caution for mercurials and similarly toxic substance shall be the word "Poison" which shall be in type size, prominently displayed on the label in red.

(*v*) the name and address of the person who offers for sale, sells or otherwise supplies the seed and who is responsible for its quality;

(*vi*) the name of the seed as notified under section 5 of the Act.

9. Manner of marking or labelling the container under clause (C) of section 7 and clause (B) of section 17.

(1) The mark or label containing the particulars of the seed as specified under clause (b) of section 6 shall appear on each container of seed or on a tag or mark or label attached to the container in a conspicuous place on the innermost container in which the seed is packed and on every other covering in which that container is packed and shall be legible.

(2) Any transparent cover or any wrapper, case or other covering used solely for the purpose of packing of transport or delivery need not be marked or labelled.

(3) Where by a provision of these rules, any particulars are required to be displayed on a label on the container, such particulars may, instead of being displayed on a label be etched, painted or otherwise indelibly marked on the container.

10. Mark or Label not to contain false or misleading statement. – The mark or label shall not contain any statement, claim, design, device, fancy name or abbreviation which is false or misleading in any particular concerning the seed contained in the container.

11. Mark or label not to contain reference to the act or rules contradictory to required particulars. – The mark or label shall not contain any reference to the Act, or any of these, rules or any comment on, or reference to, or explanation of any particulars or declaration required by the Act or any of these rules which directly or by implication contradicts, qualifies or modifies such particulars or declaration.

12. Denial of responsibility for mark or label content prohibited. – Nothing shall appear on the mark or label or in any advertisement pertaining to any seed of any notified kind or variety which shall deny responsibility for the statement required by or under the Act to appear on such mark, label or advertisement.

Part VI: Requirements

13. Requirements to be complied with by a person carrying on the Business referred to in Section 7.

    (1) No person shall sell, keep for sale, offer to sell, barter or otherwise supply any seed of any notified kind or variety, after the date recorded on the container, mark or label as the date upto which the seed may be expected to retain the germination not less than that prescribed under clause (a) of section 6 of the Act.

    (2)No person shall after, obliterate or deface any mark or label attached to the container of any seed.

    (3)Every person selling, keeping for sale, offering to sell, bartering or otherwise supplying any seed of notified kind or variety under section 7, shall keep over a period of three years a complete record of each lot of seed sold except that any seed sample may be discarded one year after the entire lot represented by such sample has been disposed of. The sample of seed kept as part of the complete record shall be as large as the size notified in the official Gazette. This sample, if required to be tested, shall be tested only for determining the purity.

14. Classes and sources of certified seed. –

    (1) There shall be three classes of certified seed, namely, foundations, registered and certified and each class shall meet the following standards for that class:

        (*a*) Foundation seed shall be the progeny of breeders' seed, or be produced from foundation seed which can be clearly traced to breeder's seed. Production shall be supervised and approved by a seed certification agency and be so handled as to maintain specific genetic purity and identify and shall be required to meet certification standards for the crop being certified.

(*b*) Registered seed shall be the progeny of foundation seed that is so handled as to maintain its genetic identity and purity according to standard specified for the particular crop being certified.

(*c*) Certified seed shall be the progeny of registered or foundation seed that is so handled to maintain genetic identity and purity according to standards specified for the particular crop being certified.

(2) At the discretion of the certification agency (when considered necessary to maintain adequate seed supplies) certified seed may be the progeny of certified seed provided this reproduction may not exceed three generations and provided further that it is determined by the seed certification agency, that the genetic purity will not be significantly altered.

Part VII: Certification of Seeds

15. Application for the grant of a certificate. – Every application for the grant of a certificate under sub-section (1) of section 9 shall be made in Form I in accordance with the procedure outlined by the certification agency for submission of applications and contain the following particulars, namely:

(*a*) the name, profession, and place of residence of the applicant;

(*b*) the name of the seed to be certified; its notified kind or variety;

(*c*) class of the seed;

(*d*) source of the seed;

(*e*) limits of germination and purity of the seed;

(*f*) mark or label of the seed.

16. Fees. – Every application under sub-section (1) of section 9 shall be accompanied by a fee of Rs. 25 in cash.

17. Certificate. – Every certificate granted under sub-section (3) of section 9 shall be in Form II and shall be granted by the certification agency, after making enquiries and satisfying itself in accordance with the provisions of the said sub-section on the following conditions, for the period to be specified by the certification agency, namely:

(*i*) The person to whom the certificate is granted under sub-section (3) of section 9 shall attach a certification tag to every container of the certified seed and shall follow the provisions in respect of marking or labelling provided by or under the Act.

(*ii*) The certification tag shall contain the following particulars, namely:

(*a*) name and address of the certification agency;

(*b*) kind and variety of the seed;

(*c*) lot number or other mark of the seed;

(*d*) name and address of the certified seed producer;

(*e*) date of issue of the certificate and of its validity;

(*f*) an appropriate sign to designate certified seed;

(*g*) an appropriate word denoting the class designation of the seed.

(*iii*) The colour of the certification tag shall be white for foundation seed, purple for registered seed and blue for certified seed.

(*iv*) The container of the certified seed shall carry a seal of such material and in such form as the certification agency may determine and no container carrying a certification tag shall be sold by the person if the tag or seal has either been tampered with or removed.

(*v*) The certification tag on the container shall specify:

(*a*) the period during which the seed shall be used for sowing or planting;

(*b*) that the use of seed after the expiry of the validity period by any person is entirely at his risk and the holder of the certificate shall not be responsible for any damage to the buyer of the seed;

(*c*) that no one should purchase the seed if the seal or the certification tag has been tampered with.

(*vi*) The holder of the certificate shall keep record of the details of each lot of the seed which is issued for sale in such form as to be available for inspection and to be easily identified by reference to the number of the lot as shown in the certification tag of each container and such records shall be retained in the case of a seed for which expiry date is fixed for a period of two years from the expiry of such date.

(*vii*) The holder of the certificate shall allow any Seed Inspector, authorized in writing by the certification agency in that behalf, to enter with or without prior notice the premises where the seeds are grown processed and sold and to inspect premises, plant and the process of processing at all reasonable hours.

(*viii*) The holder of the certificate shall allow the Seed Inspector, authorised in writing by the certification agency, to inspect all registers and records maintained under these rules and to take samples of the seeds and shall supply to the Seed Inspector such information as he may require for the purposes of ascertaining whether the conditions subject to which the certificate has been granted, have been complied with.

(*ix*) The holder of the certificate shall on request furnish to the certification agency from every lot of the seed or from such lot or lots as the said agency may from time to time specify, a sample of such quantity as the agency may consider adequate for any examination required to be made.

(*x*) If the certification agency so directs, the holder of the certificate shall not sell or offer for sale any lot in respect of which a sample is furnished under the preceding clause until the agency authorises the sale of such lot.

(*xi*) The holder of the certificate shall on being directed by the certification agency that any part of a lot has been found by the said agency not to conform to prescribed standards of quality or purity specified by or under the Act, withdraw the remainder of that lot from sale and so far as may, in the particular circumstances of the case, be practicable, recall all issues already made from that lot.

(*xii*) The holder of the certificate shall comply with the provisions of the Act and these rules and with the directions given after not less than one month's notice by the certification agency to such holder.

Part VIII: Appeals

18. The form and manner in which and the fee on payment of which the appeal may be referred.

(1) Every memorandum of appeal under sub-section (1) of section 11 shall be in writing and shall be accompanied by a copy of the decision of the certification agency against which it has been preferred and shall set forth concisely and under distinct heads the grounds of objection to such decision without any argument, or narrative.

(2) Every such memorandum of appeal shall be accompanied by a treasury receipt for a sum of Rs. 100/-/

(3) Every such memorandum of appeal may be presented either in person or through an agent duly authorised in writing in this behalf by the appellant or may be sent by the registered post.

19. Procedure to be followed by the appellate authority. – In deciding appeals under the Act the appellate authority shall exercise all the powers which a Court has and shall follow the same procedure which a Court follows in deciding appeals from the decree or order of an original Court under the Code of Civil Procedure, 1908 (5 of 1908)

Part IX: Seed Analysts and Seed Inspectors

20. Qualifications of Seed Analyst. – A person shall not be qualified for appointment as Seed Analyst unless he–

(*i*) possesses a Master's or equivalent degree in Agriculture or Agronomy or Botany or Horticulture of a University recognised for this purpose by the Government and has had not less than one year's experience in seed technology; or

(*ii*) possesses a Bachelor's degree in Agriculture or Botany of a University recognised for this purpose by the Government and has had not less than three years' experience in seed technology.

21. Duties of a Seed Analyst. – On receipt of a sample for analysis the Seed Analyst shall first ascertain that the mark and the seal or fastening as provided in clause (b) of sub-section (1) of section 15 are intact and shall note the condition of the seals thereon.

(2) The Seed Analyst shall analyse the samples according to the provisions of the Act and these rules.

(3) The Seed Analyst shall deliver the copy of the report of the result of the analysis to the persons specified in sub-section (1) of section 16.

(4) The Seed Analyst shall from time to time forward to the State Government the reports giving the result of analytical work done by him.

22. Qualifications of Seed Inspectors. – A person shall not be qualified for appointment as Seed Inspector unless he is a graduate in Agriculture of a University recognised for the purpose by the Government and has had not less than one years experience in seed production, or seed development or seed analysis or testing in a seed testing laboratory.

(*a*) inspect as frequently as may be required by certification agency all places used for growing, storage or sale of any seed of any notified kind or variety;

(*b*) satisfy himself that the conditions of the certificates are being observed;

(*c*) procure and send for analysis, if necessary, samples of any seeds, which he has reason to suspect are being produced stocked or sold or exhibited for sale in contravention of the provisions of the Act or these rules;

(*d*) investigate any complaint, which may be made to him in writing in respect of any contravention of the provisions of the Act or these rules;

(*e*) maintain a record of all inspections made and action taken by him in the performance of his duties including the taking of samples and the seizure of stocks and submit copies of such record to the Director of Agriculture or the certification agency as may be directed in this behalf;

(f) when so authorised by the State Government detain imported containers which he has reason to suspect contain seeds, import of which is prohibited except and in accordance with the provisions of the Act and these rules;

(*g*) institute prosecutions in respect of breaches of the Act and these rules; and

(*h*) perform such other duties as may be entrusted to him by the competent authority.

Part X: Sealing, Fastening, Despatch and Analysis of Samples

24. Manner of taking Samples. – Samples of any seed of any notified kind or variety for the purpose of analysis shall be taken in a clean dry container which shall be closed sufficiently tight to prevent leakage and entrance of moisture and shall be carefully sealed.

25. Containers to be labelled and addressed. – All containers containing samples for analysis shall be properly labelled and the parcels shall be properly addressed. The label on any sample of seed sent for analysis shall bear–

(*a*) serial number;

(*b*) name of the sender with official designation, if any;

(*c*) name of the person from whom the sample has been taken;

(*d*) date and place of taking the sample;

(*e*) kind or variety of the seed for analysis;

(*f*) nature and quantity of preservative, if any, added to the sample.

26. Manner of packing, fastening and sealing the samples. – All samples of seeds sent for analysis shall be packed, fastened and sealed in the following manner:
    (*a*) The stopper shall first be securely fastened so as to prevent leakage of the containers in transit.
    (*b*) The container shall then be completely wrapped in fairly strong thick paper. The ends of the paper shall be neatly folded in and affixed by means of gum or other adhesive.
    (*c*) The paper cover shall be further secured by means of strong twine or thread both above and across the container, and the twine or thread shall then be fastened on the paper cover by means of sealing wax on which there shall be at least four distinct and clear impressions of the seal of the sender, of which one shall be at the top of the packet, one at the bottom and the other two on the body of the packet. The knots of the twine or thread shall be covered by means of sealing wax bearing the impression of the seal of the sender.
27. Form of order. – The order to be given in writing by the Seed Inspector under clause (c) of sub-section (1) of section 14 shall be in Form III.
28. Form of receipt for records. – When a Seed Inspector seizes any record, register, document or any other material object under clause (d) of sub-section (1) of section 14, he shall issue a receipt in Form IV to the person concerned.
29. Samples how to be sent to the Seed Analyst. – The container of sample for analysis shall be sent to the Seed Analyst by registered post or by hand in a sealed packed enclosed together with a memorandum in Form V in an outer cover addressed to the Seed Analyst.
30. Memorandum and impression of seal to be sent separately. – A copy of the memorandum and a specimen impression of the seal used to seal the packet shall be sent to the Seed Analyst separately by registered post or delivered to him or to any person authorised by him.

31. Addition of preservatives to samples. – Any person taking a sample of seed for the purpose of analysis under the Act may add a preservative as may be specified from time to time to the sample for the purpose of maintaining it in a condition suitable for analysis.

32. Nature and quantity of the preservative to be noted on the label. – Whenever any preservative is added to a sample, the nature and quantity of the preservative added shall be clearly noted on the label to be affixed to the container.

33. Analysis of the sample. – On receipt of the packet, it shall be opened either by the Seed Analyst or by an officer authorised in writing in that behalf by the Seed Analyst, who shall record the condition of the seal on the packet. Analysis of the sample shall be carried out at the State Seed Laboratory in accordance with the procedure laid down by the Central Government.

34. Form of notice. – The notice to be given under clause (a) of sub-section (1) of section 15 to the person from whom the Seed Inspector intends to take sample shall be in Form VI.

35. Form of report. – The report of the result of the analysis under sub-section (1) or sub-section (2) of section 16 shall be delivered or sent in Form VII.

36. Fees. – The fees payable in respect of the report from the Central Seed Laboratory under sub-section (2) of section 16 shall be Rs. 10/–per sample of the seed analysed.

37. Retaining of the sample. – The sample of any seed shall, under clause (c) of sub-section (2) of section 15, be retained under a cool, dry environment to eliminate the loss of viability and insect proof or rat proof container. The containers shall be dusted with suitable insecticides and the storage room fumigated to avoid infestation of samples by insects. The sample shall be packed in good quality containers of uniform shape and size before storage.

Glossary

A line: The male sterile parent in a cross being made to produce hybrid seed (male sterile line).

Abortion of pollen grains: Shedding of pollen grains before attaining full maturity.

Acclimatization: The process whereby plants are accustomed to a new environment under human management.

Adaptation: The process by which individuals (or parts of individuals), populations or species change in form of function in such a way to better survive under given environmental conditions.

Additive gene effect: Gene action in which the effects on a genetic trait are enhanced by each additional gene, either an allele at the same locus or genes at different loci.

Adventitious: Occurring in unusual or unexpected locations such as roots on aerial stems, buds on leaves etc.

Aerial roots: Roots produced above ground level, as with those that develop on the stems of ivy.

Aestivation: The arrangement of the parts of a flower or leaves within the bud.

Air screen cleaner: It is the basic seed processing machine. Basic cleaning of all seeds is done by an air screen cleaner.

Allele: An allele is an alternative form of a gene.

Allin: A colourless, odourless, water soluble amino acid present in the uninjured bulb of garlic which on crushing, breaks down in the presence of the enzyme allinase to allicin, the principal ingredient of the odoriferous diallyl disulfide.

Allogamy: When pollen grains from flowers of one plant pollinate the flowers of other plants.

Allopolyploid: An organism in which two different species have each contributed one or more sets of chromosome.

Anemophilous: Wind pollinated.

Angiosperm: A member of the Angiospermae class in which the seeds are enclosed in an ovary. One of the main divisions of flowering plants.

Annual: Plant, which survives for one season only from germination to death.

Anther culture: The culturing of anthers *in vitro* for the purpose of generating haploid plantlets.

Anther: That part of the stamen carried at the top of the filament and bearing the pollen.

Anthesis: Full flower expansion including anther extrusion.

Antibody: Substances in a tissue or fluid of the body that acts in antagonism to a foreign substance (antigen).

Antigen: A substance, usually a protein, introduced in to a living organism that elicits antibody formation.

Apical dominance: In plants, the inhibition of lateral buds by high level of auxins, produced in the lead shoot or apical meristem.

Apical: Pertaining to the apex: situated at the tip.

Apomixis: Reproduction from an unfertilized egg or from somatic cells associated with the egg.

Aroids: A group of vegetable crops under the family Araceae where edible plant are corm and cormels e.g *Colocasia, spp, Amorphophallus spp* etc.

Artificial selection: The practice of choosing individuals from a population for reproduction, usually because these individuals possess one or more desirable traits.

Asexual reproduction: Any process of reproduction that does not involve that formation and union of gametes from the different sexes or mating types.

Asnapsis: The failure or partial failure in the pairing of homologous chromosomes during the meiotic prophase.

Aspirator: A seed processing machine in which seed separations are made by use of air, on the basis of differences in terminal velocity of seed. It has a fan, at the discharge point, which creates a vacuum or negative pressure with in the machine. Air rushing to fill in the vacuum creates a stream of air which is used to separate seeds.

Assortive mating: Mating in which the partners are chosen because they are phenotypically similar.

Asymbiotic: A plant which is hostile to certain other plants, such as horse radish (*Cochlearia armoracia*)

Autogamy: High self pollination where cross pollination is less than 5 per cent, as seen in cowpea, cluster bean, dolichos bean, pea, tomato, fenugreek, French bean etc.

Autopolyploid: A polyploid containing more than two species of the same genome *i.e.* autotriploid (3x), autotetraploid (4x), autohexaploid (6x), e.g potato, sweet potato etc.

Autoradiograph: A record or photograph prepared by labelling a substance such as DNA with a radioactive material such as tritiated thymidine and allowing the image produced by decay radiations to develop on a film over a period of time.

Autosome: A chromosome other than a sex chromosome.

Autozygote: A diploid individual in which the two genes of a locus are identical by descent from an ancestral gene.

Auxanometer: A device or measuring plant growth under laboratory conditions.

Auxotroph: A mutant microorganism (*e.g.*, bacterium or yeast) that will not grow on a minimal medium but that requires the addition of some compound such as an amino acid or a vitamin.

B–line: The fertile counterpart of a line. This line does not have fertility restorer genes and used as the male parents to maintain the A–line.

Back cross method of Breeding: It is method of breeding in which the desirable character (s) of a non-recurrent (donor) parent are added to the genetic background of recurrent (recipient) parent through subsequent generations of backcrossing and selection.

Back cross: Crossing the F_1 hybrids with either of the parents. This may be done to test the genotypic ratio of F_1 or to transfer specific gene complex from one species to another.

Bag closer: The machine used for sewing filled seed bags.

Bagger weigher: These are the small machines which, when properly mounted beneath a bin, will fill and weight a bag in a single operation.

Balanced Heterosis: It is a type of true heterosis, which occurs from the balanced combinations of genes in a hybrid.

Basic number: The number of chromosomes in ancestral diploid ancestors of polyploids, represented by x.

Basic seed: A class of seed in seed certification programme that is the last step in the intial seed multiplication and is intended for the production of certified seed. Seed stock used for the same purposes as basic seed, but not under a certification programme is refered to as "the equivalent of basic seed".

Batch: A quantity of harvested crop put into bin or container on a repetitive basis, specificllly for treatment such as drying.

Berry: Fleshy, superior (sometimes inferior), usually many seeded fruits, developing commonly from a syncarpous pistil (rarely from a single carpel) with axial and parietal placentation, e.g tomato, etc. with the growth of the fruit, the seeds separate from the placentae and lie free in the pulp.

Biennial: Plant having a two-year lifecycle, vegetative in the first season and reproductive in the second season, and this transition from vegetative to reproductive stage often requires environmental trigger such as vernalization or photoperiod. *e.g.* cabbage, onion, carrot etc.

Biometrics: The science dealing with the application of statistical methods to biological problems.

Biotechnology: The application of recombinant DNA, cell and tissue culture, and other methods used to develop new and improved plants and plant products.

Biotype: Distinct physiological race or strain within morphological species. A population of individuals with identical genetic constitution. A biotype may be made up of homozygotes, of which only the former would be expected to breed true.

Bisexuality. A system of producing both stamen and carpel in the same flower on a plant.e.g brinjal, tomato etc.

Blanching: Making plants whiter and less bitter by the partial or total exclusion of light. A process widely used in the production of celery, cauliflower, etc.

Bolting: Significant stem elongation that proceeds flowering; also includes the case of premature emergence of flower stalk. Under normal conditions all vegetables eventually flower and set seed but most are harvested and used in their immature state. Some, however, e.g lettuce, will often bolt in hot weather, producing flower spikes but no hearts.

Border rows: The recommended number of rows of the male parental lines grown on all the sides of hybrid seed field growing two different parents.

Bract: A more or less modified subtending a flower or belonging to an inflorescence.

Breeder seed: Seed increased by the originating, or sponsoring, plant breeder or institution, used as the source for production of foundation seed. It has the maximum genetic purity.

Breeder's rights: Legislation giving the equivalent of patent rights for plant cultivars. The effect is to give the developer of a cultivar sole legal possession of that cultivar and a legal basis for compensation for its use by others. Breeder's rights may be referred to as 'Plant cultivar protection'

Bucket elevator: This euipment is used to lift the seeds to the top of bins/machines for cleaning, sorting or sacking.

Bud pollination: Pollination of flowers before they attain maturity (open and shed pollen grains) with the pollen grains collected from the same plant/genotype. Artificial pollination in the bud stage; practices in the selfing of cabbage.

Bulb: A specialized underground organ consisting of a short, fleshy, usually vertical stem axis bearing at its apex a growing point or a flower primordium enclosed by thick, fleshy scales, eg. onion, etc.

Bulbil: Vegetative part that is actually the modification of flowers (s). It develops into plant directly without formation of seeds; seen in onion, garlic etc.

Bulblets: Miniature bulbs produced around the base of the mother bulb due to development of meristem in the axil of scale leaves.

Buttoning: A condition of cauliflowers resulting in a 'loose head': usually caused by lack of moisture at some stage of development.

Capsanthin: A carotenoid pigment responsible for the characteristic orange red colouration of ripe chilli.

Central Seed Committee: The Central Seed Committee constituted under sub section (1) of section 3 of the Indian Seeds Act, 1966 to advise the Central Government and the State Governments on matters arising out of the administration of the Seeds Act, 1966, and to carry out the other functions assigned to it by or under the Seed Acts, 1966.

Centrifugal divider: This equipment is also commnly used for dividing a sample in the seed testing laboratory. It makes use of centrifugal force to mix and scatter the seeds over the dividing surface.

Certified Seed: The progeny of a foundation or registered seed, which maintains the satisfactory genetic identity and purity and has been certified and approved by the certifying agency.

Chemical hybrdisig agents (CHA'S): Chemicals which cause pollen abortion and render the treated plants male sterile while not affecting the ovule fertility.

Chilling injury: Undesirable changes to some vegetables occur when the temperature is reduced below a specific optimum level and such changes are collectively known as chilling injury.

Chitting: Promoting the germination of seed before sowing. Sprouting of seed potatoes prior to planting.

Chromosomes: Darkly staining nucleoprotein bodies that are observed in cells during division. Each chromosome carries a linear array of genes.

Cleistogamy: A built in breeding mechanism where flowers remain closed at the time of pollination, which favour self-pollination, as seen in lettuce.

Clone: A group of individuals derived by a single original plant propagated by vegetative means.

Closed anther mutant: A mutated anther type which though produces viable pollen grains but doesn't shed them due to non-rupturing and causes bareness in plants particularly where self pollination is the rule. Closed anther mutants can be maintained by hand pollination with the pollens collected from such flowers by force opening at full maturity.

Clove: One of the young bulbs of shallots or garlic developed as an offshoot of a larger bulb.

Club roots: A fungus disease, which attacks cruciferous plants, including brassicas. The fungus lives in badly drained acid soils and may persist for many years. Prevention is by dressing the soil with lime before planting, and allowing a lapse of at least three years before replanting brassicas on the same piece of ground.

Co dominant genes: Alleles, each of which produce an independent effect in F_1.

Colchicine: An alkaloids extracted from seeds or corms of *Colchicum autumnale,* which induces polyploid by arresting, spindle formation during mitosis.

Collar rot: The rotting of plants and young seedlings at soil level, caused by soil borne fungi. Stem rot, foot rot, black legs and damping-off are a few of the names applied to this problem. There is no cure but some control can be obtained by the timely application of a fungicide.

Combinational heterosis: The heterosis in quantitative characters resulting from the overall combination of the favourable cumulative effects of a number of componential characters.

Combining ability: General combining ability: The comparative ability of a line or a genetic stock to combine with a tester or a group of tester.

Compatible: Capable of fertilization.

Complementary genes: Those genes which have more or less similar phenotypic expression individually but when they come together they interact to produce a new character expression. If two such genes are complementary for a dominant effect, a 9:7 ratio results in F_2; if two are complementary for a recessive effect, a 15:1 ratio results in F_2.

Composite sample: A sample formed by combining and mixing all the primary samples taken from the lot.

Composites: The advanced generations seed mixture of an interval or interracial cross.

Composite variety: A composite variety is developed by inter mating (3-4 generations) selected open pollinated parent varieties (genotypes).

Conditioning and pre cleaning: This involves removing awns or hulls, breaking up clusters, scapling off large trash and other operations, which improve the condition and flowability of the seed.

Contaminants: The factors which affect the genetic and physical quality of seeds such as off types, foreign pollen, other crop plants, weed plants, plants affected by designated diseases, weed seeds etc., are referred to as contaminants.

Convergent improvement: A system of double back crossing for the purpose of improving each of the two inbred lines without greatly modifying the yield of their F_1 cross.

Coppicing: The cutting of tree back to ground level in order to promote the growth of several new stems from one rootstock.

Corm: The swollen, solid, usually subterranean base of a stem, *e.g.* colocasia, amorphophallus etc. Tiny corms which are produced around the base of the corm are known as cormels or cormlets.

Cotyledon: A primary leaf or seed leaf. Quite different from the true leaves, which develop later, they are usually dicotyledons (two seed leaves) or monocotyledons (single seed leaves).

Cracking: Bursting of lenticels of the roots favoured by dry weather followed by wet weather.

Crop rotation: A system by which a crop is not grown in the same place in successive years. A three-year rotation is the norm but two-four-or more may be operated according to the condition of the soil and the type of crop to be planted. This method improves the fertility of the soil and, to some extent, control the population of the soil pests.

Cross–pollinated crops: An assembly of genetically heterozygous individuals under commercial cultivation which share a common gene pool and in which each individual takes new genotype generation after generation.

Cross–pollination: Transfer of pollen from the anther of one plant to the stigma of another plant. It affects the union of genetically dissimilar gametes.

Cross: The product of the mating between two or more parents of dissimilar genetic constitution. The various types of crosses utilized in heterosis breeding programmes are single, three-way, double crosses, top cross, multiple crosses etc.

Crossing: Artificial mating of two or more parents of unlike genetical constitutions.

Crown: Top stem structure of a perennial herbaceous plant or the upper branchy part of a tree above the bole.

Culm: The stem of a plant, usually hollow except at the nodes, such as the stems of grasses, bamboos.

Cultivar: A hybrid name derived from cultivated variety and applied to a plant that has originated and persisted under cultivation (as distinct from a species). A subdivision of species as "a group of plants within a species which are uniform in characters".

Curd: Edible part of cauliflower which is actually the repeatedly branched prefloral fleshy apical meristem.

Curing: Thickening of skin and healing up of the wounds of potato tubers when kept for about a week or so at room temperature.

Cytogenetic/cytonuclear male sterility: Male sterility is determined by interaction of genes and cytoplasm but none of them singly can control sterility.

Cytoplasmic sterility: Transmission of male sterility by the cytoplasm.

Cytoplasmic: Concerned with the cytoplasm in the cell.

Damping off: The collapse of seedlings at or just below soil level, usually caused by bad handling when pricking out, overcrowding or poor drainage-conditions which encourage botrytis and fungal attack. The term covers a variety of fungal diseases and some control can be obtained by using only sterilized compost and watering with a fungicide such as chestnut compound but it is best prevented by good growing practice.

Day neutral plant: Plant in which flowering is not influenced by day length *e.g.* tomato, cucumber, okra, asparagus, capsicum, snap bean etc.

Defoliation: The removal of unwanted leaves from tomato plants to expedite ripening, increases air movement and thus reduce the risk of botrytis: removal of the lower yellowing leaves helps to control disease generally. Whether defoliation actually speeds up ripening is disputed, but some growers still remove all the lower leaves, leaving only those carried on the top third of the plant.

Degreening: The process of decomposing the green pigment in fruits by applying ethylene (1000-2000 ppm) or similar metabolic inducers to give a fruit its characteristics colour as preferred by consumers, generally followed in citrus fruits but also practiced in banana, mango, tomato etc.

Dehaulming: Removal of the top portion (haulm) of potato in the seed crop to avoid the infestation of virus carrying insect vectors.

Dehiscence: Splitting open of a fruiting structure or anther.

Dehiscent fruit: Fruit whose pericarp bursts to liberate the seeds at maturity as seen in okra etc.

Determinate: Where the main stem ends in a terminal flower or truss, thereby arresting the further development of the main axis, as in bush tomatoes.

Diallele cross: A diallele cross can be defined as all possible combinations i. e; n (n-1)/2 of single crosses among n parents.

Dichogamy: The maturing of male and female gametes at different times when the male gametes matures earlier than the female it is called protandry, but when female matures first, it is called protogyny. This situation called dichogamy.

Dicliny: A situation, where plants produce unisexual flowers.

Die-back: Any of several fungus diseases which cause shoots or branches gradually to die back from the growing tips. It is a much slower process than wilt, where infected shoots collapse rapidly and entirely. Diseased shoots should be cut back to sound wood and the infected material burned.

Dimorphic: Two different forms in the same species, as in the adult and juvenile leaves of ivy.

Dioecious: The condition in which male and female sexes are produced on different individuals, as in pointed gourd etc.

Diploid: An organism having two distinct genomes i.e two sets of chromosome.

Disorder: Ailment caused by poor conditions, mineral deficiencies or a hostile environment as distinct from one emanating from attack by pets.

Distal: Farthest from the point of attachment, usually at or near the apex.

DNA: Deoxyribonucleic acid; the information–carrying genetic material that comprises the genes.

Dominance: The phenomenon in which the dominant gene has an overriding effect on its allele in such a way that heterozygote (Aa) is phenotypically indistinguishable from the dominant homozygote (AA).

Dominant gene: When two parents with contrasting characters are crossed, the character of one parent appears in F_1 generation to the exclusion of the character of the other parent even though both the genes are present in the hybrid. The gene, which expresses itself, is said to be "dominant" over the other "recessive".

Double cross hybrid: A hybrid obtained by crossing two single crosses i.e; (Ax B) x (C x D).

Double fertilization: A phenomenon in angiosperms whereby one male nucleus unites with the egg nucleus to form a zygote (2n) which develops into the embryo and the second male nucleus unites with two polar nuclei in the embryo sac to form the endosperm (3n).

Double top cross: A cross-obtained from crossing single cross with an open pollinated variety.

Drupe: A stone fruit where the seed is protected by the hard casing (stone) contained in a fleshy layer within an outer skin.

Earthing up: The process of putting the soil just near the base of stems of certain crops like potato, cassava, banana, etc to provide support and to prevent root exposure.

Electronic colour sorter: The machine used for separating seeds by differences in colour. The electronic colour sorter views each seed individually with photo-electric cells. The seed is compared with selected background, or range, according to its colour. If its colour or shade falls within the reject range, a blast of compressed air deflects the seed and sends it into the reject discharge spout.

Electronic seed separator: This machine separates the seed of different electrical properties essentially independent of differences in size, shape, weight or surface texture. It separates seeds by allowing them to fall free in space, and then using an electric field to deflect some seed from their normal flight path.

Electrophoresis: The migration of suspended particles in an electric field.

Electroporation: A process whereby cell membranes are made permeable to DNA by applying an intense electric current.

Emasculation: In bisexual flowers, the removal of stamens before they burst and shed pollen grains. Its purpose to check self–pollination and is done before effecting cross-pollination.

Embryo culture: The cutting of an immature embryo on a sterile nutrient medium.

Embryo: The portion of seed, which contains the dormant, miniature, rudimentary plant. It arises from the zygote.

Endosperm: The nutritive tissue formed inside the embryo sac in the seed. It arises from the triple fusion of sperm nuclei with polar nuclei of embryo sac.

Enzyme: A protein that accelerates a specific chemical reaction in a living system.

Epigeous germination: A pattern of germination where the hypocotyl elongates and raises the cotyledons above the ground, as seen in tomato, beans, gourds etc.

Epigyny: The phenomenon in which the thalamus completely encloses the ovary and gets fused with it, and bears the sepals, petals and stamens on the top of the ovary as seems in cumber etc.

Epistasis: The phenomenon in which a non–allelic or gene combination exert a dominant effect over another gene or combination of genes (non–allelic interaction).

European carrot: Carrot cultivars which are biennial in nature and require low temperature (4.8-10 ^{0}C) treatment for certain periods for flowering, hence do not produce seeds in plains of India, *e.g.* Nantes, Chanteny, Imperator etc.

Evolution: The process of the origin of the organisms (varieties, species, genera, families etc).

Exhausting: Removal of air by passing through exhausting box where steam circulation around it. Exhausting is done at 190 ^{o}F for about 5 to 7 minutes.

Eye: Usually an undeveloped growth bud, such as the eye of a potato. Lateral bud in the rhizomes or tuber; young shoot bud used for budding.

$F_{1.}$ An abbreviation to designate the first hybrid generation. It is composed of progenies raised by sowing the seeds obtained from cross between two genetically unlike parents.

F_2 hybrid: Second generation cross between two F_1 hybrids. Not so vigorous as F_1 hybrids and usually showing great variation.

Fasciculated roots: Swollen tuberous roots, which are developed in a cluster or fascicle at the base of the stem as in asparagus etc.

Fertility restorer gene: Usually a gene which when put into a cytoplasmic male sterile background is able to bring back the production of normal functional pollen grains.

Fertility: The ability of an organism to form viable off spring.

Fertilization: The fusion of the pollen of the male plant with the ovule of the female to form a seed.

Foreign seeds: Seeds of weeds and crops other than the kind being tested.

Foundation Seed: The seed stocks that are so handed as to most nearly maintain the specific genetic identity and purity of the original stock and provide the source for the production of called certified and registered seed.

Freeze-drying: Freeze drying (also called lyophilization) is characterized by heating (*i.e.* conduction, radiation, radio frequency, microwaves etc.) a moist frozen product in a vacuum chamber maintained at an absolute pressure below the vapour pressure of ice within the product. When the material is heated, the ice sublimes.

Freezing injury: Freezing injury refers to freezing of water present in intercellular spaces resulting in the formation of ice crystals. These ice crystals then penetrate the cells and result in their damage.

Fruit: The ripened ovary: the whole seed bearing organ whether it be pod, capsule, drupe etc.

Full slip: A harvesting index of muskmelon or cantaloupes for local market when the fruit can easily be removed (slip) with a slight pressure from the stem leaving a clean stem cavity.

Functional male sterility: Where plants produce normal pollen grains but anthers remain closed and cause male sterility.

Gamete selection: It is a type of selection for detecting and combining of desirable gemete from a genetically variable heterozygous population in to the background of an inbred line of known performance and combining ability.

Gamete: A matured sex cell, capable of fusing with another to form a zygote.

Gametocide: A chemical substance destructive to gametes.

Gametophyite: A phase of life cycle of plants, which has haploid nuclei; during it the sex cells are produced. It arises from a spore produced by meiosis from a sporophyte (diploid).

Garden for vegetable processing: Type of vegetable farming where vegetables are produced with a sole objectives of supplying them to the processing factories.

Gemete selection: It is a type of selection for detecting and combining of desirable gemete from a genetically variable heterozygous population in to the background of an inbred line of known performance and combining ability.

Gene expression: The hereditary properties of an organism as represented by the gene and are expressed in generation under a set of environmental factors.

Gene flow: The spread of genes from one breeding population to another by migration, possibly leading to allele frequency changes.

Gene frequency: Proportion of a gene or allele or its series present in a population or a sample thereof. It is usually expressed as number between 0 to1.

Gene interaction: The modification of gene action by a nonallelic gene or genes.

Gene pool: Sum total of all genes in a breeding population.

Gene: A heredity determinant of a specific biological function; a unit of inheritance (DNA) located in a fixed place on the chromosome.

General combining ability: Average performance of a line or lines in a series of hybrid combinations, which is basically due to additive effect of polygenes.

Genetic advance: The expected gain in the mean of the population for a particular quantitative character by one generation of selection of a specified per cent of the highest ranking plants.

Genetic architecture: A term used to denote the general genetic structure of a species.

Genetic breakdown: A term used to indicate the loss of vigour and often the early death of F_2 plants which lack the necessary adaptive complexes of either or both the original parents.

Genetic diversity: The genetic differences as observed between individuals, or genetic stocks with respect to individual trait or an array of traits.

Genetic engineering: Genetic manipulation that use recombinant DNA methods (gene splicing) to change the genetic make up of an organism.

Genetic equilibrium: In a random mating population, the stage in which genotype frequencies do not change from one generation to another.

Genetic erosion: Gradual disappearance of various forms of a cultivated species and of its wild relatives.

Genetic homeostasis: The ability of a random mating population to equilibrate its genetic composition so as to resist sudden environmental changes.

Genetic male sterility: Sterility of pollen grains which are governed by a single recessive gene as found in squash, pumpkin, muskmelon, brussels sprouts, cabbage, cauliflower, lettuce, sprouting broccoli etc.

Genetic purity: Trueness to type; varietal purity; plants/seeds conforimg to characteristics of the variety as described by the breeder.

Genetic shift: Change in the genetic make up of varieties, if grown over a long period in areas outside their adaptaion.

Genetics: The science of heredity and variation.

Genotype x environmental interaction: The interplay in effect of the genetic and non–genetic factors on the development of an organism.

Genotype: The hereditary properties of an organism as represented by the gene constituents. They may be expressed or latent.

Genus (pl. genera): A taxonomic category that includes groups of closely related species.

Geographical diversity: The diversity of the biological population produced by the presence of a geographical barrier such as mountains, rivers, canyons etc.

Germplasm complexes: The advanced generation of mixed seeds obtained by a purposeful intermixing of a number of genotypes or hybrids among them.

Germplasm: In plant breeding sense, germplasm is the sum total of genetic stocks of a particular crop species.

Gourd: Generally it refers to the fruit of cucurbits. Actually this epithet refers to the fruit character: hard and tough rind upon complete maturation as in bottle gourd, pumpkin or summer squash, even though the term gourd is applied to other fleshy fruits like bitter gourd or snake gourd whose skin do not become tough when ripe.

Grading: Vegetables are sorted by quality into two or more grades as per the standards available in that country.

Green pepper: Tender, semi-mature green pepper (*Piper nigrum*) spike which is used commercially in pickles.

Greening of potato: Appearance of green colour on the tuber exposed to direct sunlight due to the presence of solanin, which is slightly poisonous.

Grow-Out test: Performed to determine the genuiness of seed as to species or variety, or freedom from seed borne infecton.

Growth crack of sweet potato: Longitudinal or transverse splits and fissures to irregular on interrupted growth.

Growth inhibitors: A category of growth regulator which inhibits growth of shoots, blocks action of auxin, GA and cytokinins, sometimes stimulate fruit and leaf abscission and senescence, can substitute for chilling, delays bud break and inhibit germination, *e.g.* ABA, SADH, Alar, CCC, Phosfon-D and MH etc.

Guar gum: The mucilaginous seed flour of guar (*Cyamopsis tetragonolobus*) is valued as guar gum (*galactomannan*) used in textile, paper, cosmetic and oil industries throughout the world. It is also useful absorbent for explosives.

Gynodioecious: Having female and hermaphrodite flowers on different plants.

Gynoecious: A sex form where only pistillate flowers are produced giving rise a super female plant.

Gynomonoecious: Having male and hermaphrodite flowers on the same plants.

Half slip: A harvesting index of muskmelon or cantaloupes for distant market when netting is not complete and fruit can be separated from stem by twisting.

Half-hardy vegetables: Vegetable crops, which can thrive well in cool weather condition but cannot tolerate frost *e.g.* beet, carrot, cauliflower, lettuce, spinach etc.

Hardening: Subjecting plants to adverse conditions to hasten tissue maturation for increasing hardiness, mainly done by withholding irrigation.

Hardy vegetables: Vegetable crops which are tolerant to frost *e.g.* asparagus, broccoli, cabbage, peas, turnip, garlic, knolkhol, radish, spinach and onion etc.

Haulm: Above ground portion (stem and foliage) of potato, peas and beans etc..

Head shape index: A head character of cabbage, Chinese cabbage calculated by dividing mean head length (cm) with mean head width (cm).

Head to seed method: A method of seed production practiced in cabbage where the selected plants with fully matured heads are lifted prior to snowfall, stored and again replanted at the onset of spring for seed production.

Head: Edible portion of cabbage, Chinese cabbage and head lettuce, which is a structurally distinct, compact leafy portion, made up of numerous overlapping leaves covering the terminal bud.

Heredity: Resemblance among individuals related by descent; transmission of traits from parents to off spring.

Heritability: Degree to which a given trait is controlled by inheritance.

Hermaphrodite: An individual with both male and female reproductive organs.

Heterobeltiosis: Heterosis expressed over the better parent of the cross.

Heterochromatin: Chromatin staining darkly even during interphase, often containing repetitive DNA with few genes.

Heterosis: Hybrid vigour such that an F_1 hybrid falls outside the range of parents with respect to same characters. Usually applied to size, rate of growth or general anthocyanin absent fitness.

Heterozygosity: The phenomenon in which the homologous chromosomes of an organism possess different genes of the same allelic series.

Heterozygote: An organism with one or more heterozygous pairs of genes or unlike alleles at one or more corresponding loci. As a result of heterozygosity, the organism will not breed true.

Homologus chromosomes: Chromosomes that cover in pairs and are generally similar in size and shape, one having come from the male parent and the other from the female parent. Such chromosomes contain the same array of genes.

Homozygote: An organism with identical genes at corresponding loci on homologous chromosome.

Horticultural traits: Characters of economic importance in horticultural crops.

Hub crop: Crop, which has the greatest comparative advantage over the crops in a sequential cropping system *e.g.* vegetable crops.

Hybrid vigour: Increase in vigour, growth, yield or function of a hybrid over the parents that result from the crossing of genetically unlike organisms.

Hybrid: The progeny of a cross between two or more individual plants of unlike genetic constitution.

Hybridization: A method of crop improvement in which two or more plants of unlike genetical constitutions differing in one or more characters are crossed together.

Hypogeous germination: A pattern of germination where the lengthening of the hypocotyls does not raise the cotyledons above the ground and only the epicotyl emerges, as seen as in pea etc.

Hypogyny: The phenomenon in which the ovary occupies the highest position on the thalamus while the stamens, petals and sepals are separately and successfully inserted below the ovary, as seen as in brinjal etc.

Imbibition: Absorption of mositure by a colloidal substance such as seed coats, which is accompanied by sweeling of tissues.

In situ: From the latin, meaning "in natural place". Refers to experimental treatments performed on cells or tissues rather than on extracts from them.

In vitro: From the latin, meaning "within glass"; biological processes made to occur experimentally outside the organism in a test tube or other container.

Inbred line: A line produced by continued inbreeding. In crop breeding a nearly homozygote line usually originating from continued self-fertilization, accompanied by selection.

Inbreeding depression: The loss of vigour as a consequence of inbreeding. It is primarily due to the break down of specific gene system governing the expression of vigour, governing a particular trait or traits and is often accompanied by reduction in yield, size, fecundity etc.

Inbreeding: Mating of closely related individuals. Example-self pollinated crops.

Incompatibility: Failure to obtain fertilization and seed formation after self–pollination, usually due to failure of pollen tube to penetrate the stigma, or to reduced growth of the pollen tube in the styler tissue.

Inert matter: Inert matter includes seeds and seed like structures, namely, pieces of broken or damaged seeds which are one half the original size or less; empty glumes, lemmas, paleas, unattached sterile florets and florets with a caryopsis less than the minimum size; prescribed, and other matter, namely soil, sand, stones, chaff, stems, leaves, cone scales, wings pieces of bark, flowers, nematodes galls, fungus bodies, caryopses of graminae replaced by insect larvae; and all other matter not seeds.

Inflorescence: (1) A flower clusters (2) the arrangement and mode of development of the flowers on a flower axis.

Inter specific hybridization: Hybridization between plants belonging to two different species of a same genus.

Intercropping or companion cropping: The system of copping where two or more crops are grown together on the same land.

Internal browning of tomato: Gray brownish discolouration of internal tissues in green fruit, which extend to the surface and for lesion that remain greenish or yellow in ripe fruit; caused by water imbalance and high temperature and /or nutrient imbalance.

Introgressive hybridization: A type of hybridization in which a number of genes or gene block of one species are added to the genetic background of another species by crossing and often by backcrossing.

Inulin: A storage polysaccharide ($C_6H_{10}O_5$) found in the foots of Jerusalem artichoke, dahlia and few other plants.

Irradiation: It involves exposure to ionizing radiation, such as gamma rays from radioactive source and electron beams an X-ray generated by machines. When such rays pass through cells, they knock electrons off atoms and molecules, breaking chemical bonds and creating ions and free radicals.

Isogenic lines: Lines differing from each other genetically at one locus only *i.e.* lines identical in all traits but one.

Isolation: The condition in which individuals of common ancestry is separated into two or more mating groups that mating between or among groups is prevented.

Juvenile phase: Early or vegetative phase of growth characterized by carbohydrate utilization and before which flowering can not be induced in plant.

Kidney bean: Usually, the French dwarf bean, especially its seeds.

Kidney potato: Any long oval-shaped variety of potato tubers e.g "The Jersey Royal".

Landraces: Traditional cultivars with sufficient genetic integrity to be morphologically identifiable and differing in adaptation and to cultural practices, but genetically variable.

Leaf roll: A virus disease of potatoes. Leaves are rolled, especially when attacked by sap-sucking pests such as aphids and some types of sawfly whose leave can often be found inside the rolled leaf.

Leafyness: A physiological disorder of cauliflower in which green bracts grow out of the curd due to exposure of the curd at temperature higher than optimum for its development.

Legume: The word legume is derived from the Latin word 'Legre' (Together) because the pods have to be gathered or picked by hand. The legumes belong to family Leguminosae and produce nitrogen-fixing nodules on their roots *e.g.* peas, beans etc.

Ligase: An enzyme that joins the ends of two strands of nucleic acid.

Line breeding: A system of breeding in which a number of genotypes which have been progeny tested in respect to some character or group of characters are composite to form a variety.

Line x Tester analysis: A system of lines for combining ability in the genetic background of a number of proven testers.

Line: This term refers to a group of individuals obtained from a common ancestry. In maize programme, it is synonym to inbred line.

Linkage: Association of two or more non-allelomorphic genes, so that they tend to be passed from generation to generation as an inseparable unit and fail to show independent assortment. The parental types appear in greater frequency than expected in F_2. This is due to locating of linked factors on the same chromosome.

Lipase: An enzyme that joins the ends of two strands of nucleic acid.

Long day plant: A plant which requires a day longer than its critical day length for flowering e.g lettuce, radish, onion, cabbage, carrot, spinach, beet etc.

Luxuriance: It refers to the phenomenon in which the crossing of two parental forms brings in an excessive, accidental, unadaptable and often unbalanced expression of a number of an attribute.

Lycopene: Red pigment found in ripe tomato, which is a straight chain derivative of carotene with no vitamin activity. Its chemical composition ($C_{40} H_{56}$) is same as that of carotene.

Maintainer line: A genotype used to maintain the male sterility of cytoplasmic-genic system.

Maker genes: Common gene differences which assort independently from all other readily testable gene loci are called marker genes *e.g.* potato leaf in tomato brown seed in onion, anthocyanin pigment in cotyledons of chilli/brinjal atc.

Male sterility: An inherent inability of plant to produce viable pollen.

Marker genes: Common gene differences which assort independently from all other readily testable gene loci are called marker genes *e.g.* potato leaf in tomato brown seed in onion, anthocyanin pigment in cotyledons of chilli/brinjal etc.

Market gardening: Vegetable farming for supply of vegetables to the consumers in the local market; one of the most intensive types of vegetable gardening.

Mating system: The system in which individuals are arranged in pairs leading to sexual reproduction. There are following system of mating.

Mealy potatoes: Those potatoes in which the cells, on cooking, separate from one another so that they may be piled up in a loose flour with air between them instead of remaining in a solid mass with water in the small intercellular spaces.

Meristem culture: Culture of apical meristems, particularly shoot apical meristem, for production of shoots and plantlets.

Modified atmosphere storage (MAS): The build up of respiratory gases such as carbon dioxide inside packed vegetables for extending the shelf life of storage fresh produce is called as MAS.

Momeostatic: A genotype giving better performance even in adverse environments because of wide genetic base.

Monoecious: The condition in which both male and female sex organs are produced separately but on the same plant.

Monoecy: Where male and female flowers are unisexual but produced on the same plant.

Mucilage: A complex glutinous carbohydrate secreted by certain plants (okra, basella etc.).

Mulch: Any material such as straw, plant residues, leaves or plastic film placed on the soil surface to reduce evaporation, erosion or to protect plant roots from extremely low or high temperatures.

Mulching: Practice of covering the soil surface for conservation of soil moisture and control of weeds and soil erosion.

Multiple fruit: A collection of fruits which develop from an inflorescence where the flowers are crowed together and often fused with one another e.g jack fruit.

Multistoried cropping: Growing of different crops having varied height, canopy, orientation and rooting habits together for proper utilization of solar energy *e.g.* coconut + black pepper + cacao + elephant foot yam + pineapple.

Multivitamin green: Chekurmanis (*Sauropus androgymus*), a perennial leafy vegetable crop is called so because of the availability of various vitamins like A, B, C, D, E and K from this vegetable crop.

Mutant: Any plant, which has originated or acquired a heritable variation as a result of mutation.

Mutational heterosis: It is a type of true heterosis, which results from the occurrence of the balanced types of mutations plants.

Napiform root: Root which when swollen become spherical at the upper part and sharply tapering at the lower part, as in beet, turnip etc.

Nodulation: Development of knob like outgrowth in the roots of leguminous plants in response to the stimuli of *Rhizobium* bacteria, which is involved in fixation of atmospheric nitrogen.

Nodulose root: Slender root when becomes swollen suddenly at or near the apex as in, turmeric, arrowroot etc.

Non recurrent parent: The donor parent in the back crossing programme. The desire character of this parent is added in the genetic background of recurrent parent.

Nucleic acid: A macromolecule composed of phosphoric acid, pentose, sugar, and organic bases; DNA and RNA.

Nucleotide: A unit of DNA and RNA molecules containing a phosphate, a sugar, and an organic base.

Nucleus: The part of a eukaryotic cell that contains the chromosomes, separated from the cytoplasm by a membrane.

Nursery bed: A prepared are where seed is sown or into which transplants or cuttings are planted.

Obtuse: Blunt pointed.

Offset: A side shoots which develops naturally from the parent plant, usually at ground level and leaf axil (water hyacinth) and developing its own roots while still attached.

Off-type: Any type of plant, which is outside the acceptable limits of varietal variation.

Often cross pollination: Essentially self pollination where extent of cross pollination vary from 5-25 per cent as in lima bean, brinjal, okra, chilli etc.

Oleoresin: A natural combination of resinous substances and essential oils present in the fruits of certain crop plants like chilli.

Olericulture: A Latin term used to designate science and management of vegetable crops.

Open pollinated: The progeny of individuals where pollination took place automatically in flowers in nature.

Orbicular: Circular, round.

Organic vegetables: Vegetables which are produced by organic approach which is an integrated, living system aims to enhance and develop the biological cycles in the system involving micro-organism, soil life, plants and animals. Organic approaches are achieved by the use of sound rotations, management of manures, proper cultivation practices, biocontrol of pest and diseases and ethical treatment of livestock.

Organoleptic evaluation: Descriptive index characters of the cooked samples of vegetables for appearance, firmness, taste, mouth feel etc.

Orthodox seed: A category of seeds which can be dried to a low moisture level and would show a loss of viability with the rise in moisture content, and are usually long lived, e.g most vegetables.

Out breeding: Mating of unrelated individuals.

Out cross: A type of natural cross obtained from the crosses of a number of unknown genotypes.

Ovate: Egg shaped; broadest below center (contrast ovate, elliptic).

Over dominance: An effect the heterozygote (Aa) that is greater than the effect of homozygous dominant (AA).

Ovule: The term generally applied to the whole seed forming apparatus inside an ovary i.e; nucellus plus the integument.

Parietal placentation: When placentae bearing the ovules develop on the inner wall of the one chambered ovary corresponding to the confluent margins of carpels as seen in radish etc.

Parthenocarpy: The development of fruits without fertilization and the formation of normal seeds, *e.g.* cucumber.

Parthenogenesis: The development of an individual from the female gamete without fertilization.

Pedicels: The flower stalks, which grow from the upper branches of the peduncle.

Peduncle: The main flower stalk, often branched once or twice (or more) in its upper part and bearing pedicels near the ends of the branches.

Pentagonal: In which the lobes are rather broader than long, giving the corolla a pentagonal or 5-pointed appearance.

Pepo: Fleshy, many seeded fruit which develops from an inferior, one celled or spuriously three celled, syncarpous pistil with partieal placentation, e.g cucumber, melons, squash, gourds, etc.

Perianth: When the calyx and corolla do not differ much in shape and colour, they together are said to form the perianth as seen in onion, garlic etc.

Pericarp: The outer wall of the fruit.

Peripheral embryo: Embryo, which encloses endosperm or perisperm tissue as seen in Amaranths etc.

Petiole: The stalk of a leaflet.

Petiolulate: With petiolules.

Phenotype: The visible manifestation of the genotype produced as a consequence of growth and development.

Phenotypic assortative mating: Mating of individuals with similar phenotypes. Example–crossing between tall individuals with Aa x AA or Aa or Aa x AA and aa x Aa genotypes.

Phenotypic disassortative mating: Mating of individuals with contrasting phenotypic characters. Example–crossing between tall x dwarf plants.

Phenotypic stability: The stability of a genotype (or genotypes) over a spectrum of environmental conditions. It is also known as developmental homeostasis).

Photo dormancy: A type of physiological dormancy of seed where germination of seed is sensitive to light i.e; seeds of some plants require light to germinate whereas others require darkness. e.g lettuce seed require light and *Allium*, Amaranth etc. seed require darkness for germination.

Phytoalexin: A phenolic substance having antifungal principle, synthesized by plants in response to parasite invasion or infection by certain fungi, e.g pisatin, phaseolin, trifolirhizin, orchinol and isocumarin from pea and bean pods and carrot root, respectively.

Pie plant: Rhubarb (R*heum raponticum*), one of the oldest cultivated vegetable crops, is commonly known as " pie plant".

Pinnate: A leaf divided into terminal and lateral leaflets.

Piperin: The alkaloid present in the crops of Piperaceae family (black pepper) which gives the biting taste.

Pithiness: Development of spongy tissue from inside of the root, resulting from delayed harvest.

Planting ratio: The male and female plants when planted in a certain proportion to ensure proper pollination and fertilization *e.g.* 10:1 (female: male) ratio in pointed gourd.

Pleiotrophy: Phenomenon of a single gene affecting two or more different characters.

Pod: Dry, one chambered fruit developing from a simple pistil and dehiscing by both the margins e.g pea, bean etc.

Poi: The pressure-cooked taro (*Colocasia esculenta*) corms after being passed through strainer are allowed to ferment which gives an acidic product called ' poi'.

Polycross: An isolated group of plants or clones arranged in some fashion to facilitate random interpollination.

Polymerase chain reaction (PCR): A procedure involving multiple cycles of denaturation, and polynucleotide synthesis that amplifies a particular DNA sequence.

Polymerization: Chemical union of two or more molecules of the same kind to form a new compound having the same elements in the same proportions but a higher molecular weight and different physical properties.

Polymorphism: Two or more kinds of individuals maintained in a breeding population.

Polynucleotide: A linear sequence of joined nucleotides in DNA or RNA.

Polypeptide: A linear molecule with two or more amino acids and one or more peptide groups. They are called dipeptides, tripeptides, and so on, according to the number of amino acids present.

Polyploid: An organism with more than two sets of chromosomes (2n diploid) or genomes (*e.g.*, triploid (3n), tetraploid (4n) and so on.

Positional sterility: A type of male sterility, also called functional sterility, ehre pollens are functional but anthers fail to dehise, found in some mutants of tomato in some mutants of tomato.

Post harvest: After harvest.

Precooling: Removal of field heat as soon as feasible in order to retard ripening and deterioration of vegetables harvested during hot weather.

Presizing: For many vegetables, fruit below a certain size are eliminated manually or mechanically.

Pricking: A method of raising secondary nursery for the crops having very small seeds; in the case of high density sowing, it helps to develop a thinner stand in the nursery. In cole crops and lettuce, pricking is done when first pair of true leaves develop.

Progeny testing: The practice of ascertaining the genotype of an individual by mating it to an individual of known genotype and examining the progeny.

Progeny: The off springs of a particular mating.

Prokaryote: A number of a large group of organisms (including bacteria and blue green algae) that lack true nuclei in their cells and that do not undergo meiosis.

Pseudo–heterosis: See luxuriance.

Pseudo–dominance: The phenomenon of the apparent dominance of a recessive gene in the area opposite a chromosome deficiency.

Pseudo incompatibility: Incompatibility due to physiological and physical reasons only and is not heritable *e.g.* physiological factors like temperature, light and physical factors like heterostyly, protoandry, closed anther, etc.

Puffiness of tomato: A disorder due to high or low temperature and low soil moisture where the outer wall of the fruit continues to develop normally but remaining internal tissue growth is retarded resulting in partially filled, less firm and light weight tomato.

Pure line: The descendents obtained from self-fertilization of a single homozygous or an inbred homogenic strain.

Qualitative characters: The characters, which show discrete variation and easily identified by visual observations. Examples, flower colour, fruit shape etc.

Quantitative characters: The characters, which show continuous variation such that visual identification of individual genes segregation is not possible. These are usually governed by the cumulative effect of polygenes and are highly influenced by environmental conditions. Examples are yield per plant, fruit per plant etc.

Quercetin: A pigment, which imparts coloration to the outer skin on onion bulb.

Quick freezing: Its method of preservation of vegetables. This requires low temperature storage usually from–5 to +5 ^{0}F, depending upon the vegetable.

Raceme: Inflorescence where the main axis is elongated and bears laterally a number of flowers, which are all stalked, the lower flowers having longer stalks than the upper ones, as in radish etc.

Rajmah: Dried seed of French bean, used for the preparation of various dishes.

Random mating: Arrangements of pairs is by chance *i.e.* each individual has equal chances to mate with another. Examples– cross-pollinated crops.

Recessive gene: When parents with contrasting characters are crossed, the character of one parent does not appear in F_1 even though genes for both the factors are present in the hybrid. The gene, which does not appear, is said to be recessive.

Reciprocal hybrids: Two hybrids produced by crossing the same parents but the male of first is used as female in the another and similarly the female of first is used as male in another such as (Ax B) and (B x A).

Recurrent apomixes: A type of apomixes where embryo sac develops from egg mother cell or from some adjoining cell without complete meiosis. The embryo develops directly from the diploid egg nucleus without fertilization, seen in onion etc.

Registered seed: The progeny of foundation seed that is so handled as to most nearly maintain the satisfactory genetic purity and has been approved by certifying agency.

Relative heterosis: Usually refers to the heterosis expressed over the mid parental value of a cross.

Restriction enzyme: An endonuclease that recognizes a specific short sequence in DNA and cleaves the DNA molecule.

Restriction fragment length polymorphism (RFLP): A genetic difference among individuals that is decided by comparing DNA fragments released by digestion with one or more restriction enzyme.

Restriction map: A linear or circular diagram of a DNA molecule showing the sites that are cleaved by different restriction enzyme.

Reverse transcriptase: An enzyme that catalyzes the synthesis of DNA using an RNA template.

Reversion: Restitution of a mutant gene to the wild type condition, or at least to a form that gives the wild phenotype. More generally, the appearance of trait expressed by a remote ancestor; a throwback; atavism.

RFLP (Restriction fragment length polymorphism): A genetic difference among individuals that is decided by comparing DNA fragments released by digestion with one or more restriction enzyme.

Ribosome: Cytoplasmic organelle on which proteins are synthesized.

RNA: Ribonucleic acid. The information–carrying material in some viruses. More generally, a molecule derived from DNA by transcription that may carry information (messenger or mRNA), provide subcellular structure (ribosomal or rRNA), transport amino acids (transfer or TRNA), or facilitate the biochemical modification of itself or other RNA molecules.

Root crops: A group of vegetable crops whose swollen tap roots and in some cases hypocotyls alongwith tap root such as carrot, beet, radish, turnip, rutabaga etc. are cooked or eaten raw.

Root forking: A branching of tap roots in the roots crops, particularly in radish and carrot, due to the presence of impediment, undecomposed organic matter or plant refuse in the soil.

Root to seed method: A method of seed production in root crops where the fully matured rooted are harvested, selected and after giving proper root and shoot cuts, they are replanted for seed production.

Root tuber: The fleshy root of an herbaceous perennial plant with buds or eyes in the upper regions.e.g sweet potato.

Rouging: The act of removing undesirable individuals from varietal mixtures or seed production plots.

Salad crops: Green leafy vegetables which are usually consumed raw with oil, vinegar and various other condiments, e.g lettuce, endive, celery, chicory, parsley etc.

S-allele: A set of alleles controlling sporophytic self-compatibility.

Sauer kraut: A favorite food prepared by fermenting chopped, ground or sliced cabbage in its own juice with a little salt added to it.

Scalper: A seed-processing machine designed to remove the bulk of trash from seed so as to facilitate elevating and processing operations.

Scarification: The process of mechanically scarring or roughing the hard seed coat to make it more permeable to water.

Scooping: Removal of central portion of the cure for easier initiation of flower stalk in cauliflower.

Seed certification agency: The certification agency established under Section 8 or recognized under Section 18 of the Seed Act, 1966. Its major function is to certify seeds of any notified kind or varieties.

Seed certification: A system of maintaining the quality of seeds. The crops offered for certification are raised as per requirements for seed certification established by a seed certification agency. Several inspections are made to ensure purity and quality of seeds.

Seed disinfestations: Dipping the seeds in chemicals, soaps and fungicide to kill pathogens.

Seed parent: The female (pistillate) parent of a hybrid.

Seed potato: Potato tubers used for planting.

Seed: The matured ovule having all the essential structures to produce a new plant.

Seedless watermelon: It refers to autotriploid (3x) watermelon, which is both male and female sterile due to unequal chromosomal distribution in meiosis resulting in seedless condition in the fruit.

Segregating generations: The F_2 and onward generation of a hybrid where separation of parental from material chromosomes at meiosis and consequent separation of genes takes place leading to the possibility of recombination in the off spring.

Selection: The process in which a number of individuals with certain desirable characteristics are favoured for further reproduction.

Self–pollinated crops: An assembly of homozygous plants. These crops often have one single genotype and reproduce it as such from generation to generation.

Self–incompatibility: Inability of viable pollen to fertilize an ovule carrying an identical S-allele and inhibiting self or sib–pollination.

Self sterile: The organism which fails to fertilize and set seeds after self–pollination.

Self-blanching: Exclusion of sunlight by the inner whorl of leaves surrounding the developing curd of cauliflower for keeping the curd white. It is a varietal character.

Self-incompatibility: The inability of a functional gamete or the plant to set seed when self-pollinated, even though it forms normal zygotes when cross–pollinated.

Selfing: The process of putting the pollen of the same flower on its stigma *i.e.* enforcing self-pollination. It is extensively practiced in maize and other cross-pollinated crops to obtain inbred lines for utilization in the hybrid programme.

Self-pollination: The transfer of the pollen from the anther to the stigma of the same flower or the plant. It is also the union of genetically similar gametes.

Semi–sterility: A state of only partial fertility in plant zygotes usually associated with chromosomal translocations.

Semigamy: Abnormal fertilization in which the male gamete fertilizes the egg, but does not fuse with the egg nucleus.

Set: Onions shallots, potatoes etc used as seed. Blossom, which has been successfully fertilized. A slip or shoot for planting.

Shoot apex culture: A tissue culture procedure for eliminating virus or other pathogens from plant parts where excision and aseptic culture of the small segment of the terminal growing points is done because the terminal growing point of a plant is often free from virus and other pathogens even if the rest of the plant is infected; practiced in potato, sweet potato, cassava etc.

Shuttle vector: A plasmid capable of replicating in two different organisms such as yeast and *E. coli*.

Sib–mating: Sibbing or crossing at random the two or more individual obtained from the same parentage. It is form inbreeding and refers to brother–sister mating.

Sib pollination: Pollination between closely related biotypes.

Silencer: A DNA sequence that helps to reduce or shut off the expression of a nearby gene.

Single cross hybrid: A hybrid involving two cross-compatible lines.

Single cross: A cross between two genetically dissimilar parents.

Sinigrin: A glucoside present in all the vegetables under genus *Brassica* which on hydrolysis under the influence of enzyme myrosinase produces allyl-iso-thiocyanate which gives atypical pungent flavour.

Solar drying: The drying of vegetables using solar energy direct or indirect. In the direct solar drying commonly called open sun drying or just "sun drying" the product is placed directly on trays in open are where only the solar radiation's is used as the energy source for drying.

Somatopalstic sterility: A type of sterility, which results as a consequence of the collapse of fertilized ovules or zygote during the embryonic or early development stages due to disturbance in embryo endosperm relations.

Southern blot: The transfer of DNA fragments from an electrophoretic gel to a cellulose or nylon membrane by capillary action.

Spear: The shoot, which is the edible part of asparagus.

Species: A unit of taxonomic classification containing group of individuals enough alike so that it may be reasonably assumed that they have arisen from a common ancestor.

Specific combining ability: The deviation in the performance expected on the basis of general combining ability.

Splicing: The process that covalently joins exon sequence of RNA and eliminates the interventing intron sequences.

Splitting of carrot: A major physiological disorder of carrot where root cracks, seems to be controlled by genetic factors but a number of other factors like heavy side dressing with nitrogen fertilizer in the early stages and boron deficiency may be involved.

Sporophytic incompatibility: Failure of fertilization due to genetic abnormalities.

Stamens: Organ of flower, which produces pollens.

Standard heterosis: Heterosis expressed over the standard or check variety. It is also known as useful heterosis.

Starter solution: Refers to nutrient solution, which is generally applied in the root zone of seedlings immediately after transplanting for helping in better survival of plants and larger early yield.

Steckling: Matured root of carrot, radish, turnip etc. to be replanted after over-wintering for seed production in root to seed method.

Sterility: The phenomenon in which an organism is infertile and is incapable of reproducing.

Sterilization: Removal of unwanted organisms by heat, radiation, chemicals or by filtration.

Stolon: Slender, underground lateral stems arising from buds on the underground portion of the stem, which enlarge at its tip to produce the tuber, as in potato.

Strain: The mating group within a variety or species with distinct morphological or physiological features.

Stump method: A method of seed production in cabbage where the head after full maturity is cutoff just below the base, keeping the stem with outer whorl of leaves intact.

Succession cropping: Growing of two or more crops on the same land in one growing season. Two, three or even four crops may be grown in one year.

Sulphitation: Blanched vegetables are sulphited by placing them in 0.5 per cent solution of potassium metabisulphite for sufficient time.

Sun scald of tomato: Appearance of blistered water soaked area followed by rapid desiccation and formation sunken spots, which are grey or white in green fruits or yellowish in red fruits due to sudden exposure to sunlight, which is more serious in hot weather.

Super market on a stalk: The winged bean (*Psophocarpus tetragonolobus*) plant is described so, as six different foods are supplied by this plant: leaves like spinach, succulent shoots resembling large thin asparagus, fried flowers for making a sweet garnish, tender pod as vegetable, the seed and the underground tubers are exceptionally rich in proteins.

Synthetic varieties: These are the open–pollinated advanced generation population of a number of hybrids obtained by crossing a number of tested lines grown in isolation.

Tabasco pepper: Small fruited, very pungent chilli (*Capsicum frutescence*).

Taxa (Sing. Taxon): A general term for taxonomic classification, irrespective of rank.

Tender vegetables: Vegetable crops, which cannot withstand frost and some of them even do not thrive in cool weather *e.g.* brinjal, okra, chilli, cucurbits, sweet potato, cowpea, tomato, cassava, bean etc.

Tendrometer: An instrument by which toughness of the seed coat and firmness of pulp is determined and is mostly used to determine the seed quality of pea where high value of tendrometer indicated low quality.

Test cross: The cross of the F_1 with recessive homozygous parent i. e; Aa x aa.

Three way cross: The cross between a single cross and a inbred line. With 3 lines (A, B, C), it could be represented as (Ax B) x C. It has been genetically used in the maize-breeding programme for the production of commercial sweet maize hybrids.

Tissue culture: The development of an entire organism from plant cells and tissues *in vitro* or artificial media.

Tomatine: A steroidal glycoalkaloid present in *Lycopersicon* and *Solanum* species, which exhibits an antibiotic activity against a number of organisms.

Top cross: An out cross of selections clones or inbred, to a common pollen parent. In corn, a commonly an inbred–variety cross.

Transformation (bacteria): Genetic alteration of bacteria brought about by the incorporation of foreign DNA in the bacterial cells.

Transgenic: A term applied to organisms that have been altered by introducing DNA molecules into them.

Transgressive segregation: The appearance of individuals showing an extreme development of a character than either of the parent in F_2 or later generations. This usually occurs because of cumulative and complementary effect of genes contributed by the parents of the cross.

Transition: A mutation caused by the substitution of one purine by another purine or one pyrimidine by another pyrimidine in DNA or RNA.

Trifoliate: A compound leaf with three separate leaflets as in French bean, cowpea etc.

Triple cross: A cross between a single cross used as a female and an inbred used as a male parent *i.e.* (A x B) x C.

Truck gardening: Specialized type of vegetable farming where one or two special vegetable crops are grown usually in large quantities suited for the particular region.

Truss: A cluster of fruits or flowers growing on one stem, *e.g.* tomato.

Tuber: A special kind of swollen modified stem structure that functions as an underground storage organ as in potato, Jerusalem artichoke etc.

Tubercle: Small aerial tuber produced in the leaf axils as seen in yam (*Dioscorea alata*).

Tunic: The papery or fibrous coats covering bulbs and corms.

Umbel: A recemose inflorescence, characteristic of family Umbelliferae in which the flowers arise from the same point in the main stem and have stalks of the same length so that the flowers are seen spreading out from a common point like an umbel.

Undeveloped embryo: Partially developed torpedo-shaped embryos that may attain a size upto one half that of the seed cavity at maturity seen in carrot etc.

Urticle: Dry, one-celled, one or few seeded fruit developing from a superior monocarpellery ovary, and the fruit is provided with a thin membranous loose pericarp which is not attached to the seed coat as seen *in Basella spp., Chenopodium spp*. etc.

Utilization index: Ability of root and tuber crops for better accumulation of photosynthates in the organ, which can be judged by root: shoot ratio.

Variation: Differences amongst the individuals arising as a consequence of differences in the genetic make up; the effect of the environment or interplay of both. It is the chief characteristics of living organisms and is the basis of evolution.

Varietal blend: Mechanical mixtures of seed of two or more varieties

Variety: An agricultural variety is a population of similar individuals having common identifiable plant fruit or seed characteristics along with a good agronomic base. This is generally utilized for commercial cultivation.

Vector: An animal or insect, which transmits parasites *e.g.* white fly, is the vector of viral diseases of tomato okra, chilli, etc.

Vegetable forcing: A specialized type of vegetable farming where vegetables are grown out of their normal season. Vegetable forcing requires some special structures like glasshouse, hot bed, cold frame etc.

Vegetable oyster: It is named to salsify (*Tragapogon porrifolius*), a vegetable because of its characteristics mushroom flavor.

Vegetable: An edible plant or plant part eaten cooked or raw as a main part of meal, side dish or appetizer.

Vegetative apomixes: A type of apomixes where flowers are replaced in the plants by bulbils or buds which fall to the ground like seeds, found in *Allium cepa* var. *viviparum*, *Poa bulbosa* etc.

Vertical resistance: Resistance conferred by few genes, specific to few known races.

Vexiliary: Of the five petals, when the posterior one is the largest and almost covers the two lateral petals as in pea, dolichos bean, etc.

Viability: The capability to live and develop normally.

Waxy blister of tomato: A disorder where white to cream coloured irregular blisters, 3 to 6 mm in diameter and often more than 3 mm high occur which become light to dark brown, depressed and crack as the fruits ripen.

Whiptail: A nutritional disorder of Cole crops due to deficiency of molybdenum resulting severe reduction of the lamina leaving the large bare mid-rib in acute condition.

White heart: Whiteness at the central portion of watermelon instead of uniform development of pink colour from centre to rind, indicating poor quality.

Xenogamy: Cross-pollination between flowers borne in different plants.

Xerophytic: A plant capable of survival under conditions of extreme drought, or one adapted to arid conditions.

Xylem: Specialized cells through which water and minerals move upward inside the plant from the soil.

Yearling: One-year-old bulblets which have been formed on the scale.

Zoning of beet: Under unfavourable conditions, particularly in hot weather, beetroot show alternate white and coloured circles when sliced, called zoning.

Zygote: The cell formed by the fusion of male and female gametes.

Appendices

Appendix 1: Area and Production of Major Horticulture Crops (area-million hectare, production–million tonnes)

Crops	*2002-03*		*2003-04*		*2004-05**		*2005-06**	
	Area	*Prod-uction*	*Area*	*Prod-uction*	*Area*	*Prod-uction*	*Area*	*Prod-uction*
Fruits	4.8	49.2	5.1	49.8	5.3	52.8	5.9	54.4
Vegetables	5.9	84. 8	6.7	101.4	7.1	108.2	7.2	113.5
Spices	2.4	3.8	5.2	4.0	3.2	4.9	3.2	5.9
Plantation Crops	3.1	13.1	3.3	9.4	3.1	10.4	3.2	9.8
Flowers	0.1	0.2	0.2	0.6	0.1	0.7	0.1	0.8
Others	0.09	0.9	0.1	0.3	0.4	0.4	0.4	0.5
Total	16.4	152.0	20.6	165.5	19.2	177.4	20.0	184.9

Source: NHB, Gurgaon (2006)
* Estimated.

Appendix 2: Plot Size and Spacing of Vegetables for IET and AHT Trials

Sl.No.	*Names of Crops*	*Plot Size (m)*	*Spacing (cm)*	*Design*	*Replication*
1.	Brinjal	4x 3.6	60x45	RBD	3
2	Tomato	3.6x 3	60x45	RBD	3
3.	Chilli	4.3x 3	50x45	RBD	3
4.	Capsicum	4.3x 3	50x45	RBD	3
5.	Okra	3x 2.7	60x30	RBD	3
6.	Cucumber	4x 1.8	100x 30	RBD	3
7.	Bottle gourd	10x3	300x50	RBD	3
8.	Bitter gourd	5.5x 2.5	150x300	RBD	3
9.	Watermelon	2x 5.4	150x 90	RBD	3
10	Ridge gourd	6x 2.5	300x50	RBD	3
11.	Cauliflower	4.5x 3.2	45x40	RBD	3
12.	Cabbage	4.5x 3	45x30	RBD	3

Appendix 3: State Wise Area and Production of Horticulture Crops (area-million hectare, production–million tonnes) During 2003–04 and 2004-05

State	Area (in 000'ha)		Production (in 000'MT)	
	2003-04	2004-05	2003-04	2004-05
Andaman and Nicobar	39.1	42.4	149.2	158.1
Andhra Pradesh	1498.6	1506.1	12525.1	13668.8
Arunachal Pradesh	78.3	80.8	218.9	218.8
Assam	415.4	410.1	3158.2	3189.6
Bihar	1222.1	1192.6	16822.2	16189.0
Chandigarh	0.2	0.2	2.8	2.8
Chhattisgarh	135.0	232.2	1979.2	1636.2
Dadra and Nagar Haveli	2.2	2.2	20.6	20.6
Daman and Diu	0.0	0.1	0.1	0.2
Delhi	48.3	48.3	652.8	652.8
Goa	99.8	99.8	313.2	313.2
Gujarat	887.4	1023.3	6981.3	9402.6
Haryana	239.8	236.6	3018.8	3268.2
Himachal Pradesh	271.8	277.2	1470.3	1739.7
Jammu and Kashmir	281.3	308.8	1752.7	2170.7
Jharkhand	142.4	256.8	1493.5	3798.3
Karnataka	1491.4	1508.1	10147.9	10569.9
Kerala	1680.9	1689.1	9453.1	9323.3
Lakshadweep	3.2	3.2	54.3	54.3
Madhya Pradesh	714.6	737.6	3653.4	4129.5
Maharashtra	1926.8	2010.8	14460.7	15192.8
Manipur	79.1	79.1	417.0	417.0
Meghalaya	85.2	85.2	555.8	555.8
Mizoram	35.8	35.8	107.1	107.1
Nagaland	29.8	29.8	163.7	163.7
Orissa	1298.0	1304.9	9915.3	9997.7

Contd...

Appendix 3–Contd...

State	Area (in 000'ha)		Production (in 000'MT)	
	2003-04	2004-05	2003-04	2004-05
Pondicherry	25.8	25.9	139.9	142.6
Punjab	208.2	217.3	3249.5	3390.2
Rajasthan	2589.7	2599.2	2878.6	3019.2
Sikkim	58.2	57.4	112.7	118.3
Tamil Nadu	1026.3	1061.1	11252.2	13681.6
Tripura	74.0	76.2	871.1	908.3
Uttar Pradesh	1144.5	1173.1	17325.3	18850.6
Uttranchal	130.9	285.5	1130.0	1778.5
West Bengal	1484.8	1501.9	21115.2	20725.6
Total	19449.0	20198.9	157834.9	169828.8

Source: NHB, Gurgaon (2006).

Appendix 4(a): State-wise Area and Production of Major Vegetables (area-million hectare, production–million tonnes) During 2004–05

State/Uts	*Brinjal*		*Cabbage*		*C.Flower*		*Okra*		*Peas*		*Tomato*	
	A	*P*	*A*	*P*	*A*	*P*	*A*	*P*	*A*	*P*	*A*	*P*
Andaman Nicobar	0.0	0.0	0.0	0.0	0.0	0.0	0.0	0.0	0.0	0.0	0.0	0.0
Andhra Pradesh	24.7	493.2	8.2	98.7			24.1	301.2	0.2	1.8	69.5	1251.0
Arunachal Pradesh	0	0	0	0	0	0	0	0	0	0	0	0
Assam	13.1	171.52	24.57	485.23	16.95	251.69	9.6	119.99	25.13	14.866	14.2	330.19
Bihar	53.7	1073.0	59.7	955.2	36.5	598.8	56.2	730.2			46.0	735.8
Chandigarh	0.0	0.0	0.0	0.0	0.0	0.0	0.0	0.0	0.0	0.0	0.0	0.0
Chhatishgarh	11.2	125.1	4.6	51.4	5.8	61.8	10.3	102.4	7.6	76.1	20.7	217.6
Dadar & Nagar Haveli	0.0	0.0	0.0	0.0	0.0	0.0	0.0	0.0	0.0	0.0	0.0	0.0
Daman & Diu	0.0	0.0	0.0	0.0	0.0	0.0	0.0	0.0	0.0	0.0	0.0	0.0
Delhi	1.6	27.3	0.0	0.0	0.0	0.0	1.7	14.1	0.8	10.2	1.7	33.6
Goa	0.0	0.0	0.0	0.0	0.0	0.0	0.0	0.0	0.0	0.0	0.0	0.0
Gujarat	49.2	745.6	14.4	255.2	12.5	217.9	36.0	299.1			22.8	421.9
Haryana	10.7	174.6	10.4	165.3	17.0	268.8	12.7	100.0	8.7	80.9	13.8	219.7
Himachal Pradesh	2.0	46.9	2.9	88.9	1.8	34.4	1.4	14.5	15.4	182.1	9.0	301.0
Jammu & Kashmir	1.8	25.4	2.6	55.5	2.5	43.8	1.7	22.5	7.3	112.2	3.1	60.3
Jharkhand												
Karnataka	16.7	380.7	8.8	163.2	2.3	36.3	8.5	73.3	1.4	20.1	35.5	954.8

Contd...

Appendix 4(a)–Contd...

State/Uts	Brinjal		Cabbage		C.Flower		Okra		Peas		Tomato	
	A	P	A	P	A	P	A	P	A	P	A	P
Kerala	0.7	1.5	0.6	10.0	0	0	1.1	31.5	0.0	0.0	0.0	0.0
Lakshdweep	0.0	0.0	0.0	0.0	0.0	0.0	0.0	0.0	0.0	0.0	0.0	0.0
Madhya Pradesh	15.9	238.7	8.2	164.6	3.5	56.0	9.1	54.9	4.8	52.7	21.0	314.8
Maharashtra	31.5	473.7	14.1	348.9	12.3	284.0	27.6	160.5	6.6	22.4	33.5	884.0
Manipur	0.4	1.8	1.6	21.0	1.1	9.9	0.8	3.1	1.2	7.3	1.1	9.7
Meghalaya	0.759	9.006	1.469	30.299	1.094	17.623			0.6	3.5	1.7	22.3
Mizoram	0.4	1.5	0.5	2.3	0.1	0.3	0.204	0.699	0.088	0.2	0.2	0.5
Nagaland	0.1	0.2	0.1	0.2	0.1	0.1			1.2	0.3		
Orrisa	127.7	1852.2	33.7	931.0	45.0	637.9	71.4	619.7	4.8	41.8	100.3	1330.8
Pondicherry	0.4	10.8	0.0	0.0	0.0	0.0	0.4	6.3	0.0	0.0	0.1	2.7
Punjab	2.6	37.0	3.1	68.8	5.5	128.1	1.8	13.8	16.7	100.5	7.8	187.3
Rajasthan	5.5	23.4	8.0	25.9	0.0	0.0	4.2	9.1	12.5	28.0	14.1	54.5
Sikkim	0.0	0.0	0.0	0.0	0.0	0.0	0.0	0.0	0.0	0.0	0.0	0.0
Tamil Nadu	8.0	10.7	1.4	93.0	0.6	11.4	4.9	44.4	1.3	24.3	25.3	321.5
Tripura												
Uttar Pradesh	2.0	61.3	7.0	209.9	6.4	128.2	8.3	100.0	129.2	1033.2	4.9	194.4
Uttranchal	1.4	16.8	3.0	59.2	2.5	35.6	2.5	26.4	6.2	37.3	5.5	94.5
West Bengal	148.3	2701.7	71.3	1863.9	64.6	1685.2	63.7	677.3	25.0	122.0	46.1	694.9
Total	530.3	8703.8	290.3	6147.7	238.2	4507.9	358.3	3524.9	276.7	1971.8	497.6	8637.7

Appendix 4(b): State-wise Area and Production of Major Vegetables (area-million hectare, production–million tonnes) During 2004–05

State/Uts	Onion		Potato		S.Potato		Tapioca		Others		Total	
	A	P	A	P	A	P	A	P	A	P	A	P
Andaman Nicobar	0.0	0.0	0.0	0.0	0.5	2.6	0.8	5.0	5.9	23.3	7.1	30.8
Andhra Pradesh	35.4	566.6	3.8	60.6	0.7	14.4	18.8	375.2	73.1	699.2	258.4	3861.9
Arunachal Pradesh	0.0	0.0	4.2	27.5	0.0	0.0	0.0	0.0	16.2	51.3	20.4	78.8
Assam	7.122	16.786	73.104	589.07	8.0	28.0	2.724	13.055			194.5	2020.4
Bihar	48.8	975.2	318.05	5656.7	4.6	63.9			193.2	2560.2	816.6	13349.1
Chandigarh	0.0	0.0	0.0	0.0	0.0	0.0	0.0	0.0	0.1	1.7	0.1	1.7
Chhatishgarh	3.9	42.8	12.9	130.2	2.7	28.0			45.362	430.9	125.1	1266.3
Dadar & Nagar Haveli	0.0	0.0	0.0	0.0	0.0	0.0	0.0	0.0	1.5	13.5	1.5	13.5
Daman & Diu									0.0	0.1	0.1	0.1
Delhi	0.1	0.1	0.1	0.4	0.0	0.0	0.0	0.0	37.7	541.1	43.7	626.8
Goa	0.0	0.0	0.0	0.0	0.0	0.0	0.0	0.0	7.8	74.7	7.8	74.7
Gujarat	58.5	1340.6	43.3	978.2					94.9	609.4	331.5	4867.9
Haryana	17.2	353.0	18.4	441.7	0.0	0.0			99.0	1176.4	207.8	2980.4
Himachal Pradesh	1.3	21.7	14.1	151.9					11.1	172.1	59.1	1013.5
Jammu & Kashmir	2.6	46.0	7.1	117.4					23.6	360.0	52.1	843.0
Jharkhand							0.0	0.0	223.6	3394.9	223.6	3394.9
Karnataka	115.2	724.5	60.0	603.2	6.7	71.9	37.7	449.6	74.5	905.2	367.2	4382.9

Contd...

Appendix 4(b)–Contd...

State/Uts	Onion		Potato		S.Potato		Tapioca		Others		Total	
	A	P	A	P	A	P	A	P	A	P	A	P
Kerala	0.0	0.0	0.0	0.0	0.8	10.4	104.4	2436.8	0.0	0.0	107.6	2490.1
Lakshdweep	0.0	0.0	0.0	0.0	0.0	0.0	0.0	0.0	0.2	0.2	0.2	0.2
Madhya Pradesh	34.7	520.8	52.2	782.5	4.3	34.7			30.7	439.8	184.4	2659.6
Maharashtra	121.7	1422.3	17.4	163.3	5.0	70.0			102.5	215.4	372.2	4044.4
Manipur	0.5	1.1	5.0	26.1	0.0	0.0	0.2	0.7	1.52	5.3	13.4	86.0
Meghalaya	0.2	1.8	18.0	149.4	4.9	16.0	4.0	20.6			32.7	270.5
Mizoram	0.14	0.2	0.2	0.9	0.2	0.9	0.1	0.5	3.6	16.0	5.7	24.0
Nagaland	1.6	20.6	5.1	44.1	0.4	5.7	1.3	15.9	2.0	1.0	11.9	88.1
Orrisa	26.8	241.9	12.9	165.1	47.1	394.3			186.3	1831.0	655.9	8045.6
Pondicherry	0.1	1.0	0.0	0.0	0.0	0.0	0.7	19.5	2.8	34.4	4.5	74.7
Punjab	7.4	159.8	72.9	1470.2					40.8	511.9	158.6	2677.4
Rajasthan	42.9	347.9	3.1	48.2	2.1	6.1	0.1	0.3	31.5	106.9	124.1	650.2
Sikkim	0.0	0.0	7.5	32.6	0.0	0.0	0.0	0.0	9.5	43.9	17.0	76.5
Tamil Nadu	26.5	256.4	5.0	79.1	1.4	21.1	110.6	4563.8	30.39	792.7	215.3	6218.3
Tripura	0.1	0.2	5.3	93.5	1.0	8.4			26.3	271.3	32.7	373.4
Uttar Pradesh	21.2	245.8	440.0	9821.7	20.0	240.8			201.9	3757.5	840.9	15792.8
Uttranchal	3.8	25.1	22.1	448.5					25.6	208.3	72.8	951.8
West Bengal	16.33	183.58	320.6	7106.6	26.1	193.7			407.0	2874.3	1189.0	18103.2
Total	593.9	7515.4	1542.3	29188.6	136.5	1211.0	281.3	7900.8	2010.2	22123.7	6755.6	101433.5

Source: NHB, Gurgaon (2006).

Appendix 5: Floral Biology of Vegetable Crops

Crop	*Time of Anthesis*	*Time of Dehiscence*	*Duration of Pollen Fertility*	*Duration of Stigma Receptivity*
Lycopersicon esculentum	7.00-8.00 hr	9.00-11.00 hr	On day of the anthesis	16 hr before anthesis to the day of anthesis
Solanum tuberosum	5.00-6.00 hr	5.30-7.00 hr	On the day of anthesis to next day early morning	On the day of anthesis to next day afternoon
Capsicum annuum	5.00-6.00 hr	8.00-11.00 hr	On day of the anthesis	On day of the anthesis
Solanum melongena	6.00-11.30 hr	4.15-11.00 hr on the next morning also	On day of the anthesis to 2 days after anthesis	2-3 hr before anthesis to 2 days after anthesis
Abelmoschus esculentus	8.00-10.30 hr	6.00-11.30 hr	6.00-12.00	One day before 6 to the day of anthesis
Pisum sativum	5.45-6.30 hr	5.45-7.30 hr	One day of the anthesis to 24 hr thereafter	48 hr before and 24 hr after anthesis
Phaseolus spp.	5.00-7.00 hr	5.30-7.30 hr	On day of the anthesis	On day of the anthesis with maximum peak between 7.00-8.00 hr
Raphanus sativus	9.00-10.00 hr	9.00-10.00 hr	On the day of antheis when stored at 3ºC the pollen is viable for 60 days	Most receptive at the time of anthesis and last till 4 days after anthesis

Contd...

Appendix 5–Contd...

Crop	*Time of Anthesis*	*Time of Dehiscence*	*Duration of Pollen Fertility*	*Duration of Stigma Receptivity*
Brassica oleracea (Gongylodes Group)	8.00-10.00 hr	8.00-10.30 hr	On day of the anthesis	On day of the anthesis
Beta vulgaris var. bengalensis	7.00-17.00	8.30-18.30	On day of the anthesis	6 hr before to 8 hr after anthesis
Cucumis melo (muskmelon)	5.30-6.30 hr	5.00-6.00 hr	5.00-14.00 hr	2 hr before to 2-3 hr after anthesis
Cittrullus lanatus	6.00-7.30	5.00-6.30 hr	5.00-11.00 hr	2 hr before to 3 hr after anthesis
Laganaria siceraria	17.00-20.00 hr	13.00-14.30 hr	On the day of anthesis till next morning	36 hr before to 60 hr after anthesis
Momordica charantia	9.00-10.30 hr	7.00-8.00 hr	5.00-12.00 hr	One day before and one day after anthesis
Cucumis sativus	5.30-7.00 hr	4.30-5.00 hr	Up to 14 hr	12 hr before to 6-7 hr after anthesis

Appendix 6: Daily Allowances of Nutrients for Indians

Persons/ Category	Age (Years)	Weight (kg)	Height (cm)	Energy (Calories)	Proteins (gm)	Water Soluble Vitamins (mg)			
						Thiamine	Riboflavin	Vit. C	Niacin
Infants	0.0–1.0	9–Jun	60–71	690–945	12–19.8	0.3–0.5	0.4–0.6	35	8–Jun
Children	10–Jan	13–28	90–132	1300–2400	23–3b4	0.7–1.2	0.8–1.4	45	16–Sep
Males	18–Nov	45–66	157–176	2700–2800	45–56	1.4	1.6–1.7	50–60	18
	19–50	70	177–178	2700–2900	56	1.5	1.7	60	19
	51+	70	187	2400	56	1.2	1.4	60	16
Females	18–Nov	46–55	157–163	2100–2200	46	1.1	1.3	50–60	14–15
	19–50	55	163	2000–2100	44	1.0–1.1	1.2–1.3	60	13–14
	51+	55	163	1800	44	1	1.2	60	13
Pregnant	–	–	–	300	30	0.4	0.3	20	2
Lactating	–	–	–	500	20	0.5	0.5	40	5

Source: Food and Nutrition Board Academy of Sciences, National Research Council, Washington D. C.

Appendix 7: Vegetable Rich in Nutrients

Nutrients	*Sources*
Calories	Immature seeds of lima bean, broad bean and peas, tapioca, sweet potato, yams, colocasia (corms), potato, garlic, brussel's sprouts and onion.
Protein	Peas, double bean, winged bean, garlic, brussel's sprouts, cowpea, lima bean seeds, amaranth leaves, drumstick leaves and methi.
Precursor of vitamin-A (ß-Carotene)	Carrot, spinach, turnip green, palak, mustard green, amaranth, coriander, colocasia leaves, sweet potato (yellow), pumpkin (yellow) and tomato.
Vitamin-B complex	Peas, broad bean, lima bean, garlic, asparagus, colocasia corms and tomato.
Vitamin-C (Ascorbic acid)	Turnip green, green chillies, brussel's sprouts, mustard green and amranths, coriander, drumstick leaves, cauliflower, knolkhol, spinach, cabbage, bitter gourd and radish leaves.
Calcium	Curry leaves, amaranth leaves, drumstick leaves, methi, turnip-green, mustard green, coriander and palak.
Iron	Drumstick leaves and fruits, amaranth, methi, mint, coriander, spinach, palak and mustard green leaves.
Roughages	Spinach, lettuce, cabbage, amaranth and root vegetables.
Vegetable milk	Pea pods and cabbage leaves.

Source: Anon (1986).

Appendix 8: Quality Parameters in Vegetable Crops

Sl.No.	Name of Crops	Quality Parameters
1.	Tomato	Large round fruit with adequate firmness and shelf life, uniform fruit size, shape, red colour and freedom from external blemishes or abnormalities for fresh market. Large fruit size, high fruit quality and continuous production for home garden tomatoes. Deep uniformally red coloured tomatoes, pH below 4.4, high total soluble solids (4.5-7 per cent) and high alcohol insoluble solids (AIS) in processing tomatoes.
2.	Brinjal	Fruit shape, size and colour as per consumer preference with soft flesh. Low proportion of seed. Soft flesh. Lower olenine content
3.	Bell pepper and Chilli	Oblete or round fruit in bell pepper and long fruit in chilli. Pleasing flavour, high sugar/acid ratio, high pigment content and vitamin C in bell pepper and high capsaicin ($C_{18}H_{27}NO_3$). High Oleoresin in chilli
4.	Okra	Dark green, tender, thin, medium long, smooth 4-5 ridged pods at marketable stage. Pods should be free from conspicuous hairs and optimum seed setting ability.
5.	Pea	Long, attractive green pods with more seeds/pod. Sweetness in pods with high shelling percentage.
6.	Cauliflower	Non-ricey, compact bract free protected curds with retentive cream/white colour. Bettter seeding quality.
7.	Cabbage	Longer staying capacity in field after formation, Narrow, short and soft core with short stem.
8.	Radish	White, long/stump roots with thin tap root and non-branching habit, nonpithy roots with slow bolting habit. Pungency of roots should be as per consumer preference.
9.	Carrot	Scarlet/orange colour roots with uniformity in shape and size. Smooth root surface, thick flesh roots, and thin and self-coloured core in roots. Broad shouldered, cylindrical, uniformly tapering or stump rooted with non-branching habit. Early rooting with free from cracking. High sugar and dry matter content in roots.
10.	Turnip	Root colour as per consumer's preference, white, purple types more liked in India that golden ball types. Stump rooted hybrids with thin taproot and non-branching habit. Slow bolting habit and appropriate dry matter (8-9 per cent) in roots.
11.	Onion	High TSS is important for dehydration industry producing onion chips and powder.

Contd...

Appendix 8–Contd...

Sl.No.	*Name of Crops*	*Quality Parameters*
12.	Cucumber	Attractive green or darkgreen fruits with smooth surface and without prominent spines or prickles. Uniform long cylindrical shape without crookneck. Fruits should be free from carpel seperation without hollow spots with free from bitterness alongwith less seeds at edible maturity.
13.	Muskmelon	Fruits should be attractive round/spherical in shape with thick flesh and attractive in colour. Small seed cavity, sweet, juicy musky flavoursome fruits. TSS should not less than 10 per cent with tough netted skin of fruit.
14.	Watermelon	Tough skinned fruits for long distance transportation. TSS content not less than 10 per cent. Fruits with smaller and fewer seeds with attractive deep red flesh and firm flesh.
15.	Bottle gourd	Attractive green fruit of round/long/club shaped with sparse hairs persisting on skin and nonfibrous flesh at edible stage. Fruits should be nonbitter.
16.	Bitter gourd	Less ridged fruit surface with thick fruits particularly suitable for stuffing. Fruit size variation as per consumer's preference (Small 7.5-10 cm long, medium long 10-15 cm, long 15-20 cm and extra long 20-40 cm). Immature seeds for longer period during green edible stage.

Appendix 9: Seed Rate and Area Required for Nursery Raising in Vegetable Crops

Vegetables	*Seed rate (g)*	*Area required (m^2)*
Tomato		
Hybrid	200-250	75-100
Open pollinated	400-500	100-125
Brinjal	400-500	100-125
Chillies	500-600	75-100
Capsicum	400-600	100-150
Early cauliflower	700	100-150
Mid late cauliflower	400-500	75-100
Cabbage	450-500	75-100
Onion	8000-10,000	500

Source: Chaursia (2001).

Appendix 10: Vegetables and Approximate Days Taken in Seed Germination

Vegetables	*Seed Germination (Days)*
Tomato	5-7
Brinjal	5-6
Chilli	7-8
Cauliflower	3-4
Cabbage	3-4
Onion	7-10
Lettuce	4-5

Source: Chaurasia (2001).

Appendix 11: Suitable Age of Transplantable Seedlings of Some Vegetable Crops

Crops	*Age in weeks*
Brinjal	3-4
Cabbage	4-6
Cauliflower	3-4
Lettuce	4-5
Onion	3-4
Tomato	3-4

Source: Chaursia (2001).

Appendix 12: Bio-control Agents Recommended in Vegetables

Bioagent	*Dose*	*Target pest*
Trichogramma brassiliensis	2,50,000 parasitised Corcyra eggs/ha (Inundative release)	Okra shoot and fruit borer
	50,000-parasitised Corcyra eggs/ha (Weekly inoculative release)	Tomato fruit borer
Chrysoperla carnea	50,000 first instar larvae/ha (weekly release)	Okra aphid and cabbage aphid
HNPV	250 LE/ha (10 days interval)	Tomato fruit borer
SNPV	250 LE/ha (10 days interval)	Spodoptera litura
Bacillus thuringiensis	500 gai/ha (10 days interval)	Diamond back moth, shoot and fruit borer of brinjal and okra and tomato fruit borer.

Source: Satpathy *et al.* (1998).

Appendix 13: Selective Herbicides for Weed Control in Vegetable Crops

Herbicide	*Dose (kgha^{-1})*	*Treatment*	*Crops*
Pendimethalin	0.65-1.0	Pre-emergence	Transplanted pepper, onion, garlic, spinach, brassica crops, umbelliferous crops, legumes and potato.
Fluchloralin/ Trifluralin	1.0-1.5	Pre plant incorporation	Transplanted tomatoes, pepper, brinjal, potato, okra, brassica crops, legumes, garlic and umbelliferous crops.
Oxyfluorfen	0.24-0.36	Early post	Direct seeded and transplanted onion and potato.
Butachlor	2.0	Pre-emergence	Transplanted tomatoes, cucurbits
Metribuzin	0.2-0.35	Pre or early post emergence	Direct seeded and transplanted tomato and potato.

Source: Yaduraj and Dubey (2002).

Appendix 14: Effect of Plant Growth Regulators in Vegetable Crops

Crops	Growth Regulators	Concentrations (mg/l)	Method of Application	Attributed Affected
Tomato	NOA	25-50	Seed treatment 8 hr.	Increases fruit yield, TSS and Sugar
	GA3	5-20	Seed treatment 8 hr.	Increases fruit yield, TSS and Sugar
	CIPA	10-20	Seed treatment 8 hr.	Increases fruit yield, TSS and Sugar
	2,4-D	0.5	Seed treatment 8 hr.	Increases fruit yield, TSS and Sugar
	NAA	100	Foliar spray	Increases fruit yield
	PCPA	50	Foliar spray	Increases fruit yield
	CCC	200	Foliar spray	Increases fruit yield
	Ethephon	250	Foliar spray	Increases fruit yield
	Paclobutrazol	30-50	Foliar spray	Increases fruit yield
Brinjal	IAA	50	Seed treatment	Increases fruit yield
	NAA	0.2	Seedling treatment	Increases fruit yield
	IAA	20	Foliar spray	Increases fruit yield
	Mixtallol 2,4-D	4 2	Foliar spray 4 weeks after transplanting	Increases fruit yield
Chilli	GA	50	Foliar spray at fruit setting	Decrease flower shedding and increase fruit yield
	Planofix (NAA)	10	Double foliar spray (flowering and 5 week later)	Decrease flower shedding and increase fruit yield
	Triacontonol	2	Double foliar spray(30 days after transplanting and at bloom stage)	Decrease flower shedding and increase fruit yield

Contd...

Appendix 14–Contd...

Crops	*Growth Regulators*	*Concentrations (mg/l)*	*Method of Application*	*Attributed Affected*
Cabbage	NAA	0.1	Seedling treatment for 1/2 hr.	Improve head size and yield
	IBA	0.4		
	GA	5-10	Foliar spray	Improve head size and yield
	GA	50	Foliar spray	Increases yield
Cauliflower	IBA	0.1	Seedling treatment	Increases yield
	GA+	100	Foliar spray	Increases yield
	NAA+	120	Foliar spray	Increases yield
	MO+	0. 2%	Foliar spray	Increases yield
	GA + Urea	50 + 1%	Foliar spray	Increases yield
Okra	GA	40-150	Seed treatment	Increases yield
	IAA	100		
	NOXA	400		
	CCC	250	Seed treatment	Increases fruit yield
	NAA	15	Foliar spray at 20-25 days after sowing	Increases fruit yield
	GA	10-20	Foliar spray at 20-25 days after sowing	Increases fruit yield
	MH	800	Foliar spray at 20-25 days after sowing	Increases fruit yield
	CCC	500	Foliar spray at 20-25 days after sowing	Increases fruit yield

Contd...

Appendix 14–Contd...

Crops	*Growth Regulators*	*Concentrations (mg/l)*	*Method of Application*	*Attributed Affected*
Pea	MH	25	Foliar spray before flowering	Increases pod yield and protein content in seed
	CCC	500	Foliar spray before flowering	Increases pod yield
Cow pea	IBA	250	Seed treatment	Increases pod yield
French bean	Paclobutrazol	150	Foliar spray at 20 days after sowing	Increases pod yield
Onion	NAA	10-200	Seed treatment	Improved bulb growth and yield
	IAA	10		
	GA	40	Seedling treatment	Improved bulb growth and yield
	NAA	300	Foliar spray	Improved bulb growth and yield
	GA	60		
	MH	2500	Foliar spray at one week before bulb digging	Check the sprouting during storage
Bottle gourd	NAA	25	Foliar spray at 2 to 4 true leaf stage	Increases fruit yield
	MH	150		
	TIBA	50	Foliar spray at 2 to 4 true leaf stage	Increases fruit yield
	CCC	500		
	Ethephon	480	Foliar spray at 2 to 4 true leaf stage	Increases fruit yield
	Planofix(NAA)	50	Foliar spray at 2 to 4 true leaf stage	Increases fruit yield

Contd...

Appendix 14–Contd...

Crops	Growth Regulators	Concentrations (mg/l)	Method of Application	Attributed Affected
Ridge gourd	Ethephon	250	Foliar spray at 2 to 4 true leaf stage	Increases fruit yield
Sponge gourd	Ethephon	250	Foliar spray at 2 to 4 true leaf stage	Increases fruit yield
Bitter gourd	MH	2-33	Foliar spray at 2 to 4 true leaf stage	Increases fruit yield
Cucumber	TIBA	50	Foliar spray at 2 to 4 true leaf stage	Increases fruit yield
	GA	40		
	MH	100		
	Ethephon	150-500	Foliar spray at 2 to 4 true leaf stage	Increases fruit yield
Musk melon	NAA	25	Foliar spray at 2 to 4 true leaf stage	Increases fruit yield
	GA	10		
	MH	50-100		
Water melon	GA	10	Foliar spray at 2 to 4 true leaf stage	Increases fruit yield
	Ethrel	500	Foliar spray at 2 to 4 true leaf stage	Increases fruit yield
	MH	100		
Potato	CCC	500	Soaking of potato seed tuber	Increases tuber yield
	CCC	25	Foliar spray	Increases tuber yield
	Ethephon	400	Foliar spray	Increases tuber yield
	MH	2500	Foliar spray at 2-3 week before digging	Inhibit the sprouting during storage

Appendix 15: List of Indian Institutes Related to Vegetable/ Horticulture Science

Sl.No.	*Name of Institutes*	*Address*	*Discipline*
1.	Cardamom Research Centre	Indian Council of Agricultural Research, Appangala Coorg District, Karnataka State, Mercara–571201	Genetics, breeding, biotechnology, culture (crop improvement, crop production), crop protection, post-harvest technology and transfer of technology of cardamom
2.	Central Horticultural Experiment Station	Indian Council of Agricultural Research, PO Rarajaulatu Plandu, via Namkum Bihar State, Ranchi–834010	Culture (crop improvement, crop production), soil management, crop protection, post-harvest technology and extension of fruit, vegetables
3.	Central Institute for Arid Horticulture	Ganganagar Road, Bikaner, Rajasthan	To conduct mission oriented research for improvement in productivity of horticultural crops and development of horticulture based cropping system under arid environment; and to act as a repository of information related to arid horticulture
4.	Central Institute of Temperate Horticulture (CITH)	P.O. Sanatnagar–190005, Srinagar, J& K	Genetic resources, breeding, propagation, culture (crop improvement, crop production, seed production), crop protection and green house technology of temperate fruits, vegetables and flowers
5.	Central Plantation Crops Research Institute (CPCRI)	Kasaragod–670124, Kerala	To develop appropriate production, protection and processing technologies for coconut, arecanut and cocoa through basic and applied research
6.	Central Potato Research Institute (CPRI)	Shimla–171001, Himachal Pradesh	Breeding, biotechnology, crop improvement, crop production, physiology, nutrition, soil and water management, crop protection, engineering, post-harvest technology and extension of potato

Contd...

Appendix 15–Contd...

Sl.No.	Name of Institutes	Address	Discipline
7.	Central Tuber Crops Research Institute (CTCRI)	Sreekartiyam, Thiruvananthapuram Trivandrum–695017, Kerala	Genetics, breeding, biotechnology, crop improvement, crop production, crop protection, engineering, post-harvest technology, economics, extension and transfer of technology of tuber crops
8.	CITH Regional Station	Indian Council of Agricultural Research, Mukteshwar–263138, Uttar Pradesh	Propagation, culture (crop improvement, crop production) and crop protection of temperate fruit, vegetables and ornamental crops
9.	Department of Vegetable Crops	College of Agriculture, Punjab Agricultural University, Ludhiana	Teaching, research, and extension in vegetables
10.	Department of Vegetable Science	College of Agriculture, Govind Ballabh Pant University of Agriculture and Technology, Pantnagar–263145, Uttar Pradesh	Teaching, research, and extension in vegetables
11.	Division of Vegetable Crops, Indian Agricultural Research Institute	Indian Council of Agricultural Research, Pusa, New Delhi–110012	Breeding, culture (crop improvement, crop production), physiology, nutrition, soil and water management, crop protection, post-harvest technology, extension and transfer of technology of subtropical fruit, vegetables and ornamentals
12.	Indian Institute of Horticulture Research (IIHR)	Hassaraghatta, Bangalore–560089, Karnataka	Biotechnology, culture (crop improvement, crop production), crop protection, post-harvest technology and extension of fruit, vegetables, ornamentals, medicinal and aromatic crops

Contd...

Appendix 15–Contd...

Sl.No.	*Name of Institutes*	*Address*	*Discipline*
13.	Indian Institute of Spices Research (IISR)	P.O. No. 1701, Marikannu P.O., Calicut–673012	Genetics, breeding, biotechnology, biochemistry, culture (crop improvement, crop production), crop protection, post-harvest technology, extension and transfer of technology of spices
14.	Indian Institute of Vegetable Research (PDV)	P. B. No.01 P.O. Jakhani, (Shanshahpur), Varanasi–221305, Uttar Pradesh	Breeding, culture (crop improvement, crop production), physiology, nutrition, soil and water management, crop protection, post-harvest technology, extension and transfer of technology of vegetables
15.	National Research Center for Onion and Garlic	Rajgurunagar–410505, Pune, Maharasthra	Crop improvement, crop production, nutrition, crop protection, post-harvest technology, extension and transfer of technology of onion and garlic
16.	National Research Center for Medicinal and Aromatic Plants	Boriavi, Anand–387310, Gujarat	Biochemistry, culture (crop improvement), crop protection and post-harvest technology of medicinal and aromatic plants
17.	Yashwant Singh Parmar University of Horticulture and Forestry (YSPUH&F)	Nauni, Solan–173230, Himachal Pradesh	Crop science, genetics, breeding, culture (crop improvement, crop production, seed production green house technology), propagation, physiology, nutrition, crop protection, post-harvest technology and landscaping of fruit (temperate and subtropical), vegetables and ornamental crops

Appendix 16: List of Foreign Institutes Related to Vegetable/Horticulture Science

Sl.No.	Name of Institutes	Address	Discipline
1.	Alata Horticultural Research Institute	Turkish Ministry of Agriculture and Rural Affairs, P.O. Box 27 Erdemli, Mersin-33740, Turkey	Breeding, biotechnology, protected culture and crop protection of fruits, vegetables and ornamentals
2.	All-Russian Research Institute of Breeding and Seed Production of Vegetables	Russian Academy of Agricultural Science, Moscow region, P.O. Lesnoy Gorodok, Odintsovo–143080, Russia	Breeding and culture (seed production) of vegetables
3.	All-Russian Research Institute of Irrigated Vegetables and Melons	Russian Academy of Agricultural Science, Astrakhan region, Lubitcha str.16, Kamyziak–416306, Russia	Culture, soil and water management (irrigation) of vegetables and melons
4.	All-Russian Research Institute of Vegetable Crops	Russian Academy of Agricultural Science, Moscow region, Ramenskoe District, Vereya village, 500, 140153, Russia	Culture of vegetables
5.	ARC-Roodeplaat Vegetable and Ornamental Plant Institute	Agricultural Research Council, Private Bag X293, Pretoria–0001, South Africa	Breeding, biotechnology, culture and crop protection of potato, vegetables, indigenous vegetables, ornamentals (indigenous flowers) and medicinal plants
6.	Asian Vegetable Research and Development Center	PO Box 42, Shanhua, 741, Taiwan	Breeding, culture (production), soil management, crop protection (IPM), training and outreach of vegetables (alliums, tomato, chili, sweet peppers, leafy vegetables, mungbean, soybean)

Contd...

Appendix 16–Contd...

Sl.No.	Name of Institutes	Address	Discipline
7.	Ataturk Central Horticultural Research Institute	Turkish Ministry of Agriculture and Rural Affairs, Suleyman Bey Mah. Yalý Cad. P.O. Box 15, Yalova–77102, Turkey	Breeding, biotechnology, propagation, physiology, protected culture, crop protection, post-harvest (processing), economics and training of fruit, grape, vegetables and ornamentals
8.	Athalassa Experimental Station	Agricultural Research Institute, Agricultural Research Institute, Nicosia–1516, Cyprus	Culture of vegetables, subtropical fruits (olives, pistachio, date palm) and flowers (*i.e.* roses and greenery in soilless cultivation)
9.	Austrian Horticultural Society	Parkring 12, Vienna–1010, Austria	Interest promotion of horticultural crops
10.	Beijing Vegetable Research Center	Beijing Academy of Agricultural and Forestry Sciences, Banjingcun, Haidian District, Beijing–100089, China	Genetics, breeding, biotechnology, culture, (post-harvest) physiology, nutrition and seed testing of vegetables and flowers
11.	Biological Engineering applied to Horticulture	National Institute of Horticulture (INH), Rue Le Nôtre 2, Angers–49045, France	Genetics, germplasm preservation, molecular and cellular biology, genetic engineering, developmental physiology of vegetables and ornamental plants
12.	Bubia	National Agricultural Research Institute, PO Box 1639, Morobe Province, Lae, Papua New Guinea	Breeding, germplasm conservation, cultivar evaluation, phytopathology (entomology) and crop protection of tropical fruit, taro, yam and vegetables
13.	Centre for Agro-industrial Research and Development	Universidad de Bogota Jorge Tadeo Lozano, Apartado Aereo 140196, Chia-Cundinamarca, Colombia	Agronomy, culture, soil management (irrigation, fertilisation), crop protection and rural extension of fruit, vegetables and flowers

Contd...

Appendix 16–Contd...

Sl.No.	Name of Institutes	Address	Discipline
14.	Centre for Arboriculture and Horticulture, Les Fougères	Swiss Federal Research Station for Plant Production (Changins), Conthey, CH-1964, Switzerland	Culture (production) of fruit (berries), vegetables and medicinal plants
15.	Chiredzi Research Station	Deptt. of Research and Specialist Services, Horticultural Research Institute, P.O. Box 97, Chiredzi, Zimbabwe	Agronomy, culture, cultivar selection, physiology, plant protection of fruit (mainly tropical and subtropical), vegetables, root and tuber crops
16.	Crops Research Institute	Council for Scientific and Industrial Research, P.O. Box 3785, Kumasi, Ghana	Breeding, genetics, propagation, culture, physiology, nutrition, crop protection (IPM) and post-harvest management of fruit (citrus, mango, plantain, banana, pineapple), cashew, and vegetables (tomato, pepper, okra, eggplant, leafy vegetables, onion)
17.	Department of Medicinal and Aromatic Plants	University of Horticulture and Food Industry, Villanyi ut 29, Budapest-1118, Hungary	Crop science, breeding, culture, (post-harvest) and physiology of horticultural crops (medicinal plants)
18.	Department of Vegetable Crops	University of Applied Sciences of Weihenstephan, State Experiment Station of Horticulture, Freising-Weihenstephan–85350, Germany	Crop science, culture, computer controlled climate, irrigation and fertilization of vegetable crops
19.	Department of Vegetable Growing	Horticultural College and Research Institutes, Vienna–1131, Austria	Crop science and culture of vegetables

Contd...

Appendix 16–Contd...

Sl.No.	Name of Institutes	Address	Discipline
20.	Department of Vegetable Growing	University of Horticulture and Food Industry, Ménesiu. 44, Budapest–1118, Hungary	Crop science, breeding and culture of vegetables and mushrooms
21.	Department of Vegetable Growing	University of Horticulture and Food Industry, Ecskemét–6000, Hungary	Crop science and culture of vegetables
22.	Department of Vegetable Varieties Testing	Horticultural College and Research Institutes, Vienna-1131, Austria	Cultivar testing of vegetables
23.	Erzincan Horticultural Research Institute	Turkish Ministry of Agriculture and Rural Affairs, P.O. Box 18, Erzincan–24070, Turkey	Breeding, propagation, culture and training of fruits and vegetables
24.	Fengshan Tropical Horticultural Experiment Station	Taiwan Agricultural Research Institute, Horticultural Lane 4, Wenshan Road, Fengshan, 830, Taiwan	Crop science, biotechnology, breeding, physiology and culture of fruit and vegetables
25.	Horticultural Crop Research and Development Institute	Department of Agriculture (DOA), P.O. Box 11, Gannoruwa, Kandy, Peradeniya–KY 20400, Sri Lanka	Breeding, agronomy, culture, soil chemistry, crop protection, economics and extension of fruit, vegetables, root and tuber crops
26.	Horticultural Crops Research Institute	Xinjiang Academy of Agricultural Sciences, Nanchang Road 38, Urumchi–830000, China	Breeding, culture, micro-propagation and post-harvest physiology of fruit (apricot, pear, pomegranate, almond), grape, vegetables, sweet melon, water melon and flowers

Contd...

Appendix 16–Contd...

Sl.No.	*Name of Institutes*	*Address*	*Discipline*
27.	Horticultural Crops Research Institute	Zhejiang Academy of Agricultural Sciences, Shiqiao Road 198, Hangzhou–310021, China	Culture and post harvest (storage, processing) fruit, vegetables, edible fungi, flowers
28.	Horticultural Research	Agricultural Research Corporation (ARC), Rahad, P.O. Box 235, Sudan	Agronomy, breeding, entomology, and agricultural engineering of horticultural crops
29.	Horticulture Australia Limited	Level 1, 50 Carrington Street, Sydney, 2000	Research, development and marketing of horticultural crops
30.	Horticulture Research Centre	Dept of Research and Specialist Services, Horticultural Research Institute, P.O. Box 810, Marondera, Zimbabwe	Agronomy, breeding and selection, plant propagation, physiology, plant protection and post-harvest of flowers, deciduous fruits and vegetables
31.	Horticulture Research Institute	Pakistan Agricultural Research Council, National Agricultural Research Centre, Park Road Chak Shehzad, Islamabad–1031, Pakistan	Cultivar evaluation, seedling production, physiology, nutrition, crop protection, ecology, vinification, post-harvest, marketing and economy of fruit, grape-vine, vegetables and ornamental plants
32.	Horticulture Research Institute	Liaoning Academy of Agricultural Sciences, Dongling Road 84, Shenyang–110161, China	Breeding, biotechnology and culture of fruit, grape-vine and vegetables
33.	Horticulture Research Institute	Anhui Academy of Agricultural Sciences, Nongkeyuan Road, 40 Silihe, Hefei–230031, China	Genetics, breeding, biotechnology, culture, plant physiology, landscaping, gardening of fruit and vegetables

Contd...

Appendix 16–Contd...

Sl.No.	Name of Institutes	Address	Discipline
34.	Horticulture Research Institute	Qinghai Academy of Agricultural Sciences, Ningzhang Road 83, Xining–810016, China	Breeding and culture of fruit, vegetables and flowers
35.	Horticulture Research International	Willington Road, Kirton, Boston–PE20 1NN, United Kingdom	Agronomy, culture, crop physiology, entomology and plant physiology of field vegetables, bulbs and cut flowers
36.	Horticulture Research International	Wellesbourne, Warwick–CV35 9EF, United Kingdom	Genetics, breeding, agronomy, culture, biometrics, biochemistry, biotechnology, plant physiology, nutrition and phytopathology (entomology, pathology) of vegetables, mushrooms, protected ornamentals and protected edible crops
37.	Horticulture Research International–Stockbridge House	Cawood, Selby–YO8 0TZ, United Kingdom	Agronomy, culture, plant physiology, nutrition, phytopathology (entomology, plant pathology), crop protection of protected crops and field vegetables
38.	Hudieba Research Station	Agricultural Research Corporation, El Dammer, P.O. Box 31, Sudan	Breeding, agronomy, crop protection (entomology, virology, mycology, weed control) of horticultural crops and legumes
39.	Institute of Breeding on Vegetables and Medicinal and Spice Plants	Centre for Breeding Research on Cultivated Plants, Quedlinburg–06484, Germany	Breeding of vegetables, medicinal plants and spice plants

Contd...

Appendix 16–Contd...

Sl.No.	Name of Institutes	Address	Discipline
40.	Institute of Horticultural and Agricultural Engineering	University of Hannover, Herrenhäuser Straße, Hannover–30419, Germany	Greenhouse technology and expert systems for horticultural crops
41.	Institute of Horticultural Economics, Section Market and Policy Analysis	University of Hannover, Herrenhäuser Straße 2, Hannover–30419, Germany	Economics, marketing and communication of horticultural commodities
42.	Institute of Horticultural Economics, Section Production and Resources Economics	University of Hannover, Herrenhäuser Straße 2, Hannover–30419, Germany	Economics and marketing of horticultural commodities
43.	Institute of Post-Harvest Technology of Horticultural Products	Mendel University of Agriculture and Forestry, Valtická 337, Lednice na Morave–691 44, Czech Republic	Post-harvest physiology and technology of fruit, vine grapes, vegetables and cut flowers
44.	Institute of Vegetable and Floriculture	Mendel University of Agriculture and Forestry, Valtickå337, Lednice na Morave–691 44, Czech Republic	Agrotechnologies, cultivar testing, (seed) physiology, plant nutrition, breeding and nutritional value of vegetables, flowers, medicinal and aromatic plants
45.	Institute of Vegetable and Fruit Science	University of Hannover, Herrenhäuser Straße 2, Hannover–30419, Germany	Crop science, culture and modeling plant growth of Vegetables

Contd...

Appendix 16–Contd...

Sl.No.	Name of Institutes	Address	Discipline
46.	Institute of Vegetables and Flowers	Chinese Academy of Agricultural Sciences, Baishiqiao Road 30, Beijing–100081, China	Breeding, genetics, culture, crop protection and post-harvest physiology of vegetables and flowers
47.	Institute of Viticulture, Vegetable Crops and Floriculture	National Agricultural Research Foundation (NAGREF), Mesa Katsampas, Heraklion–71110, Greece	Culture, plant physiology (hormones), irrigation, nutrition, fertilization and wastewater management of grape-vine, vegetables and flowers
48.	International Potato Center (CIP)	La Molina, outside of Lima, Peru's	Scientific research and related activities on potato, sweet potato, other root and tuber crops, and on the improved management of natural resources in the Andes and other mountain areas
49	Inter-professional Technical Centre for fruit and Vegetables	Rue Bergère 22, Paris, 75009, France	Marketing and advertising of fruit, vegetables
50.	IRA Kebili station	Secetrtary of State of Scientific Research and Technology, Route de Douz, Jannoura, Kébili–4200, Tunisia	Fulture, soil management and crop protection of fruit trees and vegetables
51.	Jilin Vegetable and Flower Research Institute	Jilin Provincial Agricultural Bureau, Zhiyou Dalu 154, Changchun–130031, China	Breeding and culture of vegetables and flowers

Contd...

Appendix 16–Contd...

Sl.No.	Name of Institutes	Address	Discipline
52.	Lisavenko Research Institute of Horticulture for Siberia	Russian Academy of Agricultural Science, Zmeinogorsky tract 49, Barnaul-656045, Russia	Breeding, propagation, cultivar evaluation, seed production, water management, crop protection and mechanisation of fruit, berries, vegetables and ornamental plants
53.	Lithuanian Institute of Horticulture	Kauno28, Kaunas distr., Babtai–4335, Lithuania	Breeding, biotechnology, physiology, culture, crop protection and processing of fruit and vegetables
54.	Lowlands Agricultural Experiment Station, Keravat	National Agricultural Research Institute, East New Britain, Kokopo, Papua New Guinea	Cultivar evaluation, culture, phytopathology (entomology) and crop protection of sweet potato, banana, aibika (Abelmoschus manihot), vanilla and cashew nut
55.	Mediterranean Agricultural Institute–CIHEAM	Apartado 202, Zaragoza–50080, Spain	Plant resources conservation and utilization of horticultural crops
56.	National Agricultural Research Institute	NARI Main Highlands Program–Aiyura, P.O. Box 384, Eastern Highlands Province, Kainantu, Papua New Guinea	Germplasm collection and evaluation, cultivar and rootstock evaluation, culture, physiology (drought tolerance) of sweet potato, citrus and apple
57.	National Agricultural Research Institute	Ministry of Agriculture, IRESA, Rue Hédi Karray, Ariana–2049, Tunisia	Breeding, biotechnology, propagation, culture, crop protection, soil and water management of fruit, vegetables and ornamentals

Contd...

Appendix 16–Contd...

Sl.No.	Name of Institutes	Address	Discipline
58.	National Agricultural Research Institute	Co-operative Republic of Guyana, East Coast Demerara, Mon Repos	Genetic resources management, propagation, seed production, soil management, crop protection, post-harvest technology and extension of horticultural crops: fruit (pineapple, citrus, West-Indian cherry, mango), vegetables (tomato, ochro, boulanger, pepper); root and tuber crops (sweet potato, eddo, yam) and food crops (plantation, legumes)
59.	National Horticulture Research Centre	Kenya Agricultural Research Institute (KARI), PO Box 220, Kenya	Breeding, physiology, agronomy, culture, crop protection, post-harvest of fruit, vegetables and flowers
60.	National Potato Research Centre–Tigoni	Kenya Agricultural Research Institute (KARI), PO Box 338, Limuru, Kenya	Breeding, agronomy, culture, crop protection and post-harvest (food science) of potato and flowers
61.	Potato Research Centre	Agricultural Research Institute 'Serbia', Albanske Spomenice 21, Guca–32230, Yugoslavia	Breeding, selection, culture, physiology, phytopathology (virology) and crop protection of potato
62.	Pure State Horticultural Research Station	Latvia, Pure 26, Pure, Tukums–3124	Breeding, tissue culture, cultivar evaluation, culture and crop protection of fruit and vegetables
63.	Research Institute for the Dry Regions of Médenine	Secretary of State of Scientific Research and Technology, El Fjé, Médenine-4119, Tunisia	Genetic resources, physiology, culture and water management of fruit trees and vegetables
64.	Research Institute of Organic Agriculture	Research Institute of Organic Agriculture (FiBL), Ackerstrasse, Frick-CH-5070, Switzerland	Cultivar testing, quality of biological production, seedling production, physiology, soil management, pest and disease management, marketing and economy of fruit, grape-vine, vegetables and ornamentals

Contd...

Appendix 16–Contd...

Sl.No.	Name of Institutes	Address	Discipline
65.	Research Institute of Vegetable Crops	Ministry of Agriculture and Food Economy, Konstytucji 3 Maja 1/3, Skierniewice–96-100, Poland	Breeding, tissue culture, biotechnology, culture, physiology, nutrition, soil science, crop protection and post-harvest (storage, processing) of vegetables
66.	Research Station for Floriculture and Glasshouse Vegetables	Linnaeuslaan 2a, Aalsmeer–1431 JV, Netherlands	Culture (production), crop protection (biological control, IPM), soil and water management (fertilizers, amendments) of flowers and vegetables
67.	School of Higher Horticulture Studies	Ministry of Agriculture, IRESA, Chott Mériem–4042, Tunisia	Breeding, biotechnology, culture, crop protection and education of fruit, vegetables and flowers
68.	Sigatoka Research Station	Research Division of MAFF, P.O.Box 24, Sigatoka, Fiji Islands	Culture, cultivar evaluation, propagation, crop protection and package of management practice of vegetables and tropical fruit
69.	South-Urals Research Institute of Horticultural Crops and Potato	Russian Academy of Agricultural Science, Gidrostroy str. 16 P.O.Shershny, Tchelyabinsk–454902, Russia	Breeding, biotechnology, biochemistry, propagation, culture (seed production, cultural techniques) and mechanization of fruit, berry crops and potato
70.	Station for Applied Research on Arable Farming and Field Production of Vegetables	Edelhertweg 1, P.O. Box 430, Lelystad–8200 AK, Netherlands	Crop cultivation of arable crops and vegetables
71.	Swiss Federal Research Station for Fruit-Growing, Viticulture and Horticulture	P.O. Box 185, Wädenswil–H-8820, Switzerland	Breeding, cultivar and rootstock testing, culture, physiology, crop protection and post harvest of fruit (pome, stone, soft) and vegetables, grape-vine

Contd...

Appendix 16–Contd...

Sl.No.	*Name of Institutes*	*Address*	*Discipline*
72.	Teaching and Experimental Station for Horticulture Hessen	Oberzwehrenerstraße, 103, Kassel–34132, Germany	Organic growing, landscape gardening and teaching of horticultural crops (vegetables, garden plants)
73.	The Maritsa Vegetable Crops Research Institute	National Center for Agrarian Sciences of Bulgaria, Plovdiv–4003, Bulgaria	Genetics, breeding, biotechnology, biochemistry, cultivar testing, culture, seed production, physiology, nutrition, chemical analyses), crop protection, mechanization, economics and extension of vegetables.
74.	Vegetable Crops Centre	Agricultural research Institute 'Serbia', Karadordeva 71, Smederevska Palanka–1420, Yugoslavia	Genetics, breeding, culture and crop protection of vegetables
75.	Vegetable Crops Research Institute	University of Horticulture and Food Industry, Budapest, Mészöly Gy. u.6,Kecskemét–6004, Hungary	Breeding, culture (seed production, cleaning and selling) and marketing of vegetables
76.	Vegetable Research Institute	Shanxi Academy of Agricultural Sciences, Nongke Beilu 61, Taiyuan–030031, China	Breeding and culture of vegetables
77.	Vegetable Research Institute	Tianjin Academy of Agricultural Science, Tianjin high way, Tianjin–300381, China	Breeding, biotechnology, culture, physiology, storage and processing of vegetables (Chinese cabbage, cauliflower, snap bean, fruit vegetables, water melon and sweet melons, cabbage, potato)
78.	Vegetable Research Institute	Shandong Academy of Agricultural Sciences, Gongye Beilu 202, Jinan–250100, China	Breeding, biotechnology and culture of vegetables

Appendix 17: Books (Indian Authors) Related to Field of Vegetables/Horticulture Science

Sl.No.	Name of Book	Authors	Publication Agency
1.	Advances in Arid Horticulture: Present Status Vol. I	P.L. Saroj, B.B. Vashishtha and D.G. Dhandar	International Book Distributing Co., Lucknow
2.	Advances in Horticulture and Forestry Vols 7, 8	Edited by S.P. Singh	Scientific Publishers, Jodhpur
3.	Advances in Horticulture: Strategies, Production, Plant Protection and Value Addition	Edited by V.K. Sharma	Deep and Deep, New Delhi
4.	Advances in Horticulture: Vegetable Crops (Vols 5 and 6)	Ed. K. L. Chadha and G. Kalloo	Malhotra Pub. House, New Delhi
5.	Advances in Vegetable Diseases	S. Gangopadhyay	Associated Pub. Co., New Delhi
6.	Advances in Vegetable Production	N. Rai and D. S. Yadav	Researchco Book Centre, New Delhi
7.	Basic Concepts of Vegetable Science	Neeraj Partap Singh	International Book Distributing Co., Lucknow
8.	Biotechnology in Horticultural and Plantation Crops	K.L. Chadha, P.N. Ravindran and Leela Sahijram	Malhotra Publishing, New Delhi
9.	Biotechnology of Horticultural Crops	Edited by V.A. Parthasarathy, T.K. Bose, P.C. Deka and P. Das	Naya Prokash, Calcutta
10.	Breeding of Horticultural Crops: Principles and Practices	N. Kumar	New India Pub. New Delhi

Contd...

Appendix 17–Contd...

Sl.No.	*Name of Book*	*Authors*	*Publication Agency*
11.	Breeding Procedures for Cross Pollinated Vegetable Crops	Vishnu Swarup	ICAR, New Delhi
12.	Commercial Production of Horticultural Crops	H.N. Samaddar	Naya Udyog, Kolkata
13.	Comprehensive Micropropagation of Horticultural Crops	Ramesh Chandra and Maneesh Mishra	International Book Distributing Co., Lucknow
14.	Cultivation of Spice Crops	A.A. Farooqi, B.S. Sreeramu and Srinivasappa	Universities Press, Hyderabad
15.	Dictionary of Horticulture	Dinesh Arora	Biotech Books, New Delhi
16.	Diseases of Vegetables Crops	V. K. Gupta	Kalyani Publishers, New Delhi
17.	Diseases of Vegetables Crops	R. S. Singh	Oxford and IBH Publishing Co. Pvt. Ltd, New Delhi
18.	Elements of Horticulture	D.K. Singh and S.K. Singh	Agrotech Publshing Academy, Udaipur
19.	Emerging Scenario in Vegetable Research and Development	Edited by G. Kalloo and Kirti Singh	Research Periodicals and Book Publishing House, New Delhi
20.	Exploitation of Hybrid Vigour in Vegetables	Harbhajan Singh, Vishnu Swarup and Brijendra Singh	ICAR, New Delhi
21.	Flowers and Vegetables of India	Romesh Kumar Sud and Sudhir Kumar	Scientific, Jodhpur
22.	Fruit and Vegetable Preservation	R.C. Bhutani	Biotech Books, New Delhi

Contd...

Appendix 17–Contd...

Sl.No.	Name of Book	Authors	Publication Agency
23.	Genetics and Breeding of Vegetables	K. V. Peter	ICAR, New Delhi
24.	Gobivargeya Sabjiyon	Ramphal	ICAR, New Delhi
25.	Hand Book of Horticulture	Ed. K. L. Chadha	ICAR, New Delhi
26.	Handbook of Aromatic Plants	Supriya Kumar Bhattacharjee	Pointer Pub., Jaipur
27.	Heterosis Breeding in Vegetable Crops	N. Rai and Mathura Rai	New India Publishing Agencey, New Delhi
28.	Hi-Tech Horticulture	D.K. Singh	Agrotech Publshing Academy, Udaipur
29.	Horticulture for Sustainable Income and Environmental Protection	Edited by V.B. Singh, K. Akali Sema, Pauline Alila, Y.Y. Kikon and C.S. Maiti	Concept, New Delhi
30.	Horticulture Technology: Vision 2000 and Beyond	Edited by V.K. Sharma and K.C. Azad	Deep and Deep, Delhi
31.	Innovative Pest and Disease Management	Edited by S. Narasimhan,	SPIC Science Foundation, Chennai
	in Horticultural and Plantation Crops	G. Suresh and S. Daniel Wesley	
32.	Kheervargeya Sabjiyon	Nampal Singh and Pramod Joshi	ICAR, New Delhi
33.	Laboratory Manual of Analytical Techniques in Horticulture	R.S. Saini, K.D. Sharma, O.P. Dhankhar and R.A. Kaushik	Agrobios, Jodhpur
34.	Major Spices of India-Crop Management and Post Harvest Technology	J. S. Pruthi	ICAR, New Delhi

Contd...

Appendix 17–Contd...

Sl.No.	*Name of Book*	*Authors*	*Publication Agency*
35.	Minor Spices and Condiments Crop Management and Post-Harvest Technology	J. S. Pruthi	ICAR, New Delhi
36.	Nutrition of Vegetable Crops	S. K. Mitra	Naya Prokash, Calcutta
37.	Patt Wali Sabjyion	A. K. Pandey	Researchco Book Centre, New Delhi
38.	Pest and Soil Management of Horticultural Crops	S.K. Singh and D.K. Singh	Agrotech Pub., Udaipur
39.	Pest Management in Horticulture Crops: Principles and Practices	Edited by L.R. Verma, A.K. Verma and D.C. Gautam	Asiatech Pub., New Delhi
40.	Pest management in Vegetables, Vol I and II	K. P.Srivastava	Res. Periodical and Book, New Delhi
41.	Plant Growth Regulators in Agriculture and Horticulture: Their Role and Commercial Uses	Amarjit S. Basra.	International Book, Lucknow
42.	Preservation of Fruits and Vegetables	Girdhari Lal, G.S. Siddappaa and G. L. Tandon	ICAR, New Delhi
43.	Principles and Methods of Vegetable Cultivation	P. P. Sharma and R.C. Pandey	Ashiyan Publishers, New Delhi
44.	Principles of Vegetable Production	S. P. Singh	Agro Tech Publishing, Udaipur
45.	Producing Seed of Binnial Vegetables in Temperate Region	T.S. Verma, S. C. Sharma and P.V. Sane	ICAR, New Delhi
46.	Production Technology of Vegetable crops	K. G. Shanmugavelu	Oxford and IBH Publishing Co. Pvt. Ltd, New Delhi

Contd...

Appendix 17–Contd...

Sl.No.	*Name of Book*	*Authors*	*Publication Agency*
47.	Propagation of Tropical and Subtropical Horticultural Crops, Vol. I	Edited by T.K. Bose, S.K. Mitra, M.K. Sadhu, P. Das, D. Sanyal and V.A. Parthasarathy	Naya Udyog, Kolkata
48.	Sabji Ottpadan ke Siddhant Avam Vidhiya	P.C. Sharma	Asian Publishers, Muzzafarpur
49.	Sabji Vigyan	Pren Nath, Vishambarsharan Brahmchari and Dharmpal Singh	ICAR, New Delhi
50.	Scientific Horticulture: Vols. 2, 3,4, 5,6, 7, 8	Edited by S.P. Singh.	Scientific, Jodhpur
51.	Techniques of Developing Hybrids in Vegetable Crops	J. C. Kumar	Agro Botanical Pub., Bikaner
52.	Technology for Vegetable Production and Improvement	Hazra and M.G. Som	Naya Prokash, Calcutta
53.	Terminology of Horticulture	N. P. Singh	International Book Distributing Co., Lucknow
54.	The Potato	R. P. Pandey	Kalyani Publishers, New Delhi
55.	The Potato Production and Utilization	SM Paul Khurana	Mehta Publishers, New Delhi
56.	Tomato	G. Kalloo	Allied Publishers Pvt. Ltd, New Delhi
57.	Treatise on Indian Medicinal Plants Vols. 1, 2, 3, 4	Edited by Asima Chatterjee	Naya Prokash, Calcutta
58.	Tropical Horticulture, Vol. 1	Edited by T.K. Bose, S.K. Mitra, A.A. Farooqi and M.K. Sadhu.	Naya Prokash, Calcutta

Contd...

Appendix 17–Contd...

Sl.No.	Name of Book	Authors	Publication Agency
59.	Tropical Horticulture, Volume 2	Edited by T.K. Bose, J. Kabir, P. Das and P.P. Joy	Naya Prokash, Calcutta
60.	Varieties of Horticultural Crops	U. Thapa, A.B. Sharangi and A.K. Pal	Agrotech, Udaipur
61.	Vegetable Breeding	G. Kalloo	Panima Educational Book Agency, New Delhi
62.	Vegetable Breeding Principles and Practices	Hari Har Ram	Kalyani Publisher, New Delhi
63.	Vegetable Breeding, Production and Seed Production	Prem Singh Arya	Kalyani Publishers, Ludhiana
64.	Vegetable Crops of India	K. S. Yawalkar	Agri-Horticultural Publishing House, Nagpur
65.	Vegetable crops of India	P. C. Das	Kalyani Publishers, Ludhiana
66.	Vegetable crops Vol. I, II and III	Ed. T. K. Bose, J. Kabir, T.K. Maity, V.A. Parthasarathy and M.G. Som	Naya Prokash, Calcutta
67.	Vegetable Crops: Production Technology	M.S. Fageria	Kalyani Publishers, Ludhiana
68.	Vegetable Growing	S.C. Dey	Agrobios, Jodhpur
69.	Vegetable Growing in India	S. L. Katyal	Oxford and IBH Publishing Co. Pvt. Ltd, New Delhi

Contd...

Appendix 17–Contd...

Sl.No.	*Name of Book*	*Authors*	*Publication Agency*
70.	Vegetable Growing in India	Sasanka Barooach	Kalyani Publishers, Ludhiana
71.	Vegetable Production in India	D.V. S. Chauhan	Ram Prasad and Sons, Agra
72.	Vegetable Seed Production Principles	Prem Singh Arya	Kalyani Publisher, New Delhi
73.	Vegetables	B. Choudhary	National Book Trust of India, New Delhi

Appendix 18: Books (Foreign Authors) Related to Field of Vegetables/Horticulture Science

Sl.No.	*Name of Book*	*Authors*	*Publication Agency*
1.	A Colour Atlas of Cucurbit Diseases	D. Blancard, H. Lecoq and M. Pitrat	CPL Press Tall Gables, The Sydings Speen, Newbury, Berks RG14 1RZ, UK
2.	A Colour Atlas of Tomato Diseases	D. Blancard	CPL Press Tall Gables, The Sydings Speen, Newbury, Berks RG14 1RZ, UK
3	Advances in Potato Pest Biology and Management	Edited by Geoffrey W. Zehnder, Mary L. Powelson and Richard K. Jansson	CPL Press Tall Gables, The Sydings Speen, Newbury, Berks RG14 1RZ, UK
4.	Compendium of Tomato Diseases	Edited by J.B. Jones, John Paul Jones, R.E. Stall and T.A. Žitter	CPL Press Tall Gables, The Sydings Speen, Newbury, Berks RG14 1RZ, UK
5.	Cruciferous Vegetables, Isothiocyanates and Indoles IARC Handbooks of Cancer Prevention Volume 9	Harri Vainio, Franca Bianchini	Oxford University Press Bookshop, 116 High Street Oxford OX1 4BZ. UK
6.	Diseases of Fruits and Vegetables: Diagnosis and Management	Edited by S. Naqvi	Springer Publishing Agency
7.	Diseases of Vegetable Crops	Edited by Alfred Steferud	Biotech Books, New Delhi
8.	Diseases of Vegetable Crops	John Charles Walker	Greenworld, Lucknow
9.	Diseases of Vegetables Crops	John Charles Walker	McGraw Hill Book Company, New York

Contd...

Appendix 18–Contd...

Sl.No.	Name of Book	Authors	Publication Agency
10.	Fruit and vegetable biotechnology	Edited by Victoriano Valpuesta	CPL Press Tall Gables, The Sydings Speen, Newbury, Berks RG14 1RZ, UK
11.	Fruit and Vegetable Diseases	Edited by Mukerji, K.G	Springer Publishing Agency
12.	Fruit and Vegetable Processing	Wim Jonger	CPL Scientific Publishing Services Limited, The Sydings Speen, Newbury, Berks RG14 1RZ, UK
13.	Fruit and Vegetable Processing	Mircea Enachescu Dauthy	International Book Distributing Co., Lucknow
14.	Fruit and Vegetable Processing: Improving Quality	Edited by W Jongen	CPL Press Tall Gables, The Sydings Speen, Newbury, Berks RG14 1RZ, UK
15.	Fruit and Vegetables IARC Handbooks of Cancer Prevention Volume 8	Harri Vainio, Franca Bianchini	Oxford University Press Bookshop, 116 High Street, Oxford OX1 4BZ. UK
16.	Handbook of Horticulture	G.H. Preston	Reprint Pub., Dehradun
17.	Handbook of Vegetable Pests	John L. Capinera	Academic Press, London
18.	Handbook of Vegetable Preservation and Processing	Edited by Y. H. Hui, Sue Ghazala, Dee M. Graham, K. D. Murrell, Wai-Kit Nip	Marcel Dekker

Contd...

Appendix 18–Contd...

Sl.No.	Name of Book	Authors	Publication Agency
19.	Horticultural Pests: Detection and Control	G. Fox Wilson	Biotech Books, Delhi
20.	Indian Vegetable Plantation	W. Gollan	Akashdeep Publishing House, New Delhi
21.	Introductory Horticulture	E.P. Christopher	Biotech Books, Delhi
22.	Manual of Vegetable Plant Diseases	Charles Chapp	Discovery Publishing House, New Delhi
23.	Onions and Other Vegetable Alliums	J.L. Brewster	CAB International, Cambridge, UK
24.	Pests and Diseases of Vegetable Crops	B.A. Gerasimov and E.A. Osnitskaya	State Publishing House for Agricultural Literature, Moscow
25.	Phytochemistry of Fruits and Vegetables	F.A. Tomás-Barberán, R.J. Robins	Oxford University Press Bookshop, 116 High Street, Oxford OX1 4BZ. UK
26.	Postharvest Oxidative Stress in Horticultural Crops	Edited by D. Mark Hodges	International Book Distributing Co., Lucknow
27.	Postharvest Physiology and Pathology of Vegetables	Edited by J.A. Bartz and J.K. Brecht	Marcel Dekker
28.	Producing Vegetable Crops	John M Swiader	International Book Distributing Co., Lucknow
29.	The new Vegetable Grower's handbook	Arthur J. Simons	Penguin Books Ltd. Middlesex.

Contd...

Appendix 18–Contd...

Sl.No.	Name of Book	Authors	Publication Agency
30.	Vegetable Crops	Homer C. Thomson	McGraw Hill Book Company, New York
31.	Vegetable Gardening	W.E. Shewell-Cooper	Biotech Books, Delhi
32.	Vegetable Gums and Resins	F.N. Howes	Scientific, Jodhpur
33.	Vegetable Production	John H. MacGillivray	McGraw hill Book Company, Newyork
34.	Vegetable Production and Marketing	Paul Work	John Wiley and Sons, New York
35.	Vegetable Seedling Production Manual	David W. Doolan, C. Leonardi and W.O. Baudoin	Daya Publishing House, Delhi
36.	Viruses of Potatoes and Seed-Potato Production	J.A. De. Bokx	Dehra Dun

Appendix 19: List of Indian Magazines Related to Vegetable/Horticulture Science

Sl.No.	*Name of Magazine*	*Subscription*	*Publication Agency*
1.	Agriculture Today	Monthly (English)	Concept Agrotech Consultants Limited, C-17, Lajpat Nagar, 3rd Floor, New Delhi–110024
2.	Agrobios News Letter	Bimonthly (English)	Behind Nasrani Cinema, Chopasani Road, Jodhpur–342003
3.	Ahinshak Kheti	Monthly (Hindi)	Kisan-Kisani Vkas Trust, Basant Mension, RNT Marg, Indore
4.	Chotanagpur Horticulture	Bimonthly (Hindi and English)	Horticultural Scoiety of Chotangapur, C/o HARP, Plandu, Ranchi–834010
5.	ENVIS Bulletin: Himalayan Ecology	Half yearly (English)	G. B. Pant Institute of Himalayan Environment and Development, Kosi–Katarmal, Almora–263643 (Uttranchal)
6.	India Horticulture	Quarterly (English)	Agriculture Information and Publication Directorate, ICAR, Krishi Anusandhan Bhavan, Pusa, New Delhi–110012
7.	Indian Farming	Monthly (English)	Agriculture Information and Publication Directorate, ICAR, Krishi Anusandhan Bhavan, Pusa, New Delhi–110012
8.	Intensive Agriculture	Quarterly (English)	Krishi Vistar Bhavan, Directorate of Extension, Ministry of Agriculture, IASRI Campus, Pusa, New Delhi–110012
9.	Journal of North Eastern Council	Quartely, (English)	Director, Information and Public Relations, North Eastern Council, Shillong–793001 (Meghalaya)
10.	Kheti	Monthly (Hindi)	Agriculture Information and Publication Directorate, ICAR, Krishi Anusandhan Bhavan, Pusa, New Delhi–110012
11.	Krishak Samridhi	Monthly (Hindi)	A-1, Sai Nagar, Devendra Nagar Road, Raipur

Contd...

Appendix 19–Contd...

Sl.No.	*Name of Magazine*	*Subscription*	*Publication Agency*
12.	Krishi Chayanika	Quarterly (Hindi)	Agriculture Information and Publication Directorate, ICAR, Krishi Anusandhan Bhavan, Pusa, New Delhi–110012
13.	LEISA India	Quartely (English)	AME Foundation, D/No. 204, 100 Feet Ring Road, Banashankari III Stage, Bangalore–560085, Karnataka
14.	Phal–Phool	Quarterly (Hindi)	Agriculture Information and Publication Directorate, ICAR, Krishi Anusandhan Bhavan, Pusa, New Delhi–110012
15.	Spice India	Monthly (English)	Ministry of Commerce and Industry, Government of India, Sugandha Bhavan, Palarivattom, Cochin–682025 (Kerala)
16.	Unnat Krishi	Bimonthly (Hindi)	Krishi Vistar Bhavan, Directorate of Agriculture, Ministry of Agriculture, Pusa, New Delhi–110012
17.	Vishwa Krishi Sanchar	Monthly (English)	Vishwa Agro Marketing and Communication, II floor, Mohsin Manjil, 252, Shopping Centre, Kota–324007 (Raj)

Appendix 20: List of Indian Journals Related to Vegetable/Horticulture Science

Sl.No.	Name of Journals	Subscription	Publication Agency
1.	Agriculture Reviews	Quarterly (English)	Agricultural Research Communication Centre, 1130 Sadar Bazaar, Karnal–132 001, Harayana
2.	Agricultural Science Digest	Quarterly (English)	Agricultural Research Communication Centre, 1130 Sadar Bazaar, Karnal–132 001, Harayana
3.	Bhartiya Krishi Anusandhan Patrika	Quarterly (Hindi)	Bhartiya Krishi Anisandhan Samiti, F-5, Krisahi Nagar, Gwalior–474002
4.	Haryana Journal of Horticultural Sciences	Quarterly, (English)	Horticultural Society of Haryana, Department of Vegetable Crops, HAU, Hissar–125004 (Haryana)
5.	Indian Journal of Agricultural Sciences	Half yearly (English)	ICAR, New Delhi
6.	Indian Journal of Horticulture	Bimonthly, (English)	The Horticultural Society of India, Indian Agricultural Research Institute, New Delhi–110012.
7.	Journal of Applied Horticulture	Half yearly (English)	Society for the Advancement of Horticulture, A-859, Indiranagar, Lucknow–226016
8.	Journal of Assam Science Society	Half yearly (English)	The Assam Science Society, Lamb Road, Latasil, Guwahati–781001
9.	Orissa Journal of Horticulture	Half yearly (English)	Journal of Orissa Horticultural Society, Department of Horticulture, OUAT, Bhubaneswar–751003 (Orissa)
10.	Legume Research	Quarterly, (English)	Agricultural Research Communication Centre, 1130 Sadar Bazaar, Karnal–132 001, Harayana

Contd...

Appendix 20–Contd...

Sl.No.	Name of Journals	Subscription	Publication Agency
11.	Plant Horti-Tech	Bimonthly (English)	Institute of Plantation and Horticulture Management, Bangalore
12.	Potato Journal	Quarterly, (English)	Indian Potato Association, Central Potato Research Institute, Shimla–171001 (HP)
13.	Progressive Horticulture	Quarterly, (English)	Horticultural Experiments and Training Centre, CHAUBATTIA–263651, Ranikhet, Almora, (Uttaranchal) India
14.	The Horticultural Journal	Thrice issues per year (English)	Society of the Advancement of Horticulture, Faculty of Horticulture, BCKV, Mohanpur, WB
15.	Vegetable Science	Half yearly (English)	Indian Institute of Vegetable Research (IIVR), P.O. Jakhani (Shanshahpur), Varanasi–221305 (Uttar Pradesh)

Appendix 21: List of Foreign Journals Related to Vegetable/Horticulture Science

Sl.No.	Name of Journals	Subscription	Publication Agency
1.	Acta Horticulture	Irregular, (English)	ISHS Secretariat, PO Box 500, 3001 Leuven 1, Belgium
2.	Biological Agriculture and Horticulture	Quarterly (English)	AB Academic Publishers, P.O. Box 42, Bicester, OX6 7NW, U.K
3.	Canadian Journal of Plant Science	Quarterly (English)	Canadian Society of Agronomy (CSA), Agricultural Institute of Canada, 280 Albert St., Ottawa, ON, Canada K1P 5G8
4.	Field Crops Research	(English)	Elsevier, P.O. Box 1270, 1000 BG Amsterdam, The Netherlands
5.	Hort Science	Monthly (English)	American Society for Horticultural Sciences (ASHS), 113 South West Street, Suite 200, Alexandria, VA 22314-2851
6.	Hort Technology	Quarterly (English)	American Society for Horticultural Science, 113 South West Street, Suite 200, Alexandria, VA 22314-2851
7.	Horticultura Brasileira	Quarterly (English)	Sociedade de Olericultura do Brasil
8.	Horticultural Science Abstracts	Monthly (English)	CAB International Wallingford, Oxfordshire, OX10 8DE, UK
9.	HORTUS	Quarterly (English)	Smith Settle, Otley, West Yorkshire
10.	Journal of American Society for Horticultural Sciences	Bimonthly, (English)	American Society for Horticultural Sciences (ASHS), 113 South West Street, Suite 200, Alexandria, VA 22314
11.	Journal of Environmental Horticulture (JEH)	Quartely, (English)	Horticultural Research Institute, 1000 Vermont Avenue, N.W., Suite 300 Washington, D.C. 20005

Contd...

Appendix 21–Contd...

Sl.No.	*Name of Journals*	*Subscription*	*Publication Agency*
12.	Journal of Heredity	Bimonthly (English)	American Genetic Association, Oxford University Press, Evans Road, Cary NC–27513, USA
13.	Journal of Therapeutic Horticulture	Three issues/year (English)	American Horticultural Therapy Association (AHTA), 3570 East 12th Ave., Suite 206–Denver, CO 80206
14.	Journal of Vegetable Science	Quarterly, (English)	The Lane Research Center, Lane, Oklahoma
15.	New Zealand Journal of Crop and Horticultural Science	Quarterly, (English)	The Royal Society of New Zealand
16.	Nigerian Journal of Horticultural Science	Quarterly, (English)	Dept. of Agronomy, University of Ibadan, Ibadan, Nigeria
17.	Scientia Horticulture	Quarterly, (English)	Elsevier, P.O. Box 1270, 1000 BG Amsterdam, The Netherlands
18.	The Horticulturist	Half yearly (English)	The Journal of the Institute of Horticulture, Institute of Horticulture, 14/15 Belgrave Square, London SW1X 8PS.
19.	The Journal of Horticultural Science and Biotechnology	Bimonthly (English)	Department of Biological Sciences, University of Warwick Coventry, CV4 7AL, UK

Appendix 22: Vegetables Seed Companies in India

Sl.No.	*Seed Company*
1.	Agri Gold Group of Companies 40-1-21/21, 4tth Floor, Catholic Complex, M.G. Road, Vijayawada–520010, Andhra Pradesh, India
2.	Ajinkya Seeds 2, Sairaj Apartments, Survey no.14, Wadgaon Bk., Anandnagar, Pune–411 051
3.	Ankur Seeds Pvt. Ltd., 27, New Cotton Market Layout, Nagpur–440 018 (M.S.)
4.	Asgrow, USA C/o. Maharashtra Hybrid Seeds Co. Ltd. Raheja Towers, 4th Floor, West Wing, M.G.Road, Bangalore–1
5.	Bejo Sheetal Hybrid Seeds Co. Ltd. P.B. No. 72, A-2, Mama Chowk, Jalna–431203 (M.S)
6.	Century Seeds Pvt. Ltd. Lusa Tower, BA, 22-24, Mangolpuri Industrial Area Phase II, New Delhi–110034
7.	Daftari Agro Private Limited Seloo Post, Wardha–442104
8.	Golden Seeds Private Limited B-22A, Block-B, Brigade NM, R.Cross, Yediyur, Bangalore–560 082
9.	Indo-American Hybrid Seeds (India) Pvt. Ltd. PO Box No. 7099, 17th Cross, 2nd 'A' Main, K.R.Rd, BSK 2nd Stage, Bangalore–560 070
10.	J.K.Agri-Genetics (A division of J K Industries Ltd) 20, Paigah Colony, Behind Anand Theatre, S.P. Road, R.V.P. (HS) Secunderabad–3
11.	Mahyco Vegetable Seeds Ltd. Plot No. B-4, Industrial Estate, Jalna–431203
12.	Namdeo Umaji & Company 1-1-167, Dr. Ambedkar Road, Byculla, Mumbai–400 027
13.	Namdhari Seeds (Pvt) Ltd. 119, Arasappa Complex, 9th Main Rd, Ideal House, Raj Rajeshwari Nagar, Bangalore–39
14.	Nath Seeds Ltd. Seeds House Nath Res., P.B. No. 318, Aurangabad–431005 (M.S.)
15.	Nav Maharashtra Chakan Oil Mills Ltd. Nav Maharashtra House, 43, Shaniwar Peth, Opp. Shaniwar wada, Pune–411 030

Contd...

Appendix 22–Contd...

Sl.No.	Seed Company
16.	Nodai Seeds India Pvt. Ltd. 38th Mile Stone, Nh-8, Behrampur Road, Mandeep Estate, Gurgaon–122001, Haryana, India
17.	Novartis India–Seeds Division Seeds House, 1170/27, Revenue Colony, Shivaji Nagar, Pune– 411005
18.	Nunhems Proagro Pvt. Ltd. Registered Office: 10-1-127/1 Masab Tank, Hyderabad–500 028
19.	Proagro Seed Company Ltd. A-311, Ansal Chamber No.1, 3, Bhikaji Cama Place, New Delhi–110 066
20.	Richa International 2nd Floor, Krishna Bhuvan, 75, Kazi Sayed Street, Masjid Bunder, Mumbai–400003, Maharashtra, India
21.	Sartaj Seeds Farms P B No. 19, Hajipur, N. Bihar
22.	Sukanya Overseas Satyam, Sant Kabir Road, Behind Sadguru Sanidhaya Complex, Kabirvan, Society Main Road, Rajkot–360003, Gujarat, India
23.	Sun seeds India Private Limited 10-1-127/1, Masab Tank, Hyderabad–500 028
24.	Sungro Seeds Ltd., 3rd Floor, Sungro Chamber B.N. Block, Local Shopping Centre, Shalimar Bagh, Delhi–110052
25.	Sutton & Sons Pvt. Ltd. K-26, Connaught Circus Opp. Plaza Cinema, New Delhi–110001
26.	Swastha Organic Foods 114, New Ashiana Appt., Plot No-10, Sect.-6, Dwarka, New Delhi–110075
27.	Syngenta India Ltd., Seeds Sector, Waogholi, Gate No. 2347, Nagar Road, Taluk Haveli, Pune
28.	Vijay Seeds Co. (P) Ltd. A9/17, New Midc, Opp. Sham Lodge, Jalna–431203

References

Anonymous (1986). In Nutrition Garden. *Extension Bulletin 22, IIHR, Bangalore.*

Bonnierbale M, Plaisted RL and Tanksley SD (1988). *Genetics,* 120:1095-1103.

Borthwick HA, Phillips M and Robbins WW (1931). *American Journal of Botany,* 18:784-797.

Chandra S and Bhatnagar SP (1975). *Acta Botanica India,* 3:104-113.

Chaursia SNS (2001). *Compendium of Lectures in National Training course on Vegetable Production Technology,* 20-27 November, pp 7-15.

Choudhury B and Pathak SC (1959). *Indian Journal of Horticulture,* 18: 212-21.

Churchill GA and Doerge RW (1994). *Tech Rpt Biometric Unit Cornell University, Ithica, NY.*

Deshpande AA, Bankapus VM and Nalawadi UG (1978). *Current Research University of Agricultural Science, Bangalore,*7 (10): 174.

Deshpande AA, Venkatasubbaiah K, Bankapur VM and Nalawadi VG (1979). *Mysore Journal of Agricultural Sciences,* 13: 156-159.

Does and MP and Corenelissen BJC (1998). *Proceedings of 2nd International Crop Science*. Oxford IBH Publication Co. Pvt Ltd, New Delhi, pp233-244.

Doolan, David W *et al.* (2005). *Vegetable Seedling Production Manual/ FAO*. Daya Publishing House, New Delhi.

Dubey RS and Singh SP (1968). *Indian Journal of Horticulture*, 25(3-4): 210-203.

Edison S and Jhony Kallupurackal (1991). *Indian Cocoa, Arecanut and Spices Journal*, 14(4):143.

Ericksan EH, Garment MB and Peterson C (1982). *Journal of American Society of Horticulture Science*, 107: 698.

Fang Z, Sun P, Liu Y, Wang X, Hou A and Bcan C (1997). *Euphytica*, 97:265-268.

FAO (2005). *Restoring Farmers Seed Systems in Disaster Situations*. Daya Publishing House, New Delhi.

FAO (1997). *Quality Declared Seed*. Daya Publishing House, New Delhi.

Federoff NV, Futerk DB and Nelson OE (1984). *Proceeding of National Academic Science*, USA, 81: 3825-2829.

George RAT (1980). *Technical Guidelines of Vegetable Seed Technology, Food and Agriculture Organization of the United Nations*, Rome.

Haley CS and Knot SA (1992). *Heredity*, 69: 315-324

Herrera Estrella L and Simpson J (1995). *World Journal of Microbiology and Botany*, 11:383-392.

Herrington ME (1981). *Queen Journal Agriculture Animal Science*, 38: 127-132.

Hilder VA, Gatehouse AMR, Sheerman SE, Barker RF and Boulter D (1987). *Nature*, 330; 160-163.

Honne G, Melchers LS, Vleehouwers VGGAA, Van Roekel JSC and de Wit PJGM (1995). *Plant Molecular Biology*, 29: 909-920.

Iwasaki F and Takeda Y (1977). *Journal of Japanese Society of Horticulture Science*, 46:193-200.

James C (1998). *Asian Seed*, 5: 18-21.

Jansen RC and Stam P (1994). *Genetics*, 136:1447-1455.

Jones DF (1918). *Connecticut. Agriculture Experiment Station Bulletin*, 207, p. 100.

Kadam BS (1942). *Indian Journal of Genetics*. 2: 159-172.

Kalloo G and Singh Kirti (2001). *In : Souviner Emerging scenerio in vegetable research and development*. Research Periodical and Book Publishing House, New Delhi; pp.104-117.

Knapp SJ, Bridges WC and Brikes D (1990). *Theoretical Applied Genetics*, 79: 583-592.

Kumar Sanjeet, Banerjee MK and Kalloo G (2000). *Vegetable Science*, 27: 1-24.

Lysenko, T D (2005). *Agrobiology: Essays on Problems of Genetics Plant Breeding and Seed Growing*. Biotech Books, New Delhi.

Malik VS (1989). *Advances in Applied Microbiology*, 34; 263-306.

Martin GB, Ganal MW and Tanksley SD (1992). *Mol Gen Breeding Science*, 43: 327-333.

Mathur, S B & H K Manandhar (1997). *Quarantine for Seed/FAO*. Daya Publishing House, New Delhi.

Mohr HC (1956). *Proceedings of Association of South Agriculture*, Work, 53:174.

Mullar CH (1940a*). US Dept Agric Onisc Publ* 382.

Nair PKK, Kapoor SK and Verma D (1974). *New Botanist*, 1: 134-141.

Nandpuri KS and Brar JS (1966). *Punjab Agriculture University Journal Research*, 3: 395.

Nath P (1969). *Indian Journal of Horticulture*, 26: 136-139.

Nelson RS, McCormick SM, Delannay X, Dubre P, Layton J, Anderson EJ, Kanieska M, Proksch RK, Horsch RB, Rogers SG, Fraley RT and Beachy RN (1988). *Biotechnology*, 6:403-409.

NHB (2006). *National Horticulture Board*, Gurgaon.

Nieuwhof M (1961). *Euphytica*, 10: 351-356

Nieuwhof M and Garretsen F (1975). *Euphytica*, 24: 87.

Odland ML and Knoll CJ (1950). *Proceeding of American Society of Horticultural Science*, 55: 391-402.

Pal AB (1975). *Progressive Horticulture*, 7: 62-68.

Patil AV and Patil SJ (1974). *Research Journal of Mahatma Phule Agriculture University*, 5 (1): 57-58.

Prasad, D ed (2005). *Crop Protection: Management Strategies*. Daya Publishing House, New Delhi.

Purewal SS and Randhawa GS (1947). *Indian Journal of Agricultural Sciences* 42 (3A): 437-438.

Purohit SP and Nath P (1970). *Madras Agriculture Journal*, 57:99-105.

Rai Mathura and Pandey AK (2004). *Hybrid vegetables: Meeting Global Trade.* The Hindu, Survey of Indian Agriculture, pp–129-131.

Rick CM and Fobes JF (1977). *Rep Tomato Genet Coop*, 24:25.

Robinson RW, Munger HM, Whitaker TW and Bohn GW (1976). *Hort Science*, 11: 554-568.

Rotino GL, Perri E, Zottini M, Sommer and Spena A (1997). *Nature Biotechnology*, 15:1398-1401.

Satpathy S, Rai Samarjit and Kapoor KS (1998). *Souvenir. AICRP on Vegetable Crops*, IIVR, Varanasi, p 123-130.

Sedgley M (1981). *Journal of Cell Science*, 48: 137.

Sedgley M (1982). *Journal of Cell Science*, 54: 341.

Seshadri (1998) (In.): *Memorial Lectures in Silver Jubilee National Symposium on Emerging Scenario in Vegetable Research and Development*, 12-14, December, 1998. pp. 34-38.

Sharma RP and Kumar PA (1995). *In: Sustaining crop and animal productivity. In:* Deb DL (ed). The Challenge of the Decade, Associated Publishing Co, New Delhi, pp. 147-156.

Shull (1909). Report of American Breeders Association, 5: 51-59.

Sidhu AS, Kalloo G and Pandita ML (1980). *Haryana Journal of Horticultural Science*, 9.

Singh Major, S. Chakraborty, Kiriti Srivastava and Kallo G (2001). *In: Emerging Scenario in Vegetable Research and Development* (eds) Kalloo G and Singh Kirti. Research Periodical and Book Publishing House, pp.58-84.

Smartt J (1976). *Tropical Pulses*, Longman, London, pp 348.

Srivastava D (1982). *Advances in Pollen Spore Research* vol 9 pp 18-44. Today and Tomorrows Printers and Publishers, New Delhi.

Strittmatter G Janssens J Opsomer C and Botterman J (1995). *Biotechnology*, 13: 1085-1089.

Tanksley SD Ganal MW, Prince JP, de Vicente MC, Boinerable MW, Broun P, Fulton TM, Giovanonni Wing RA, Wu W and Young ND (1992). *Genetics*, 132: 1141-1160.

Tanskley SD, Bernatzky R Lapitan NL and Prince JP (1988). *Proceedings of National Academic Science* USA, 6419-6423.

Thomas BR and Pratt D (1981). *Theoretical Applied Genetics*, 59:215-219.

Thompson KF (1959). *Breeding better kales. Agricultur*e, 65:487-491.

Thompson RC (1951). *US Department Agriculture Farmers Bulletin*.

Thomson JR (1979). *An Introduction to Seed Technology*, Chapter 12, pp 157-67. Leonard Hill, London.

Tinker NA and Mather DE (1995). *JQTL*, 1:1-22.

Tomassoli L, Iiardi V Barba M and Kaniewski W (1999). *Molecular Breeding*, 5: 121-130.

Vijay OP, Singh DP and Yadav IS (1979). *Vegetable Science*, 6: 5-10.

Ward GM (1964*). Cannadian Journal of Plant Sciences*, 44: 492-494.

Whitaker TW and Davis GN (1962). *Cucurbits*. Inter-Science Publishers, INC. New York. p. 13.

Yoneyama K and Anzai H (1991). *Journal of Pest Science*, 16:291.

Young ND and Tanksley SD (1989). *Theor. Applied Genetics*, 77:353-359.

Zeng ZB (1993). *Proceeding of National Academic Science*, USA Vol. 10972-10976.

Index

A

A. caudatus 83
A. ficuleneus 179
A. graveolens 15
A. manihot 179, 183, 184
A. moschatus 179
A. spinosus 83
Abelmoschus esculentus 14, 178, 183, 184
Abiotic and biotic stress 97, 106
Abscisic acid 316
Abscission 41
Absorption 312, 314, 322
ACC oxidase 347, 358
ACC synthase 347, 358
Acid scarification 318
Aculeate 37, 39
Acuminate 37, 39, 41, 42, 70
Acute 37, 39, 42, 44, 64, 70, 71, 77, 78
Adenosine triphosphate 312
Adoptive trials 109
Advanced hybrid trials 107-108
Adventitious 40, 47, 62
Agencies in seed production 23-31
  Project Directorate of Vegetable Research 23-24
  National Bureau of Plant Genetic Resources 25
  Statutory bodies and agencies 25
  Financing organization 25-26
  Societies 26

Universities 26
Institutes 26-27
Semi-government organizations 27
State government pilot projects 27
Private sector 27
ICAR funded centres 28
Voluntary centres 28
National Seeds Corporation 28-31

Agonoscelis nubilis 258
Agricultural Development Officer 309
Agriculture–Horticultural Society 20
Agro-climatic situations 23
Agro-climatic zones 108
Agronomic principles 89
Agrotis spp 172
Air drying 322
Air-tight container 9
Aizoaceae 12
Albizia lophantha 315
Albugo candida 188, 212
All India Coordinated Research Project 22, 23, 25, 86
All India Coordinated Vegetable Improvement Project 23, 106-108, 113, 116, 117, 126
Alliaceae 12, 57
All-India Coordinated Maize Improvement Project 24
Allium cepa 12, 83, 215
Allium spp. 12
Allium ascalonicum 12
Allium cepa var. *aggregatum* 12
Allium cepa var. *viviparum* 12
Allium fistulosum 12
Allium porum 12
Allium sativum 12
Allium schoenoprasum 12
Allogamy 50, 81
Alphanaphthalene acetic acid 272
Alpinia galangal 12
Alternaria blight 188, 206
Alternaria brassicae 188, 206
Alternaria dauci 264
Alternaria porri 219
*Alternaria solani*143
Althalia proxima 188, 206
Amaranthus spp. 34, 82
Amomum subulatum 12
Amrasca biguttula 130, 181
Amrasca kerri 246
Anaemophily 82
Anatrosperm 42
Androecium 54
Andromonoecious 73, 75, 101, 326
Anethum sowa 15
Angelica archangelica 15
Angiospermae 11
Angular leaf spot 239, 241, 276
Annual 17, 35, 44

Anther culture 341, 342
Anthesis 40, 44, 50
Anthracnose 156, 159, 160, 237, 239, 240, 244, 246, 272, 276, 279, 285
Anthriscus cerefolium 15
Anti-auxins 272
Antifungal protein genes 355
Antigens RNA 344, 357
Antisense ACC oxidase 348
Antisense RNA 347, 357
Ants 252
APEDA 4
Aphid 130, 158, 188, 191, 212, 239, 246, 257, 259, 272, 279, 285, 289, 295, 290
Aphis craccivora 239
Aphis gossypii 130, 158, 295
Aphis spp. 228, 246, 272, 279, 285, 289, 290
Apium graveolens 14, 15
Appeal 384, 395
Appellate Authority 25
Appellate board 380
Aqconus calamus 12
Arabidopsis 352
Araceae 12
Area for certification 377
Armoracia rusticana 13
Artemisia dracunculus 12
Artificial pollination 101-102
  Hand emasculation 101
  Wind pollination 101
  Insect-pollination 101
  Self-incompatibility 102
  Male sterility 102
Ascochyta blight 237, 244
Asexual 18, 80, 81
Ashy stem blight 244
Asparagus officinalis 12
Aspergillus niger 345
Associated Agriculture Development Organizations 26
Association of Official Seed Analysis 25
Association of Official Seed Certification Agencies 85
ATP 312
Aulacophora foveicollis 272, 279, 284, 289
Autogamy 81
Auxins 272
Avr genes 356
Awakening or Activation 312-314
  Imbibition of water 312
  Synthesis of enzyme 312
  Cell elongation and emergence of the radicle 313-314

B

Bacillus thuringensis 345, 346
Back cross 248
Backcross-pedigree selection 151
Bacterial blight 228, 237

Bacterial resistance 344-345

Bacterial wilt 131, 134, 135, 138, 139, 143, 149, 153, 154, 159, 168, 173

Bactericide 344

Bagging 7, 8, 49, 58, 326, 332

Bagrada cruciferarum 206

Balsam apple 287, 288

Barbarea vulgaris 13

Barnase gene 356

Barnase-Bar Star 348

Batisak 35

Bean aphids 239

Bean mosaic 237

Bean thrips 239

Bejo Sheetal Seed Co Ltd 100

Bemisia tabaci 143, 159

Bervicoryne coriandri 257

Beta vulgaris 12, 83, 249-250

  Beta vulgaris rubra 250

  Beta vulgaris var. *bengalensis* 249

  Beta vulgaris var. *cicla* 250

Bhat karela 287, 288

Biennial 17, 35, 203

Biochemical markers 304

Biological diversity bill 362

Biotechnology 337-339, 354, 360, 369

Bisexuality 81

Bitter gourd mosaic 290

Bitter gourd witches broom 290

Black leg 190, 195, 198, 202, 206, 208, 210

Black rot 190, 195, 198, 199, 202, 207, 208, 210

Black scurf 170, 173

Black spot 206

Bolting 66, 189, 193, 213, 268

Bordeaux mixture 130

Botanical classification 11-16

  Monocotyledonae 12

  Dicotyledonae 12-16

Botany and floral biology 37-79

  Solanaceae 37-49

  Malvaceae 49-51

  Alliaceae 57-58

  Leguminosae 58-65

  Chenopodiaceae 65-66

  Umbelliferae 66-70

  Cucurbitaceae 70-79

  Compositae 79

Bovine pancreatic trypsin inhibitor 346

Brassica alba 14

Brassica campestris 13, 209, 342

Brassica caulorapa 13

Brassica chinensis 13, 35

Brassica juncea 14

Brassica napus 13

Brassica nigra 13

Brassica oleracea var. *acephala* 13

Brassica oleracea var. *botrytis* 13, 83, 194

Brassica oleracea var. *capitata* 13, 83, 201

Brassica oleracea var. *caulorapa* 83

Brassica oleracea var. *gemmifera* 13, 83
Brassica oleracea var. *italica* 13
Breaking dormancy 49
Breeder seed 24, 85-87, 308, 363, 376, 377
Breeding systems 18
Bremia lactucae 296
Brevicory brassicae 188, 212
Brinjal little leaf 131
Brown rot 166, 167, 168, 169
Buckeye rot 143, 155
Bulb rot 219
Bulb to seed method 217, 218

C

C. melo 73
Cabbage borer 206
Cabbage butterfly 206
Cabbage looper 295
Caging 7, 326
Campanulate 39, 42, 45, 71, 77
Campylotropus 54
Cantaloupe 73
Capitulum 79
Capsaicin 111, 161, 162, 163, 165
Capsicum annum 14, 155, 349
Capsicum fruitescence 14, 155
Carotene 266, 267
Carrot rust fly 264
Carrot yellows 264
Carum bylbocastanum 15
Carum carvi 15
Cell elongation 313
Central Seed Certification Board 25, 365
Central Seed Committee 25, 365, 382, 391
Central Seed Testing Laboratory 25, 369
Central Subcommittee on Crop Standards and Release of Varieties 120-126
Central Vegetable Breeding Station 21
Century Seeds 27, 100
Cercospora carotae 264
Cercospora leaf blight 254, 264
Cercospora leaf spots 246
Cercospora spp. 246
Certification agency 383
Certification of seeds 375-380
- Purpose of seed certification 375
- Certification standards 375
- Requirements for certification of crop varieties 376
- Classes and sources of seed 376-377
- Seed standards for genetic purity 377
- Unit area for certification 377
- Rejection of seed farms and seed lots 378

Certification stages 378-380
- Registration of seed programme 378-379

Seed certification fee 379
Field inspection 379
Processing/sampling 379
Field testing 379
Revalidation 379
Certification tags 379-380
Appellate board 380
Certification standards 375-376
Minimum seed certification standards 375
Seed certification procedure 375
Certification tag 86, 87, 376, 377, 379-380
Certified foundation seed 86
Certified seed 28, 87, 88, 92, 308, 363, 375-378
Certified seed production 24, 28, 29
Cetral Seed Laboratory 381, 391
Charcoal rot 170
Chemical dormancy 315-316
Chenopodiaceae 12, 18, 65
Chips 175
Chitinase 343, 356
Choanephora cucurbitarum 246
Chymotrypsin inhibitor 345
Cichorum endivia 12
Cichorum intybus 12
Cinnamomum aromaticum 16
Cinnamomus zeylanicum 16
Citrullus colocynthis 283
Citrullus lanatus 14, 83, 281
Cladosposium fulvum 351, 356
Classification of vegetables 11-19
Botanical classification 11-16
Nature of pollination 16
Planting response 16
Nature of vegetables 17
Area of seed production 17
Hardiness in hills 17
Methods of seed production 18
Breeding systems 18
Growing season 18-19
Soil reaction (pH) 19
Cleistogamy 82
Climatic factors 33-36
Photoperiod 34
Temperature 34-35
Rainfall 35
Wind 35-36
Colchicines 254
Cole crops 16, 18, 82, 90
Collar rot 139
Colletotrichum capsici 159
Colletotrichum lindemuthianu 246
Colletrotrichum lagenarianum 272, 279, 285, 290
Colocasia esculenta 12, 17, 18
Combining ability 102, 103, 199
Commercial hybrids 101
Single crosses 101
Double crosses 101

Triple crosses 101
All types of hybrids 101
Commercial Seed Analysts Association 25
Common bean mosaic 239, 241
Common methi 235
Common scab 166, 170, 173
Compositae 12, 79
Composits 87
Container 381, 393
Control of seed source 85-87
Convolvulaceae 14
Coreless 68
Coriandrum sativum 15, 255
Cost benefit analysis 331, 335
Cost of hybrid seed production 323-326
Cost on raising of healthy crop 323-324
Cost of emasculation and pollination 324-325
Cost of post harvest operations 325-326
Cotton bollworm 184
Coumarin 316
Cowpea mosaic 244, 246, 249
Cowpea trypsin inhibitor 346
Crocus sativus 12
Crop rotation 87, 88
Cross compatibility 83
Cross pollinated fertilizing species 301
Crossability 11
Cross-compatible 88
Cross-pollinated 2, 5, 7, 16, 32, 36, 51, 58, 66, 73, 82, 87, 88, 89, 102, 339
Cross-pollination 7, 16, 39, 42, 46, 51, 80, 81, 88, 144, 272, 290
Cruciferae 14, 51-57
Cryopreservation 341
Cucumber mosaic 272, 343, 344, 357
Cucumis hardwickii 269, 270
Cucumis melo 14, 83, 276
Cucumis melo var. *conomon* 277
Cucumis melo var. *momordica* 277
Cucumis melo var. *utilissimus* 277
Cucumis melo var. *agrestis* 277
Cucumis prophetarum 277
Cucumis sativus 14, 83, 269
Cucurbita maxima 14
Cucurbita moschata 14, 83
Cucurbita pepo 14, 83
Cucurbitaceae 13, 70-78
Cucurbits 18, 34, 83
Cuminum cyminum 15
Curcuma longa 12
Cut worms 172
Cyamopsis tetragonolaba 14
Cynara scalymus 12
Cyst nematodes 166, 176

D

Dacus cucurbitae 279, 290
Dacus dorsalis 272, 279, 290

Damping off 130, 142, 143, 158, 159, 295
Dark-color jars 9
Daucus carota 14, 83, 261
Day length 34
Day light 34, 40
Day-neutral plants 34
Decliny 82
Defence Research Laboratory 27, 28
Defence Vegetable Farm 27
Degree of aliveness of seed 306
Dehiscence 40, 42, 44-46, 50, 51, 66, 70, 76, 78
Dehumidifiers 326
Dehydration 112, 229
Delayed ripening 358
Detasselling 101
Determinate 142, 148, 150-154, 165
Detoxification 344, 358
Diadelphous 59
Diamond back moth 206
Dichogamy 82
Dicotomous 41
Dicotyledonae 11
Dicotyledonous 44
Die back 156, 162
Dietary requirement 10
Digestion and translocation 313
Dimorphic 73
Dioecious 18, 104, 325, 332
Dioscoraceae 12
Dioscorea alata 12
Directorate of Plant Protection Advisor 23
Disease resistance 116
Disease resistant hybrids 122-123
Diseases and insects test 109-110
Distichous 70
District Consumer Forum 310
DNA based markers 303
Dolichos lablab 14
Dormancy 311, 314, 320
Double cross 101
Downy mildew 199, 206, 212, 219, 234, 272, 275, 279, 285, 290, 291, 296
Drosophila 352
Dry matter 112, 163, 175, 222
Dry rot 170
Drying 325
Durga Seeds 26

E

E. coli 342
E. insulanta 181
Earias vittella 181
Early blight 140, 143, 144, 151, 153, 154, 173, 174, 176, 177
Economics of Hybrids Seed Production 323-336
  Cost of hybrid seed production 323-326
  Methods of hybrid seed production and cost analysis 326-336

Cost benefit analysis 335-336
Electrophorosis 304
Elettaria cardamomum 12
Ellipsoid 71
Emasculation 44, 54, 64, 65, 101, 105, 323, 324, 325, 326, 327, 329, 330, 332, 334
Embryo culture technique 339
Embryo rescue 341
English vegetables 20
Entomophily 82
Environment Protection Act 370
Environmental factors 34
Epicalyx 49
Epidemic 110
Epigeous germination 314
Epigynous 69, 72
Epilachna beetle 172
Epiphytotic 110
Erwinia carotovora 345
Erwinia rphapontici 188
Erysiphe cichoracearum 181, 279
Erysiphe heraclei 356
Erysiphe polygoni 228, 234, 258
Etherl 99, 272, 333, 334
Ethylene biosynthesis 347
Eublemma olivacea 130
Eugenia carryophyllus 15
Euphorbiaceae 14
Euxoa spp 172
Evaluation 106-112
Station trials 106-107
Multilocation trials 107-108
National trials 108-109
Adoptive trials 109
Minikit trials 109
Trials on disease and insects 109-110
Trials or quality test 110-112
Exalbuminous 79
EXIM policy 371
Export of seeds 4
Eyes 47

F

Fermentation 144-145, 273
Fertilization 51, 81
Ferula asafetida 15
Field inspection 379
Field testing 379
Financing organization 25
Flat bed method 171
Flea beetle 188
Foeniculum vulgare 15
Foil envelope 9
Foot-rot 130
Formaldehyde 142, 159
Formalin 141
Foundation seed 29, 86-88, 94, 308, 363, 369, 376, 377
Foundation seed stage-I 86-87
Foundation seed stage-II 86-87
Freezing 320
French Fries 175
Fruit cracking 150, 152, 154

Fruit fly 143, 164
Fruit rot 143, 164
Fruit set 1, 2, 40, 44-46, 75
Functional male sterility 330
Fungal resistance 343-344, 355
Fusarium caeruleum 170
Fusarium oxysporum 131, 143, 219, 341, 356
Fusarium wilt 131, 138, 143, 150, 153, 159, 208, 228, 231, 260

G

G. palida 176
Galangal 12
Galerucid beetle 246
Gameticides 325
Gametophyte 313
Gamma irradiation
Gamopetalous 39, 42, 45
Garciana indica 16
Garden beet 250
Geitonogamy 81
Gene cloning 351-353
Gene mapping 352
Gene tagging 349
General and specific combining abilities 103
Genetic and physical purity 31
Genetic contamination 88, 299
Genetic engineering 337, 342-349, 346
  Virus resistance 343
  Fungal resistance 343-344
  Bacterial resistance 344-345
  Insect-pest resistance 345-346
  Parthenocarpic plants 346-347
  Post harvest traits 347
  Male sterility 347-348
  Herbicide tolerance 348-349
  Salt tolerance 348-349
Genetic identity 86-87
Genetic principles 84-89
Genetic purity 86, 87, 89, 299, 301-304, 363
Genetic purity test 376
Genetic shifts 85
Genetic transformation 339-340
Genotype x environment interaction 299
Geographical region 11
Germination 42, 52, 60, 62, 63, 78, 92, 93, 145, 146, 212, 305-307, 371
Germination inhibitors 315
Germination percentage 306, 308, 318
Germination process 311
Gibberellic acid 49, 99, 272, 316
Glass jar 9
Globodera rostochiensis 176
Glucanase 343, 356
Glucose oxidase 345
Glycine max 14
Golden nematode 353
Graminae 12

Grant of certificate 376
Green house 1, 2, 92
Grow out test 89, 301, 302, 304
Growth chamber test 302
Guttuferae 16
Gymnospermae 11
Gynoecious 73, 99, 333, 334

H

Hairy caterpillers 239
Half-hardy 17
Hand emasculation 324, 326, 327, 332-334
Hand pollination 7, 8, 64, 326, 327, 330, 333, 340
Haploid plant 342
Haulms 173
Head intact method 204
Head to seed method 195
Heliothis armigera 143, 219, 246
Hellula undalis 206
Herbicide tolerance 348
Hermaphrodite 60, 66, 72, 73, 76, 81, 82, 104, 105
Hermitasonemus latus 158
Heteromorphic flower 39
Heterosis 102
Heterostyly 39, 324, 325
Heterozygosity 102
High temperature scarification 319
Hispidulous 70, 71
Hoechst India Ltd 100
Homestead gardens 10
Homogamy 82
Homozygosity 102, 103
Homozygous 340
Honey bees 56, 58, 73, 75, 213
Hot water scarification 317-318
Houseflies 58
Humidity 8, 40, 49, 66
Hyalopterus pruni 257
Hybrid 5, 10, 23, 24, 92, 97-99, 101, 103, 106, 107, 116, 119, 120, 126
Hybrid seed 5, 22, 83, 99, 101, 103, 105, 120, 323-327, 330, 332, 335, 340
Hybrid seed companies 98, 99, 102, 120
Hybrid seed production 323-327, 330-332, 334-335, 340
Hybrid testing centres in India 113-116
Hybrid varieties testing 105-126
    Evaluation 106-112
    Identification 116-120
    Release and notification 120-126
Hybrid vigour 103, 340
Hybridization 54, 65, 149, 208, 214-215, 254
Hybrids 23, 31
Hybrids developed by private sector 120
Hybrids identified/released for cultivation 117-119
Hydration 312
Hydrochloric acid 145-146, 274

Hylemia antiqua 219
Hypocotyl 51, 56
Hypogeous germination 314
Hypogynous 42
Hyssopus officinalis 15

I

Identification 25
Identification of hybrids for release 116-120
Imperfect flowers 6
Impermeability 315
Import and export of seeds 374
Inbred lines 102
Inbreeding 80, 102, 103
Indeterminate 138, 142, 149, 150, 153-155, 165, 242, 249
Indian Council of Agricultural Research 25, 120
Indian minimum seed certification standards 375, 377
Indian Society of Seed Technology 24, 26
Indian Society of Vegetable Science 26
Indo American Hybrid Seed 100
Indo-German project 27
Inflorescence 41, 48, 49, 51, 53-58, 66, 68, 73, 79
Initial evaluation trials 107-108
Insect pollination 101, 332-334
Insect-pest resistance 345-346
Insect-pollinated 16
Internal dormancy 317
International seed testing association 25
International trade 21
Interspecific 184, 348
Introduction 153, 154, 201, 242, 243, 248
Ipomoea batatas 13
Iridaceae 12
Isariopsis griseola 239
Isozymes 303

J

Jassid 130, 135, 172, 184
Juniperous 11

K

Kasuri methi 235
Kidney bean 34
Kitchen garden 92

L

L. hirsutum 151
L. pennelli 349
L. peruvianum 341, 349
Labeling 325, 392
Labiateae 15
Lactuca sativa 12, 82, 292
Lactuca scariola 14, 293-294
Lanceolate 42, 44, 55, 56, 71, 77, 78
Late blight 143, 156, 160, 170, 173-177
Lauraceae 16
Laurus nobilis 16
Leaf curl virus 151-153, 159, 161, 164

Leaf eating caterpillars 172, 234
Leaf hopper 131, 181, 252, 264
Leaf minor 143
Leaf roller 130
Leaf spot 140, 194, 236, 252
Leaf webber 252
Leguminosae 14, 58
Leguminous vegetables 90
Lenticels 47
Lettuce drop 296
Leucinodes orbonalis 130
Leveillula taurica 234
Levistcum officinale 15
Light intensity 34
Liliaceae 12
Linkage 349, 352, 353
Lipaphis erysimi 206, 295
Little leaf 131, 133, 136, 138
Long melon 277
Long styled 39-40
Long-day plants 34
Luffa acutangula 13
Luffa cylindrica 13
Lycopersicon esculentum 14, 82, 139, 151
Lysozyme 345

M

M. arenaria 143
M. cochinchinensis 287
M. dioica 287
M. javanica 143
Macrophomina phaseoli 170
Madhurasia obbscurella 246
Maharashtra Hybrid Seed Company Ltd 27, 100
Maintenance of genetic purity 98-89
Majorana hortensis 15
Male fertile 69
Male sterile 99, 101, 325-327, 330, 335, 339, 340, 352, 354
Male sterility 64, 83, 99 102, 105, 324-326, 330, 334, 339, 340, 347-348
Malformation 214
Malic hydrazide 272
Malvaceae 14, 49-51
Manihot esculenta 14
Manila envelopes 9
Marketing of hybrids seeds 323, 325
Marking 392
Mechanical damage 307
Mechanical dormancy 315
Mechanical scarification 317
Medium styled 39-40
Melilotus spp. 233
Mellisso officinalis 15
Meloidogyne incognita 143
Meloidogyne spp. 130, 172
Mentha arvensis 15
Mentha piperita 15
Meristem culture 341
Methods of hybrid seed production and cost analysis 326-336

Hand emasculation and hand pollination 326-329
Hand pollination without emasculation 330-331
Emasculation and wind pollination 332
Hand emasculation and use of insect pollination 332-334
Insect or wind pollination with out emasculation 334-335

Micro propagation 338, 340-342
Somatic embryogenesis 341
Ovule culture 341
Meristem culture 341
Embryo rescue 341
Protoplast fusion 341
Cryopreservation 341
Somaclonal variation 341

Micronutrients 90
Microsatellites 303
Minikit trials 109
Minisatellite 303
Mites 158, 181, 259
Modes of reproduction 80-83
Methods of reproduction 80-81
Modes of pollination 81-83
Moisture content 8, 307, 312, 321, 322
Moisture of seed packaged 322
Mokha 287, 288
Molecular markers 303, 339, 349-358
Gene tagging 349
QTL mapping 350-351
Gene cloning 351-353
Transpoon tagging 354
Transgenic 354-358

Molybdenum 272
Molybdenum deficiency 197
Momordica balsamina 287
Momordica charantia 14, 286
Momordica dioica 14
Monocotyledonae 11
Monoecious 71, 73, 75, 77, 78, 99 104, 105, 325, 332
Morphological dormancy 316
Rudimentary embryo 316
Underdeveloped embryo 316

Mosaic virus 294, 296
Mulching 94, 130
Multilocation trials 106, 107-108, 117
Murraya koenigii 15
Muskmelon mosaic 279
Mustard aphids 206
Mustard sawfly 188, 191, 206, 212
Myristica fragrans 15
Myristicaceae 15
Myrtaceae 15
Myzus persicae 158, 172, 188, 212, 295

N

NABARD 26, 373
NAFED 26
Namdhari Seeds Pvt. Ltd 100

Nath Seed Co Ltd 100
National Agricultural Cooperative and Marketing Federation Ltd 25
National Agricultural Research Projects 25
National Bank for Agriculture and Rural Development 26
National Bureau of Plant Genetic Resources 25, 366
National Commission on Agriculture 24
National Cooperative Development Corporation 26
National Horticultural Board 25
National Horticultural Research and Development Foundation 222
National Seed Corporation 5, 85-87, 92, 275
National Seed Map 364
National Seed Programme 23, 24
National seed register 366
National Seed Research and Training Center 368-369
National Seeds Board 365, 367
National Seeds Corporation 24, 26, 28, 363, 375, 376
National Seed Policy 359-374
  Aims and objectives 360
  Varietal development and plant variety protection 360-362
  Seed production 362-365
  Quality assurance 365-367
  Seed distribution and marketing 367-368
  Infrastructure facilities 368-369
  Transgenic plant varieties 369-371
  Import of seeds and planting material 371-372
  Export of seeds 372
  Promotion of domestic seed industry 372-373
  Strengthening of monitoring system 373-374
National trials 108-109
Natural crossing 85
Nature of pollination 16
  Highly cross-pollinated 16
  Often cross-pollinated 16
  Highly self-pollinated 16
NBPGR 25
Nematode 149, 150
NERAMAC 26
Nezara viridula 258
Nigella sativa 15
Nipping 258
Nitrogen-fixing bacteria 337
Non-transparent containers 9
Notification of the hybrids/varieties 124-125
Nucleus seed 363
Nylon net tunnels 8

O

Ocimum basilicum 15

Often cross-pollinated 16, 51, 83, 87
Oleoresin 111, 163
Onion maggot 219
Onion thrips 219
Open pollinated 99, 333, 335
Ophiomyia phaseoli 239
Orchidaceae 12
Oriental picking melon 277
Origanum vulgare 15
Ovule culture 341

P

Packing 325
Painted bug 206
Paniculiform 55
Paper bag 9
Parenchyma 54
Parthenocarpic plants 346-347
Pastinaca sativa 14, 82
Pea aphids 228
Pea mosaic 228
Pea stem fly 228
Pectinesterase 355
Pedigree 85, 89, 138, 242, 249, 254
Pendulous 43, 45, 71
Pentagonous 39
Pentamerous 41
Perennial 35, 44
Perenospora parasitica 206
Perfect flowers 6, 7
Perigynous 66, 72
Peronospora destructor 219
Peronospora trigonella 234
Peroxidase test 300
Petroselinum crispum 14
Phaseolotoxin insensitive OCTase 344, 345
Phaseolus lunatus 14
Phaseolus vulgaris 14, 82, 236
Phenol test 300
Phenols 175, 316
Phoma leaf spot 177
Phoma lingum 206
Phomopsis blight 128, 130, 131, 135, 136, 137
Phomopsis vexans 130
Photo thermo sensitive 248
Photoperiod 33
Photoperiodism 34
Photosensitive 242
Photosynthesis 34, 314
Phthorimaea operculella 172
Phyllody 198
Phylloreta sp 188
Physical dormancy 314
Physical purity 305, 308, 377
Phythium spp 130, 143, 159, 239
Phytoene synthase gene 348
Phytophthora blight 159
Phytophthora infestans 143, 170, 351, 356
Phytophthora nicotianae 130
Phytophthora parasitica 143
Pieris brassicae 206
Pimenta officinalis 15

Pimpinella anisum 15
Pinaceae 11
Pinching 325, 330, 334
Pioneer Seed Co Ltd 100
Piper longum 15
Piper nigrum 15
Piperaceae 15
Pistillate 72, 73, 75, 78, 101
Pisum sativum 14, 82, 224
Pithiness 193
Plant Varieties Protection 360-362
Plant Variety Right 5
Planting value 307
Plastic bag 8
Plutella xylostella 206
Pocha's Seed Co. 27
Pod borers 246
Pollen culture 342
Pollen dehiscence 40
Pollen fertility 51, 72, 73
Pollen viability 65, 78
Pollination 7, 16, 35, 44, 48-49, 51, 58, 69, 73, 76, 79-83, 101-103, 213-214, 323-327, 329, 332, 334, 340
Polycotomous 41
Polygalacturonase 347, 358
Polygonaceae 14
Polymerase chain reaction 339, 343
Polymorphous 37, 55, 61
Polysiphonous 51
Potassium hydroxide 300
Potassium nitrate 316
Potato breeding 22
Potato Growers Cooperative Marketing Society Ltd. 26
Potato tuber moth 172
Potato virus X 344
Potato virus Y 344
Powdery mildew 162, 181, 228-230, 234-236, 239, 241, 246, 258, 268, 272, 276, 279, 281, 285, 286, 291
Principles of vegetable seed production 84-96
  Genetic principles 84-89
  Maintenance of genetic purity 85
  Control of seed source 85
  Breeder seed 85-86
  Foundation seed 86-87
  Registered seed 87
  Certified seed 87
  Crop rotation 87-88
  Isolation 88
  Isolation by time 88
  Isolation by distance 88
  Roguing of seed crop 88
  Seed certification 89
  Grow-out tests 89
  Agronomic principles 89-96
PR–protein genes 356
Prduncles 54
Private seed companies 87, 98

Private seed sector 5, 10, 27, 29, 121, 360, 368
Procedures of hybrid seed production 102-126
  Inbreeding and production of inbred lines 102
  Testing the combining ability of inbred line 103-104
  Production of hybrid seed 104-105
  Hybrid varieties testing 105-126
Project Directorate on Vegetable Research 23-25
Promising hybrids at various locations 121
Promotion of Hybrid Research in Vegetable Crops 23
Prostrate 139
Protein polymorphism 304
Protogynous 54
Protomyces macrosporus 258
Protoplasmic plug 51
Protoplast culture 339
Protoplast fusion 341
Pruinose ovary 55
Pseudo short styled 39, 40
Pseudomonas solanacearum 131, 143, 159, 166, 168
Pseudomonas syringae 344
Pseudoperenospora cubensis 279
Psila rosae 264
Psophocarpus tetragonolobus 14
Puberulent 63
Pubescent 45, 47, 64, 77
Public sector 23, 25, 362, 363
Pure line 165, 183-184, 235-236, 242-243, 248
Pure live seed 307
Purple blotch 219, 223

Q

QTL mapping 350-351
Quality control 23, 29, 32
Quality parameters 111-112
Quality of seeds 299-310
  Trueness to variety 299-303
  Testing for genetic purity 299-303
  Sample size 303-304
  Physical purity 305
  Viability 305-306
  Vigour 306-307
  Moisture content 307
  Mechanical damage 307
  Size and appearance 307
  Seed health 307
  Planting value 307
Quality seed 3, 4, 22, 33, 99, 110-111, 299, 359, 360, 364, 375
Quarantine 4, 23, 371

R

Racemiform 55
Radish mosaic 188
Raphanus caudatus 186
Raphanus sativus 14, 83, 184
Ratoon crop 138
Rat-tail radish 186

Receptacle 71, 79
Recombinant DNA 342, 347
Recombination 338
Red pumpkin beetle 272, 279, 284, 289
Reducing sugars 175
Refrigerated stratification 319
Registered seed 87
Regression coeffcinets 350
Rejection of seed farms and seed lots 378
Relative humidity 65, 321-322
Release of hybrids 126
Reniform nematode 133
Replanting method 18
Replum 54
Reproductive phase 33
Revalidation 379
Revocation of certificate 384
Rheum rhaponticum 14
Rhizobium 234, 337
Rhizoctonia solani 143, 159, 170, 207, 239, 295
Rhizoctonia spp. 130
Ribonuclease 348
Ribosome inactivating protein 343-344
Riceyness 198, 199
Ridge and furrow method 171
Ripe rot 156
Ripening inhibitor 353
Root knot nematode 130, 133, 143, 151-153, 172, 349, 353
Root rot 143, 188, 235, 236, 295
Root to seed method 187, 211
Rosemarinus officinalis 15
Rosette 65
Rouging 88, 89, 95, 131, 144, 159, 173, 182, 189-190, 207, 214, 220, 228, 235, 240, 246, 253, 258, 272, 279, 285, 290, 296
Rudimentary embryo 316
Rumex vesicarius 14
Rust 229, 240
Rutaceae 15

S

Sale of seeds 308-309
Salt tolerance 348-349, 353
Salvia officinalis 15
Sandoz (India) Ltd 100
Satellite RNA's 343, 344, 357
Satureia hortensis 15
Scarification 317-319
    Mechanical scarification 317
    Hot water scarification 317-318
    Acid scarification 318
    Warm moist scarification 318-319
    High temperature scarification 319
Schizocarp 69
Scirtothrips dorsalis 257
Sclerotinia sclerotiorum 296
Sclerotium rolfsii 131, 170
Scooping 197

Screw-top jar 9
Scritothrips dosalis 158
Sealing 396-397
Sechium edule 14
Seed Act 308, 375, 381-388
- Short title, extent and commencement 381
- Definitions 381-382
- Central seed committee 382
- Central seed laboratory and state seed laboratory 383
- Power to notify kinds or varieties of seeds 383
- Power to specify minimum limits of germination and purity 383
- Regulation of sale of seeds of notified kinds or varieties 383
- Certification agency 383
- Grant of certificate by certification agency 383-384
- Revocation of certificate 384
- Appeal 384
- Seed analysts 384
- Seed inspectors 384
- Powers of seed inspector 384-385
- Procedure to be followed by seed inspectors 385-386
- Report of seed analyst 386
- Restriction on export and import of seeds of notified kinds or varieties 386
- Recognition of seed certification agencies of foreign countries 386
- Penalty 387
- Forfeiture of property 387
- Offences by companies 387
- Protection of action taken in good faith 387
- Power to give directions 387
- Exemption 388
- Power to make rules 388

Seed analyst 382, 384, 386, 396
Seed banks 365
Seed borne diseases 31, 156, 178, 195, 202, 210, 214, 237, 244, 255, 277, 294
Seed certification 367
Seed Certification Agency 129, 141, 157, 171, 180, 187, 205, 212, 227, 233, 238, 245, 251, 256, 263, 270, 278, 288, 294, 309, 375-378, 392
Seed certification fee 379
Seed certification procedure 375
Seed certification standards 325, 375, 376
Seed companies 3, 99, 362
Seed control 24
Seed crop insurance 365
Seed dormancy 314-319
- Physical dormancy 314-315
- Mechanical dormancy 315
- Chemical dormancy 315-316

Morphological dormancy 316
Internal dormancy 317
Seed drying 148
Seed extraction 49, 144, 159, 273, 280, 323, 325
Seed extraction in tomato 144-146
Fermentation 144-145
Acid treatment method 145-146
Alkali treatment method 146
Mechanical seed extraction 146
Seed farms 28
Seed germination 311-314
Germination process 311
Stages of germination 312
Seed health 307, 376
Seed import 3-4
Seed industry 3-5, 21, 24, 28, 29, 31-32, 99, 360, 361, 372-374
Seed inspector 309, 382, 384-386, 396
Seed moisture 322
Seed packing 325
Seed plot techniques 22
Seed policy 5, 24
Seed production 1, 2, 3, 5, 17, 21, 22, 24, 25, 29, 30-33, 65, 80, 84-86, 88-90, 92, 95, 99, 103, 105, 196, 203, 262, 217, 265, 327, 330, 336, 362-365
Seeds pure 7
Isolation in distance 7
Bagging 7
Caging 7
Seed quality 28, 307, 305, 321
Seed requirement 100
Seed ripening 35
Seed rules 376, 390-398
Preliminary 390-391
Central seed committee 391
Central seed laboratory 391
Seed certification agency 392
Marking or labelling 392-393
Requirements 393-394
Certification of seeds 394-95
Appeals 395
Seed analysts and seed inspectors 396
Sealing, fastening, despatch and analysis of samples 396-398
Seed saving 5
Seed sector 3, 4, 10, 99, 359, 363, 368
Seed set 10, 46, 91, 93
Seed technology 23, 24, 31
Seed to seed method 18, 94, 195, 211, 217
Seed trade 4, 99
Seed viability 35, 144, 320
Seed vigour 306
Seed Village Scheme 364
Seed weight 306

Seed-extracting machine 146
Seedling growth 313
Seeds (amendment) Act 388-390
Seeds Act 24, 359, 365, 381-388
Seeds Control Order 359
Seeds export promotion zones 372
Seeds Rules 31, 302, 390-398
Self–fertile species 103, 105, 301
Self–incompatibility 80, 82, 99, 102, 105, 324-326, 334, 353
Self–pollination 42, 79
Self-blanching 198
Self-incompatible lines 339, 340
Selfing 58, 102, 103
Self-pollinated 16, 39, 48, 51, 73, 87, 102
Self-pollination 5, 7, 81, 82
Semi determinate 149, 150, 152, 249
Semi-government organizations 27
Semipermeability 317
Sex expression 105, 271, 324, 326, 353
Sex ratio 271
Sexual 33, 80, 81
SFCI 5
SFDA 27
Shattering 36
Shelf life 162
Shoot and fruit borer 130, 133, 135-137, 181
Short styled 39-40
Short-day plants 34
Silver nitrate 99
Simple recurrent selection 200
Society of commercial seed technologist 25
Sodium chloride 349
Sodium hydroxide 300
Soft rot 195, 202
Solanaceae 14, 37-49
Solanum melongena 14, 127, 128,
Solanum tuberosum 166
Somaclonal variation 341
Somatic embryogenesis 340
Somatic hybrids 341
Spermatophyta 11
Sphareotheca fuliginea 279
Spinacea oleracea 12, 83
Splitting and forking 193
Spodoptera litena 257
Spontaneous mutation 254
Stages of germination 312-314
    Awakening or activation 312
    Digestion and translocation 313
    Seedling growth 313-314
Staking 142, 206
Staminate 66, 72 73, 75
State seed certification agencies 25
State seed certification authority 86
State Seed Corporation 26
State seed laboratory 382, 383

State seed testing laboratories 25, 367
State varieties release committee 105-106, 125
Station trials 106-107
Steckling 18, 188, 189, 213
Stem gall 258, 260, 261
Stem weevil 252
Stigma receptivity 40, 66, 73
Stipules 58, 62
Stock beet 250
Stolon 47
Storage 148, 320-323, 325
Store seeds 8-9
  Humidity 8
  Darkness 9
  Temperature 9
Stratification 319
Streptomyces scabies 166, 170, 173
Stringless 242, 243
Stump method 203-204
Sugar extraction 22
Sulfuric acid 318
Sun drying 95
Sungrow Seeds Co. 27, 100
Sunscalding 227
Sutton and Sons 20
Symbiosis 337
Synchytrium endobioticum 166
Synthetic seed 87, 208, 340

T

T. corniculata 231
T4 bacteriophage 345
Tabtoxin 344
Tagging 332, 376
Takmek 277
Tannic acid 300
Tapetum 99
Tap root 56, 58, 66-68
Tarai Development Corporation 24, 26
Techniques for hybrid seed production 97-102
  Hybrid seed production system 99-100
  Types of hybrids 101
  Types of crossing 101-102
Temperate 21, 22, 34, 44
Testing for genetic purity 299-303
  Morphological test 300
  KOH-Bleach test 300
  NaOH test 300
  Peroxidase test 300
  Phenol test 300-301
  Grow out test 301-302
  Greenhouse/growth chamber test 302-303
  Molecular markers-proteins/isozymes/DNA based markers 303
Tetragonia expansa 12
Tetranychus spp. 181
Thionin 343
Thrips 158, 257
Thrips tabaci 257
Thymus vulgaris 15

Tissue culture 338, 340, 341, 343, 348

Tobacco anionic peroxidase 346

Tobacco mosaic virus 140, 143

Tomato fruit worm 143

Tomato leaf curl virus 143

Tomato mosaic virus 146, 159, 161, 343, 344, 357

Tomato proteinase inhibitor 346

Tomato spotted wilt virus 344, 357

Tomato yellow leaf curl virus 344, 357

Tomentum 37

Top cross 101, 103

Top-dressing 129, 218, 252

Trachyspermum ammi 15

Trailing 73

Transgenic 338, 342, 343, 345, 354-358, 370

  Transgenic with antifungal protein genes 355-356

  Transgenic through pathogen induced cell death 356

  Transgenic with R and Avr genes 356-357

  Transgenic resistant to viral disease 357-358

  Transgenic with delayed ripening 358

Transpoon tagging 354

Treatments to overcome seed dormancy 317

  Scarification 317-319

  Harvesting immature fruits 319

  Stratification 319

Tribal area development programmes 27

Trichomes 41

Trichosanthes anguina 14

Trichosanthes dioica 14

Trigonella foenum graecum 14, 231

Triploidy 105

Tropical crop species 34

True to type 3

Trueness to variety 299

TSS 222

Tubercles 72

Types of crossing 101-102

  Emasculation 101

  Pollination 101

  Andromonoecious 101

  Artificial pollination 101-102

U

Unsatisfactory quality of seeds 307-308

Umbel 58, 68, 69, 70

Umbelliferae 14, 66

Underdeveloped embryo 316

Unisexuality 82

Urocystis cepulae 219

V

Value for cultivation and usage (VCU) 366

Vapour proof containers 326

Variability 46

Vegetable industry 4, 10
Vegetable production 10
Vegetative phase 33
Vegetative reproduction 80, 81
Vegetatively propagated 46, 87
Vernalization 33
Verticillum wilt 153
Viability 49, 76, 305, 320, 322
Viable 3
Vicia faba 14
Vigna unguiculata 14, 243, 248
Viral coat protein 343, 344
Virus resistance 343, 354, 355, 357
Vitality 320
Vitalline 71
Vitamin A 112
Vitamin C 111

W

Warm moist scarification 318-319
Wart disease 166, 173, 177
Watermelon mosaic 285
Weed melon 277
Wet rot 170
White fly 143, 181, 259
White rust 188, 191, 212, 252
Wild lettuce 293
Wild okra 179
Wild water melon 283
Wind pollination 16, 66, 88, 101, 332, 334
Wind-blown pollen 7, 35
Wire stem 207
Wrinkled-seeded 229, 231

X

X ray irridation 241
Xanthomonas campestris 207, 353

Y

Yellow vein mosaic 181, 182, 183, 184
Yellowing 209

Z

Zea mays var. *rugosa* 12
Ziad vegetable 19
Zingiber officinale 12
Zingiberaceae 12
Zygomorphic 62